LEGUMES OF THE GREAT PLAINS

LEGUMES of the

GREAT PLAINS

An Illustrated Guide

JAMES STUBBENDIECK
AND JESSICA L. MILBY

Illustrated by
BELLAMY PARKS JANSEN,
REGINA O. HUGHES,
and KEITH WESTOVER

UNIVERSITY OF NEBRASKA PRESS   LINCOLN

Library of Congress
Cataloging-in-Publication Data
Names: Stubbendieck, James L., author. |
Milby, Jessica L., author. | Jansen, Bellamy
Parks, illustrator. | Hughes, Regina O., 1895–
1993, illustrator. |
Westover, Keith, illustrator.
Title: Legumes of the Great Plains:
an illustrated guide / James Stubbendieck
and Jessica L. Milby, illustrated by
Bellamy Parks Jansen, Regina O. Hughes,
and Keith Westover.
Description: Lincoln: University of Nebraska
Press, [2021] | Includes bibliographical
references and index.
Identifiers: LCCN 2020011872
ISBN 9781496217752 (hardback)
ISBN 9781496224569 (epub)
ISBN 9781496224576 (mobi)
ISBN 9781496224583 (pdf)
Subjects: LCSH: Legumes—
Great Plains—Identification. |
Legumes—Great Plains—Pictorial works.
Classification: LCC QK495.L52 S925 2021 |
DDC 583/.63—dc23
LC record available at
https://lccn.loc.gov/2020011872

Designed and set in Adobe Text Pro
by L. Auten.

Dedicated to
Dr. Lowell E. Moser: scientist, teacher, mentor, and friend

CONTENTS

ACKNOWLEDGMENTS

Elverne C. Conard was a coauthor of *Common Legumes of the Great Plains* (University of Nebraska Press, 1989), and many of his contributions carry forward to this work. We express appreciation to Jerry Volesky, University of Nebraska West Central Research and Extension Center, North Platte, without whose help this project would not have been possible.

We acknowledge John Guretzky, Daren Redfearn, Bruce Anderson, Francis Haskins, and Walter Schacht, Department of Agronomy and Horticulture, University of Nebraska–Lincoln, for answering specific technical questions and being willing to take time to discuss problems encountered during the preparation of this manuscript. Cheryl Dunn, Department of Agronomy and Horticulture, University of Nebraska–Lincoln, provided guidance for the preparation of the maps and figures. Robert B. Kaul, School of Biological Sciences and State Museum, University of Nebraska–Lincoln, and David M. Sutherland, Biology Department, University of Nebraska–Omaha, answered many taxonomic questions.

Appreciation is expressed to the University of Nebraska Press for permission to print the map of the Great Plains (reprinted from *Atlas of the Great Plains* by Stephen J. Lavin, Fred M. Shelley, and J. Clark Archer, 2011). Texas A&M University Press is acknowledged for permission to print three illustrations (reprinted from *Texas Range Plants* by Stephan L. Hatch and Jennifer Pluhar, 1993). An illustration by Regina O. Hughes was reproduced from *Selected Weeds of the United States* (Agricultural Research Service Handbook 366, 1970).

Appreciation is expressed to the Department of Agronomy and Horticulture, West Central Research and Extension Center, and Institute of Agriculture and Natural Resources, University of Nebraska–Lincoln, for fostering a work environment that makes projects such as *Legumes of the Great Plains* possible.

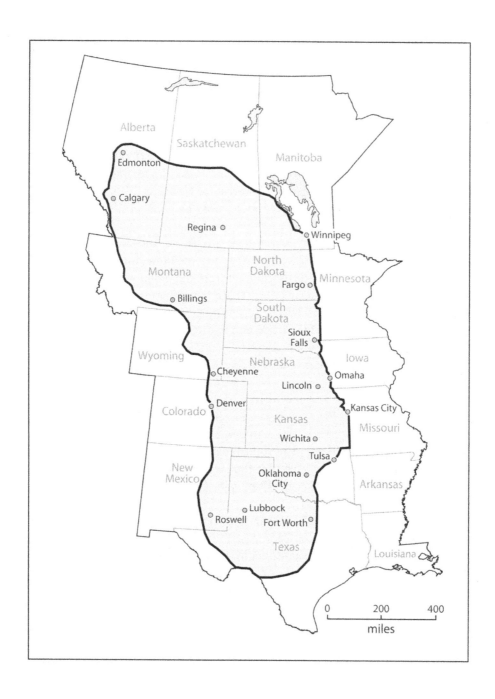

FIG. 1. The Great Plains. Reproduced from *Atlas of the Great Plains* by Stephen J. Lavin, Fred M. Shelley, and J. Clark Archer, by permission of the University of Nebraska Press. Copyright 2011 by the Board of Regents of the University of Nebraska.

The Great Plains region occupies a vast area in central North America (fig. 1). It contains all or parts of ten states of the United States and three Canadian provinces (Wishart 2004; Lavin et al. 2011). Total land area is about 62 million acres, which equals more than 30 percent of the area of the continental United States (Lavin et al. 2011). Many terms used to describe the area, such as *treeless, unwatered, flatlands, forgotten,* and *Great American Desert,* have negative connotations. None are correct. The Great Plains is a sparsely populated region with wide-open spaces, but it is far more than the Buffalo Commons, Dust Bowl, or flyover country (Stubbendieck et al. 2017). Climate and soils vary across the region, resulting in diverse plants and complex plant communities. Water is the most limiting factor for growth of most plants in the region, and annual precipitation is mainly responsible for plant species distributions and patterns of plant communities (Stubbendieck et al. 2017).

Climate

The Great Plains climate is classified as continental (Lawson et al. 1977). Temperatures can vary from -45°F (-43°C) in winter in the north to 110°F (43°C) in summer in the south. Winds can be strong in winter and spring. This semiarid to semihumid region is in the rain shadow of the Rocky Mountains. Precipitation varies from about 10 inches (250 millimeters) in the west to more than 40 inches (1,000 millimeters) in the southeast (Lavin et al. 2011). Dry westerly winds flow onto the Great Plains from the Rocky Mountains and collide with the moisture-laden air masses moving north from the Gulf of Mexico, resulting in increasing precipitation from west to east. Much of the precipitation falls during spring and early summer. Violent thunderstorms are common (Dewey 2019). Drought is also common. Winter snowfall can exceed 50 inches (125 cm) in the north. Snowfall in the southern part of the region is uncommon.

Geology

It is impossible to separate the geologic development of the region from the development of the earth as a whole (Diffendal 1991). However, the geology of the Great Plains is not as complicated as the geology of most regions (Trimble 1980).

While the earth is much older, the last amalgamation of the continental material was complete by about 1 billion years ago (Swinehart 2004). Beginning about 540 million years ago, shallow seas covered the region. These seas rose and fell numerous times during the next 480 million years. Much of the sediment deposited during this extended period consolidated into rock. By about 66 million years ago, tectonic plate motion caused the Rocky Mountains to uplift, and the inland sea began to disappear (Diffendal 2017).

The mountains eroded and huge quantities of sediment were moved to the east by rivers (Howard 1958). Volcanic activity increased west of the region 37 to 17 million years ago causing tremendous quantities of ash to be deposited in many areas of the future Great Plains. In some places these volcanic ash deposits are more than three hundred meters deep. Renewed uplift of the mountains 15 to 2.5 million years ago resulted in continued erosion and sediment transport to the east (Diffendal 2017). Intermittent deposits of ash combined with water- and wind-borne particles continued and formed the mineral base of the Great Plains (Howard 1958).

Beginning about 2.6 million years ago, the northern part of the region was subjected to several periods of glaciation. Tremendous amounts of sediment were moved by the glaciers and left behind as the ice melted. Strong winds and active rivers during the dry intervals between periods of glaciation shifted and redeposited this material (Plaster 1996). Some soil development occurred during the relatively long interglacial periods. The last glacier retreated from the region twelve thousand to ten thousand years ago (Swinehart 2004).

Soils

Soils are a product of the parent minerals, vegetative cover, and climate, primarily precipitation and temperature (Peterson and Cole 1995). Because of the variability of these factors, many different soils orders developed. Alfisols formed under hardwood forest cover near the eastern edge of the Great Plains, aridisols developed under arid and semiarid conditions in the west and south, and mollisols developed in semiarid to semihumid areas under a cover of grasses and forbs in the central and northern parts of the region (Brady and Weil 2002). Entisols are soils without profile development other than the A horizon and are scattered throughout the region.

Great Plains soils are frequently reworked by organisms such as earthworms and ants (Miller and Gardner 2001). Little water is stored in the soil from year to year. When combined with low rainfall, this results in mostly dry subsoil with little loss of soil nutrients through leaching (Brady and Weil 2002). Therefore, many Great Plains soils are fertile and have been important for cultivated crop production for more than 150 years.

Prairies

Historically, the soils and vegetation of the Great Plains have been strongly influenced by fire (Stubbendieck et al. 2017). The growing points of forbs and grasses are near the soil surface, making them better adapted to fire than shrubs and trees with elevated growing points. This resulted in prairies, interrupted by a few relatively small areas of forest, extending across most of the Great Plains before settlement by European Americans. Prairie vegetation is composed of grasses, grasslike plants, shrubs, and forbs (including most legumes). The major types of prairies were the tallgrass prairie in the eastern part of the region, receiving larger amounts of precipitation; shortgrass prairie in the western Great Plains with the lowest amounts of precipitation; and the mixed-grass prairie in the central area. Many subtypes of prairie have been described.

Plants composing the natural vegetation, including native legumes, evolved to grow and reproduce in the extremes of climate, variable soils, and frequent prairie fires. They developed many strategies to withstand herbivory by a range of organisms, from microscopic nematodes to large, hooved ungulates such as the bison.

Legumes are one of the largest and most important groups of plants in the Great Plains. They are second to grasses (POACEAE family) in economic value and third in abundance following grasses and composites (ASTERACEAE family). More than 50 genera and 250 species of legumes may be found in the region. Many are uncommon and will rarely be encountered, while others are found in nearly every county. This work includes descriptions of 44 genera and 217 species, which account for more than 98 percent of the abundance of legumes in the Great Plains.

Legumes are important in the diets of humans in most parts of the world. Numerous types of peas and beans are harvested for food while green or upon maturity. This book does not include the legumes grown and eaten by humans, but many of the legumes included here are eaten by domestic livestock and wildlife. The most important legume hay crops are alfalfa and various clovers. However, some species of legumes are among the most toxic plants found in nature. Some can be managed to be valuable forage producers but are poisonous when managed improperly. Each legume species with toxic properties is identified, and the poisonous principle is discussed.

Some legumes are important for prevention and control of soil erosion. The foliage intercepts the raindrops and reduces their impact on the soil. Roots bind the soil and the foliage covers it, protecting it from the erosive forces of both wind and water.

Many legumes improve soil nitrogen fertility through a symbiotic relationship with certain bacteria (*Rhizobium* spp.). Individual species of *Rhizobium* are often specific for an individual species of legume. The rhizobia form nodules on the roots of the legume plants. The legume plants furnish the food energy that enables the rhizobia to change atmospheric nitrogen into a form plants can use. This process is called nitrogen fixation. Surplus nitrogen increases soil fertility and is used by other plants.

Flowers and inflorescences of some legumes are large and showy, whereas those of others are small and delicate. Flower and foliage color and growth form are varied, making many legumes valuable for landscaping.

Many legumes are important for honey production and serve as hosts for pollinators. Some furnish food for adult butterflies and larvae. Seeds are eaten by many species of birds and small mammals. Some legume trees furnish excellent firewood, and some wood products are used for construction. Food, medicinal, spiritual, and other uses of legumes by Great Plains Indians are described in this book.

Keys

Keys to genera and species are based on floral or vegetative characteristics. Fruit characteristics have been used when others are not particularly diagnostic. Keys were constructed using both living and pressed plant specimens. Most of the keys were new in *Common Legumes of the Great Plains* (1989), but a few were previously published and were extensively modified to fit material from the Great Plains. Modifications were made to the keys used in this book to incorporate additional distinguishing characteristics and the expanded list of species. The keys are artificial. No attempt was made to key together closely related genera or species. Dichotomous choices are identified by the same number.

Taxonomy

Nomenclature generally follows the Missouri Botanical Garden (https://www.tropicos .org). Legumes are separated into three subfamilies: CAESALPINIACEAE (Caesalpinia Family), FABACEAE (Pea Family), and MIMOSACEAE (Mimosa Family). In the Great Plains, only six genera are in the Caesalpinia Family, five are in the Mimosa Family, and 33 are in the Pea Family. Relatively recent changes, such as removing species from *Psoralea* and placing them into *Pediomelum* and *Psoralidium* or moving species from *Oxytropis* to *Orophaca*, may be challenging for individuals with knowledge of botanical names. Extensive synonymy, alternative common names, and plant descriptions may help minimize confusion.

Meanings of botanical names are included. Where necessary, the names have been divided into two parts, and the meaning for each is given. Abbreviations for the source of names are Gk. (Greek), Lat. (Latin), Ar. (Arabic), and Tr. (Tartar). Names from other sources, such as the name of an individual or geographic location, are noted.

Authorities for botanical names, including synonyms, are included. Names, birth and death dates, and occupations are found in the section titled "Abbreviations for Nomenclature Authorities."

A single common name has been selected for each species on the basis of frequency of use in the region. Other common names used in the Great Plains are given in the text. Common names are restricted to two words, and many hyphens have been eliminated from previously published names. Exceptions to two words are those names used by Great Plains Indians. All common names may be found in the index.

Illustrations

Illustrations of the 114 primary legumes are included. These detailed pen and ink drawings illustrate the plant in flower and fruit. Individual features, such as flower structure, are often greatly enlarged to show the detail. Illustrations were prepared from living plant material or herbarium specimens.

Maps

Distribution maps are based on information from multiple publications from states and provinces and websites such as the Biota of North America Program (https://www.bonap .net/napa) and the Natural Resources Conservation Service Plants Database (https:// www.plants.usda.gov). The shaded area on the maps delineates where the species is most common in the Great Plains and is meant to provide "at a glance" information on where the species are most likely to occur. The maps are not meant to imply that the plant distributions are uniform or equally abundant across the area. Also, these species may be found outside of the areas designated on the maps.

Descriptions of Individual Species

Descriptions of individual genera and species are provided. Descriptions of stems, leaves, inflorescences, flowers, fruits, and seeds follow the growth form, flowering dates, methods of reproduction, origin, and height of each species. Ranges of measurements of plant parts are provided. These measurements will include at least 90 percent of the plants encountered, but extreme environmental conditions may cause a plant part to be either larger or smaller. Technical terms are defined in the glossary.

Chromosome numbers are provided. These numbers were taken from the literature, and no chromosome numbers were determined specifically for this book.

Habitat in which the plant grows follows the botanical description of the species. Uses and values including relative palatability and value to domestic livestock and wildlife are provided. Responses to grazing and toxic principles are included where applicable. Associated pollinators and other insects are identified. Potential use for erosion control and landscaping are discussed. Ethnobotany includes uses of the plants by Great Plains Indians, as well as medicinal and other uses.

Botanical varieties within a species are given. Where appropriate, differences between the varieties are outlined. An extensive list of synonyms and other common names is recorded for each species, and each is included in the index.

Characteristics of Legumes

Most legumes have papilionaceous flowers (fig. 2). The corolla of these irregular flowers is made up of five petals. The banner, or standard, is the single upper petal. The two lateral petals are the wings, and the two lower petals comprise the keel. The keel petals are often fused.

Other types of legume flowers are presented in figure 3. The *Chamaecrista* flower has five nearly regular petals. The petals of *Dalea* flowers are modified stamens. The *Amorpha* flower has a single petal. The *Prosopis* flower is very small and is nearly regular with five petals.

Leaf arrangements are presented in figure 4. Most legumes have compound leaves with leaflets arranged either palmately or pinnately. Examples of odd-pinnately, even-pinnately, and bipinnately compound leaves are illustrated. Leaf characteristics are important for the identification of many legumes. Leaf shapes, apices, bases, and margins are illustrated in figure 5. This figure is a comprehensive illustration of these characteristics, and not all are represented in the legumes described in this book.

Fruit characteristics are necessary for the identification of some species. The fruit of leguminous plants is highly variable. Most are legumes (sometimes called a pod or seedpod) or loments. Several types of legumes and two types of loments are illustrated in figure 6.

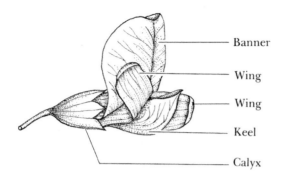

Banner

Wing

Wing

Keel

Calyx

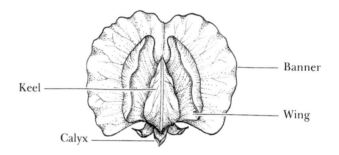

Keel

Banner

Wing

Calyx

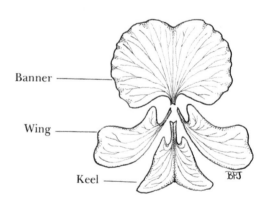

Banner

Wing

Keel

FIG. 2. The papilionaceous flower from different perspectives

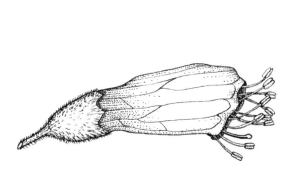

FABACEAE (*Amorpha*)

CAESALPINIACEAE (*Senna*)

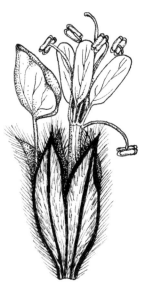

FABACEAE (*Dalea*)

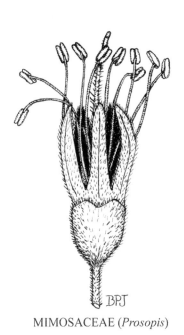

MIMOSACEAE (*Prosopis*)

FIG. 3. Other legume flowers

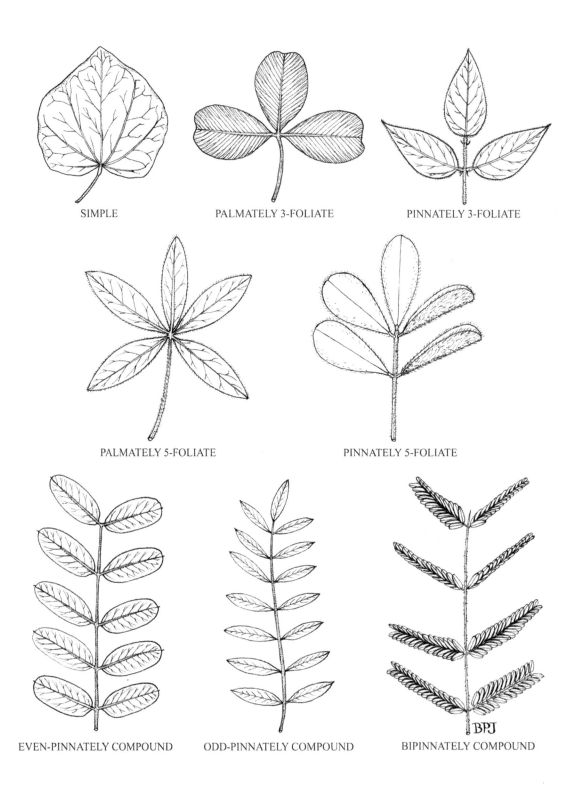

SIMPLE

PALMATELY 3-FOLIATE

PINNATELY 3-FOLIATE

PALMATELY 5-FOLIATE

PINNATELY 5-FOLIATE

EVEN-PINNATELY COMPOUND

ODD-PINNATELY COMPOUND

BIPINNATELY COMPOUND

FIG. 4. Legume leaves

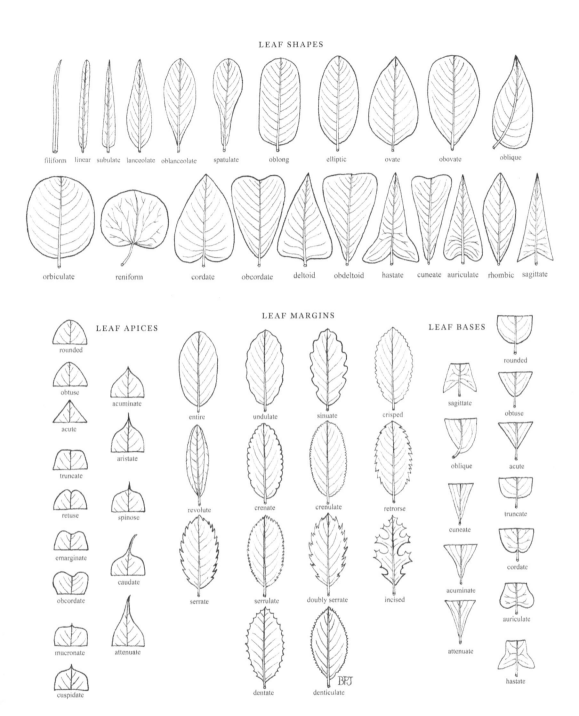

FIG. 5. Legume leaf and leaflet shapes and margins

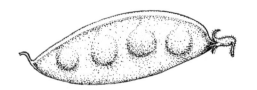

SEVERAL-SEEDED LEGUME

BILOCULAR LEGUME

ONE-SEEDED LEGUME

COILED LEGUME

INFLATED LEGUME

DEHISCENT LEGUME

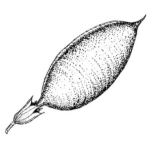

INDEHISCENT LEGUME

COILED AFTER DEHISCENCE

TWO-JOINTED LOMENT

FOUR-JOINTED LOMENT

FIG. 6. Legume fruits

Taxonomy of Legumes

The group of plants comprising the legumes has varied historically. Earlier, most taxonomists included all legumes in one family, the LEGUMINOSAE or FABACEAE. This large family was subdivided into the subfamilies CAESALPINIOIDEAE, PAPILIONOIDEAE, and MIMOSOIDEAE. This division was made primarily on the basis of floral differences. Other taxonomists elevated these subfamilies to full family ranking in accordance with the 1978 International Legume Conference held at the Royal Botanic Gardens, Kew, England. As a result, most authors now recognize three families of legumes; CAESALPINIACEAE (Caesalpinia Family), FABACEAE (Pea Family), and MIMOSACEAE (Mimosa Family).

Key to the Families

1. Flowers regular or nearly so, small; inflorescences densely flowered globose heads, spikes, umbels, or racemes; petals five (rarely four), inconspicuous, valvate in bud; stamens strongly exserted; leaves bipinnately compound or uncommonly pinnately compound. .... MIMOSACEAE

1. Flowers more or less irregular, papilionaceous or sometimes imperfectly papilionaceous; inflorescences usually racemes or panicles; petals imbricate in bud; stamens generally not exserted; leaves mostly other than bipinnately compound.

    2. Corolla irregular or imperfectly papilionaceous or not at all irregular; banner enclosed by the lateral petals in bud; stamens free; leaves bipinnately compound or pinnately compound, or rarely simple. .... CAESALPINIACEAE

    2. Corolla usually papilionaceous (mostly 5 petals, 1 in *Amorpha*, and 1 plus 4 petal-like staminoids in some *Dalea* species) with the keel petals more or less united; banner enclosing the lateral petals in bud; stamens commonly diadelphous or monadelphous; leaves mostly palmately or pinnately compound, not bipinnately compound. .... FABACEAE

LEGUMES OF THE GREAT PLAINS

I. *CAESALPINIACEAE* R. BR.

Trees, shrubs, or perennial (rarely annual) herbs; leaves alternate, bipinnately compound (rarely simple) or pinnately compound, sometimes glandular-punctate, usually with stipules and lacking stipels; inflorescences usually showy racemes or panicles; flowers irregular or imperfectly papilionaceous or not at all irregular, perfect or unisexual; calyx polysepalous or gamosepalous, lobes 5, imbricate or separate; corollas usually white or yellow or pink to rose; petals usually 5, sometimes rudimentary or absent, imbricate, not valvate; banner distinctive and innermost in bud; stamens (3) 5–10, included to exserted, mostly free or joined; ovary sessile; fruits legumes, sometimes fleshy, occasionally with winged sutures, sometimes indehiscent; seeds usually with abundant endosperm.

This family contains about 150 genera and about 2,200 species worldwide. The species occur mainly in tropical and subtropical regions. It is represented in the Great Plains by 6 genera and 13 species.

Key to the Genera

1. Trees or tall shrubs
    2. Leaves simple, cordiform, entire; flowers imperfectly papilionaceous, appearing before the leaves; corollas rose to purplish-pink. .... *Cercis*
    2. Leaves compound; flowers not papilionaceous, appearing with the leaves or after the leaves; corolla greenish-yellow or greenish- to pinkish-white
        3. Stems usually armed; staminate flowers in catkinlike racemes; pinnate leaflets 4 cm long or less, 1 cm wide or less, bipinnate leaflets smaller; entire to crenulate; legumes large, compressed, coriaceous. .... *Gleditsia*
        3. Stems unarmed; flowers in terminal racemes or panicles; leaflets 4 cm long or more, 2 cm wide or more, entire; legumes thick and woody. .... *Gymnocladus*
1. Herbs
    4. Leaves bipinnately compound. .... *Hoffmannseggia*
    4. Leaves even-pinnately compound
        5. Annuals; leaflets 1–2 cm long. .... *Chamaecrista*
        5. Perennials; leaflets 2–8 cm long. .... *Senna*

Cercis L.

kerkis (Gk.): ancient name of a tree, possibly *Cercis siliquastrum* L., judas tree.

Small trees or shrubs; leaves simple, cordiform, or broadly obovate, margins entire with a shallow or deep sinus at the base, palmately veined; inflorescences umbel-like, flowers fascicled, appearing before the leaves; calyx irregular; lobes 5, shallow; corolla irregular, imperfectly papilionaceous; petals rose (occasionally white) to purplish-pink, banner petal smallest, keel petals largest; stamens 10, free; hypanthium short, hemispheric, bearing the perianth and stamens on its margins; fruits legumes, narrowly oblong, thin, compressed, strongly margined along upper surface, valves papery, persistent, slowly dehiscent; seeds several. $x=7$.

Seven species have been described in North America. Others are found in Eurasia. One species is common in the southeastern Great Plains.

Eastern redbud
Cercis canadensis L.

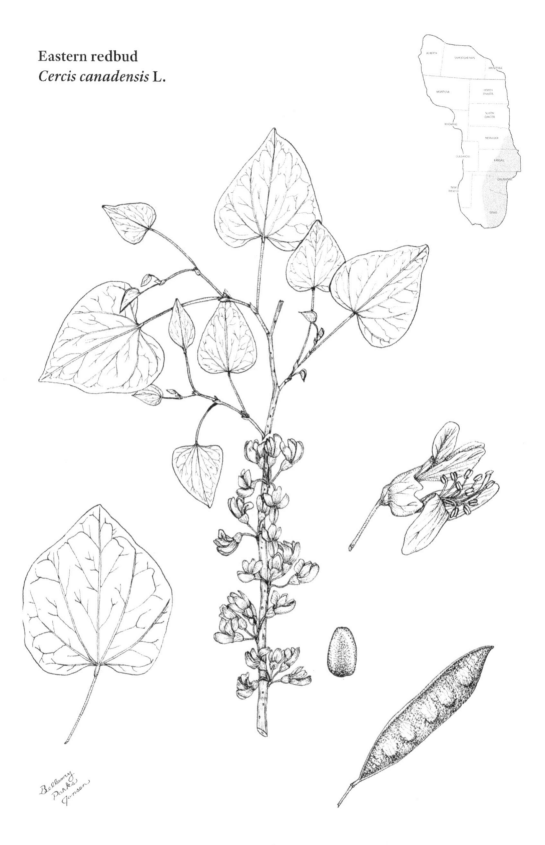

Species Plantarum 1:374. 1753.
canadensis: of or from Canada.

GROWTH FORM: small tree with a flat or rounded crown, flowers March to May, reproduces from seeds. LIFE-SPAN: perennial. ORIGIN: native. HEIGHT: to 12 m. TWIGS: zigzag; low branching or multi-trunked, lenticels white to dark brown. TRUNKS: to 30 cm in diameter, usually straight, often leaning with age; bark becoming scaly with age. LEAVES: alternate, simple, broadly cordate to broadly ovate (8–14 cm long, 5–12 cm wide); palmately veined, usually with 5 principal veins; apex acute or short-acuminate; base cordate to truncate; margins entire; upper surface dark green, glabrous; lower surface lighter in color with axillary tufts of hair, few hairs on the veins; petiole enlarged at the summit and base (4–10 cm long); stipules connate, caducous. INFLORESCENCES: umbel-like clusters from lateral buds on old wood. FLOWERS: imperfectly papilionaceous, showy; calyx tube irregularly campanulate (2–2.5 mm long, 3–3.5 mm wide), gamosepalous, base oblique, lobes 5; lobes broadly triangular to rounded, shallow, purplish; corollas appearing before the leaves, irregular, purplish-pink to rose, occasionally white in horticultural cultivars; banner 7–7.5 mm long, 3–3.5 mm wide; wings 7–7.8 mm long, strongly reflexed, enclosing the banner in bud; keel distinct (8–9 mm long); stamens 10, free; pedicels slender (8–17 mm long), glabrous, reddish. FRUITS: legumes clustered, linear to elliptic (8–10 cm long, 1.2–1.5 cm wide), pointed at both ends, compressed, slightly stipitate, bilocular; valves papery, brown, small ridge on both sides of the upper suture, glabrous; fruits remaining on the tree into early winter; seeds usually 5–12. SEEDS: oval to reniform (4–5 mm long, 4–4.5 mm wide, about 2 mm thick), first green, then light to dark brown with maturity, semilustrous, indurate. $2n=14$.

HABITAT: Eastern redbud is infrequent to common in open woodlands and along the edges of woodlands. It may occur as an understory plant. It is found on hillsides, limestone glades, and rocky stream banks. Eastern redbud does not grow well in poorly drained soils. USES AND VALUES: Cattle and deer browse the young trees, and squirrels eat the legumes and seeds. Songbirds and ground-foraging birds eat the seeds. It contains a toxic saponin, but poisoning of livestock and wildlife has not been documented. It serves as an excellent source of nectar for honey. It is pollinated mostly by bees, and caterpillars of butterflies and moths eat its foliage. It has little value for commercial lumber and is not used for erosion control. Eastern redbud is a popular plant for landscaping and is grown far beyond its natural range. ETHNOBOTANY: Some Plains Indians made tea from the inner bark to treat dysentery and diarrhea. Flowers were fried or pickled and used in salads. Immature legumes were eaten raw. Twigs were used on a limited basis for baskets.

VARIETIES: Var. *canadensis* L. is the variety most common in the Great Plains. Var. *texensis* (S. Watson) M. Hopkins grows in Texas and Oklahoma, and var. *mexicana* (Rose) M. Hopkins has limited distribution in Texas and New Mexico. SYNONYM: *Cercis occidentalis* Torr. *ex* A. Gray. OTHER COMMON NAMES: American judas-tree, American redbud, judas tree, redbud, salad tree.

Chamaecrista (L.) Moench

chamai (Gk.): on the ground; + *crista* (Gk.): crest, referring to the low growth and the inflorescence.

Annual or sometimes perennial herbs; leaves even-pinnately compound, leaflets several to numerous (1–2 cm long), asymmetrical, margins entire; petiole bearing a gland between or below the leaflets; stipules striate, persistent or caducous; inflorescences racemes, terminal or axillary, reduced; flowers 1 to few, irregular; corolla yellow, generally reddish- or purplish-spotted at the base; petals equal or 1 larger than the others; functional stamens 5–10, variable in length; ovary pubescent; fruits legumes, linear to oblong, laterally compressed, erect or pendant; seeds few to many. $x=14$.

About 260 species with pantropical distribution. Two grow in the Great Plains, and one is common in the east-central and southeastern parts of the region.

Partridgepea
Chamaecrista fasciculata (Michx.) Greene

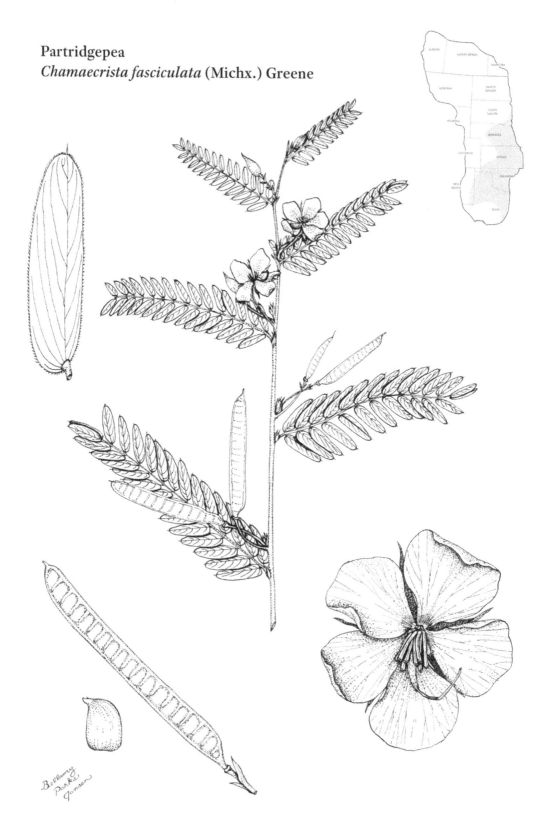

Pittonia 3 (17C): 242. 1897.

fasciculus (Lat.): clustered or growing together in groups, in reference to the inflorescences.

GROWTH FORM: forb, flowers July to September, reproduces from seeds. LIFE-SPAN: annual. ORIGIN: native. HEIGHT: 0.5–2 m. STEMS: erect to ascending, 1 to few from a woody caudex, glabrous or puberulent to conspicuously hirsute. LEAVES: alternate, even-pinnately compound (10–30 cm long), leaflets 8–24; leaflets elliptic to oblong (1–2 cm long, 5–30 mm wide), asymmetrical; apex obtuse to rounded, mucronate; base obtuse to rounded; margins entire, sparingly hairy; petiolar gland ovoid, near the base of the petiole; stipules linear-lanceolate (7–10 mm long, about 1 mm wide), striate, persistent. INFLORESCENCES: racemes terminal and upper axillary; flowers 1 to few. FLOWERS: perfect, irregular; sepals oblong to ovoid (5–8 mm long, 2.5–4 mm wide); corolla yellow, becoming cream-colored, veins brown when dry, petals 5; petals obovate to obtriangular (8–11 mm long), apices rounded or slightly emarginate, upper 4 petals with a reddish-purple spot at the base; stamens 10, dark red, upper 3 erect and sterile, middle 4 fertile and projecting forward, lower 3 longer than the others; pistil curved upward, silky; ovary appressed-pubescent; hypanthium short; pedicels 1–2.5 cm long. FRUITS: legumes linear (4–10 cm long, 8–12 mm wide); straight or slightly curved, laterally compressed, tipped by a persistent, curved style, coiling after dehiscence, septate between the seeds; seeds 10–25. SEEDS: rhomboidal to rectangular (3.5–4.5 mm long), flattened, brownish-black, with longitudinal rows of minute black dots or pits. $2n=28$.

HABITAT: Partridgepea grows in prairies and rangelands and on hillsides and stream banks. It is most frequent in sandy soils. USES AND VALUES: It produces good to fair forage for livestock and deer. Leaves contain a cathartic substance and consumption of large quantities may cause distress in cattle, but death is rare. Seeds are valuable for birds and small mammals. It is pollinated by various bees. The petiolar glands and flowers furnish nectar for honey. It is not used in landscaping. ETHNOBOTANY: The cathartic value of leaves and legumes, gathered after ripening, equals that of East Indian sennas imported for the pharmaceutical trade.

SYNONYMS: *Cassia brachiata* (Pollard) J.F. Macbr., *C. chamaecrista* L., *C. depressa* Pollard, *C. fasciculata* Michx., *C. littoralis* (Pollard) Cory, *C. robusta* (Pollard) Pollard, *C. rostrata* (Wooton & Standl.) Tidestr., *Chamaecrista bellula* Pollard, *C. brachiata* Pollard, *C. camporum* Greene, *C. chamaecrista* (L.) Britton, *C. depressa* (Pollard) Greene, *C. littoralis* Pollard, *C. mississippiensis* (Pollard) Pollard *ex* A. Heller, *C. puberula* Greene, *Grimaldia chamaecrista* (L.) Schrank *ex* Link, *Xamacrista triflora* Raf., and about 12 additional synonyms. OTHER COMMON NAMES: bundled cassia, cassia, dwarf partridgepea, dwarf senna, prairie senna, sensitive senna, showy partridgepea, wild senna.

SIMILAR SPECIES: *Chamaecrista nictitans* (L.) Moench, sensitive partridgepea, grows in the southeastern Great Plains from eastern Texas to southeastern Kansas. It has only 5 stamens. Its petals are strongly unequal and smaller than those of *Chamaecrista fasciculata*, and its pedicels are shorter (1–3 mm long).

Gleditsia J. Clayton

Gleditsia: named for Johann Gottlieb Gleditsch (1714–86), German physician and botanist.

Trees or tall shrubs, usually armed with simple or branched thorns; leaves alternate, 2 kinds, pinnately compound leaves from short spurs, bipinnately compound leaves on new growth; leaflet margins crenulate to entire, deciduous; stipules obsolete; staminate inflorescences racemose, catkinlike, pendant, appearing after leaf growth; fertile inflorescences with perfect or pistillate flowers; perianth regular, of 2 similar series, pubescent; flowers not papilionaceous, ascending, small, greenish-yellow; sepals 3–5, almost alike, inserted in the summit of the obconic hypanthium; petals 3–5, broader and slightly longer than the sepals; stamens usually 3–10, inserted at the base of the perianth; strongly pedicelled; fruits legumes, elongate to oval, compressed, large, stipitate, pulpy between the seeds, coriaceous, scarcely dehiscent; seeds few to several. x=14.

Fourteen species have been described in North and South America, Africa, and Asia. Most are in Asia. Two grow in the United States, and one is common in the Great Plains.

Honeylocust
Gleditsia triacanthos L.

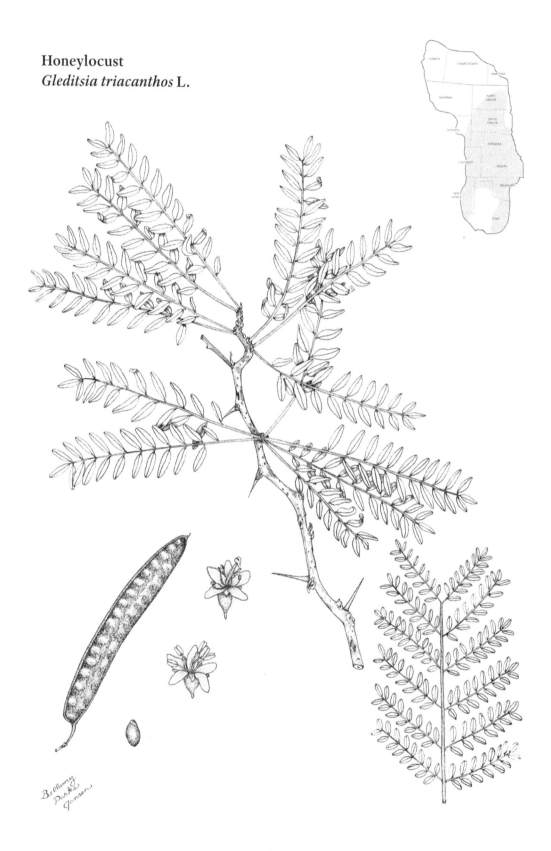

Species Plantarum 2:1056–57. 1753.

tri (Lat.): three; + *akanthikos* (Gk.): thorns; referring to the occasionally branched thorns.

GROWTH FORM: tree, flowers May to June, reproduces from seeds. LIFE-SPAN: perennial. ORIGIN: native. HEIGHT: tree to 25 m. TWIGS: glabrous, reddish-brown to grayish-brown, zigzag; thorns simple or branched (4–20 cm long), stout, terete, flattened at the base, shiny, reddish-brown; occasionally thornless. TRUNKS: usually straight (mature trees usually 60–90 cm in diameter, to 1.4 m). LEAVES: alternate, even-pinnately and bipinnately compound on the same tree; pinnate leaves 3–6 (15–30 cm long); fascicled, from old wood; leaflets 18–32; leaflets lanceolate to ovate (1–4 cm long, to 1 cm wide), larger than the bipinnate leaflets; apex obtuse, often mucronate; base acute; upper surface glabrous; lower surface pubescent on the midvein; margins obscurely crenulate or entire; sessile or with a petiolule (about 1 mm long); bipinnate leaves from new growth (2–4 cm long), 4–16 pinnae, 10–24 leaflets per pinna; leaflets usually elliptic-oblong or lanceolate to narrowly ovate (1–2.5 cm long, 4–8 mm wide); petioles 3–5 cm long, abruptly enlarged at the base, grooved above; stipules obsolete. INFLORESCENCES: racemes axillary, usually solitary, appearing after the leaves; staminate racemes catkinlike (4–10 cm long), dense, flowers many; pistillate racemes less common, shorter, with fewer flowers. FLOWERS: sweetly fragrant; staminate flowers in 3s, center flower opening first; sepals 3–5, linear (about 2 mm long); petals 3–5, obovate (2–2.2 mm long); stamens 3–10, distinct; pistillate flowers pedicellate; sepals 3–5, slightly smaller than the petals; petals 3–5, greenish-yellow; hypanthium short-campanulate. FRUITS: legumes pendant, elongate (10–45 cm long, 2.5–3 cm wide), laterally compressed, curved, twisted; pulpy between the seeds; apex acuminate or acute; base abruptly narrowed, stipitate, valves velvety pubescent when young, becoming glabrous, coriaceous; indehiscent or tardily dehiscent, pulpy between the seeds, persisting on the tree until spring; seeds 6–27. SEEDS: oval (9–12 mm long, 5–8 mm wide), brown, dull, smooth, surfaces fractured, indurate. $2n=28$.

HABITAT: Honeylocust grows in rich bottomlands, hillsides, fencerows, pastures, and planted landscapes. Its distribution has greatly increased in a northwesterly direction because of extensive planting. The thornless type, forma *inermis* Zabel, is most commonly planted as an ornamental and shade tree. USES AND VALUES: Cattle and deer browse the young plants. Deer frequently strip and eat the soft bark of young trees in winter. Squirrels, other small mammals, and many kinds of birds eat the seeds. It is visited by bees and is a good source of nectar for honey. It is planted in windbreaks and produces good firewood. It spreads rapidly in pastures and has become a pernicious weed.

SYNONYMS: *Caesalpiniodes triacanthum* (L.) Kuntze, *Gleditsia brachycarpa* (Michx.) Pursh, *G. bujotii* Neumann, *G. elegans* Salisb., *G. hebecarpa* S. McCoy, *G. heterophylla* Raf., *G. horrida* Salisb., *G. meliloba* Walter, *G. polysperma* Stokes, *G. spinosa* Marsh, *Melilobus heterophyla* Raf., and about 12 additional synonyms. OTHER COMMON NAMES: common honeylocust, honey shuck, largethorn acacia, locust thorn, sweet bean, sweet locust, thorn locust, threethorned locust.

Gymnocladus Lam.

gymnós (Gk.): naked; + *klados* (Gk.): branch, referring to the absence of small branchlets.

Trees, tall, unarmed, without small twigs; bark rough, deeply fissured; leaves opposite or irregularly arranged, large, compound; inflorescences racemose or paniculate, terminal, perfect or partly perfect, regular; sepals 3–5, nearly like the petals, inserted in a single series at the summit of the tubular hypanthium; petals 3–5, oblong greenish-white to pinkish-white; stamens 10, distinct, alternating long and short; ovary sessile; fruits legumes, asymmetrically oblong, thick, compressed, woody when mature, containing a few large seeds separated by pulp, tardily dehiscent, remaining on the tree until spring. $x=14$.

Only three species have been described. Two grow in eastern Asia. The other is native to the eastern Great Plains.

Kentucky coffeetree
Gymnocladus dioicus (L.) K. Koch

Dendrologie 1:5. 1869.
di (Gk.): two; + *oikos* (Gk.): house, in reference to some plants being dioecious.

GROWTH FORM: tree, dioecious or polygamous, flowers May to June, reproduces from seeds and basal shoots, sometimes forming colonies. LIFE-SPAN: perennial. ORIGIN: native. HEIGHT: to 30 m, with a narrow shape and rounded crown. TWIGS: freely branching; branches stout, rigid, light brown, unarmed. TRUNKS: to 90 cm in diameter. LEAVES: opposite or irregularly arranged; bipinnately compound (30–90 cm long, 30–60 cm wide); pinnae 6–14 (10–40 cm long), or a few pinnae replaced by a single large leaflet; usually 10–14 leaflets per pinna; leaflets ovate to ovate-acuminate (4–7 cm long, 2–4 cm wide); apex acute to acuminate; base cuneate to obtuse, uneven; margins entire; upper surface pubescent near the margin and on the midrib; lower surface sparsely pubescent; petioles 5–6 cm long; petiolules 2–3 mm long; stipules small, caducous. INFLORESCENCES: staminate inflorescences in racemes or panicles of racemes (7–11 cm long); pistillate inflorescences in racemes (10–30 cm long), lax, terminal; peduncles stout (2–7 cm long). FLOWERS: imperfect or partly perfect, regular, citrus fragrance; sepals 3–5, subequal; corolla greenish-white to pinkish-white; petals 3–5, obovate to oblong (4–5 mm long), white pubescence on the outer side; stamens 10, distinct, alternating long and short; hypanthium tubular-obconic (6–10 mm long). FRUITS: legumes oblong to linear (5–25 cm long, 2–6 cm wide), straight or slightly curved, slightly beaked, laterally compressed, walls thick and woody, containing pulp around the seeds, purplish-brown to reddish-brown; tardily dehiscent; seeds 1–8. SEEDS: nearly circular (1.2–2.1 cm in diameter), compressed, dark olive-brown, smooth, indurate. $2n=28$.

HABITAT: Kentucky coffeetree is scattered to locally common on rich alluvial soils of bottomlands. It occasionally grows on rocky hillsides and was commonly planted on farmsteads. USES AND VALUES: Kentucky coffeetree has essentially no value for forage. The leaves, seeds, and pulp of the legume contain a quinolizidine alkaloid which has caused poisoning of cattle, horses, and sheep. Sprouts eaten in the spring and legumes and seeds eaten in the autumn or winter are toxic. Death is infrequent but can come within one day of ingestion. Sprouts growing on mowed areas may particularly be a problem. Human poisoning has occurred following ingestion of the pulp between the seeds. It produces durable, valuable wood. It has been used for erosion control in gullies, but there are other plants better for this purpose. ETHNOBOTANY: Plains Indians used this tree for a variety of purposes. Members of several tribes used a mixture of the pulverized bark and water for enemas. Seeds were roasted, ground, and used to make a coffeelike drink. Snuffed pulverized bark caused sneezing, which was thought to relieve the pain of headaches. Lakota used the root to make a poor-quality dye. Some Winnebago used the seeds as counters in gambling.

SYNONYMS: *Guilandina dioica* L., *Gymnocladus canadensis* Lam. OTHER COMMON NAMES: American coffeetree, American coffeebean, Canadian bonduc, chicot, coffeetree, Kentucky mahogany, nantita (Omaha-Ponca), nicker tree, stump tree, tohuts (Pawnee), wah'nah'na (Dakota).

Hoffmannseggia Cav.

Hoffmannseggia: named for Johann Centurius, Count of Hoffmannsegg (1766–1849), German nobleman and botanist.

Perennial, unarmed herbaceous or suffrutescent plants, from spreading roots or a woody caudex surmounting a taproot; subscapose to caulescent; leaves bipinnately compound, with 3 to several pinnae; leaflets mostly small, smooth, not glandular-punctate; stipules ovate to ovate-deltate, persistent; inflorescences racemes, terminal, ascending, open; flowers perfect, nearly regular; calyx tube short, lobes 5; petals 5, corolla yellow to orangish-red, conspicuous; stamens 10, distinct; fruits legumes, oblong, asymmetric, straight or falcate, often glandular, tardily dehiscent or indehiscent; seeds few to several. $x=12$.

A genus of about 25 species occurring mostly in dry areas of subtropical North America, South America, and southern Africa. Several have extended into the southern United States from Mexico. Two grow in the Great Plains, and one is common in the southwest portion of the region.

Indian rushpea
Hoffmannseggia glauca (Ortega) Eifert

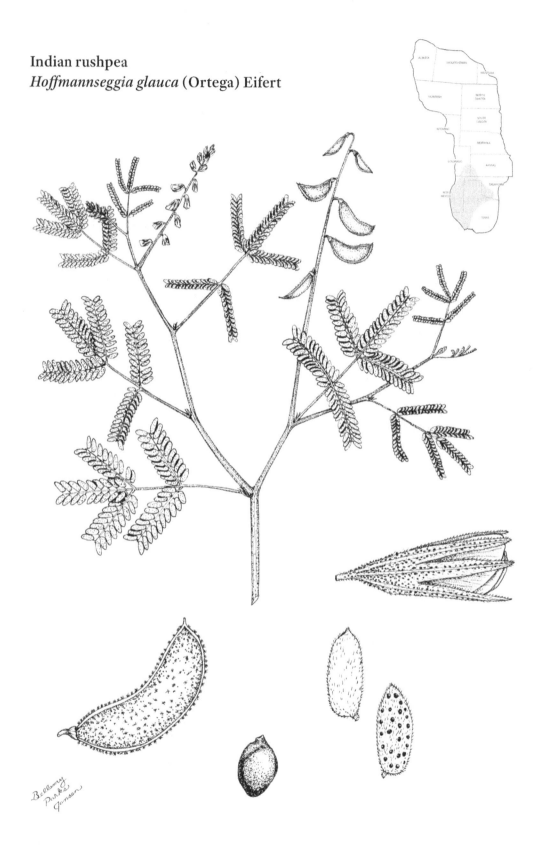

Sida 5(1):43. 1972.

glauca (Gk.): silvery, in reference to the pubescence.

GROWTH FORM: forb (suffrutescent), flowers May to September, reproduces from tubers and seeds. LIFE-SPAN: perennial. ORIGIN: native. HEIGHT: 5–35 cm. STEMS: bases woody; terminal portion herbaceous, erect to spreading, low-growing, weak-stemmed, simple or branched at the base from a caudex and laterally spreading roots with fleshy tubers; stipitate-glandular, pubescent to glabrate; tubers spheroid (up to 1 cm in diameter), brownish-black. LEAVES: mostly basal or partly cauline, alternate, bipinnately compound (6–13 cm long); pinnae 3–11; leaflets 8–24 per pinna; leaflets oblong to elliptic (2–8 mm long, 2–3 mm wide), sessile; apex rounded to obtuse or acute; base rounded; margins entire; lower surface sparsely pilose, glandular; upper surface minutely pubescent to glabrate; petioles 2–7 cm long, equaling or exceeding the rachis; stipules ovate-deltoid, ciliate, persistent. INFLORESCENCES: racemes (10–20 cm long) usually elevated above the leaves, terminal, stipitate-glandular, pubescent; peduncles 1–19 cm long; flowers 4–17. FLOWERS: perfect, nearly regular (small gap between the 2 lowermost petals); sepals glandular, pubescent; calyx lobes 5 (5.5–8 mm long); corolla yellowish-orange (1–1.3 cm long), fading to reddish, often with red spots near the base of the banner, claws glandular, connate basally; stamens 10, not exceeding the petals, filaments red; pedicels glandular (2–5 mm long), pubescent. FRUITS: legumes oblong to lunate (2–4.5 cm long, 5–8 mm wide), straight or slightly falcate, laterally compressed, surfaces shiny, stipe absent, indehiscent, persistent; seeds 2–8. SEEDS: ovate (3.5–4.5 mm in diameter), compressed, brown to black, smooth, indurate. $2n=24$.

HABITAT: Indian rushpea is scattered to locally common in sandy or rocky prairies, rangelands, abandoned fields, and roadsides. It is most common in alkaline soils. It can form large patches or colonies on disturbed sites. It is often categorized as a weed. USES AND VALUES: It is palatable to livestock but not extensively grazed by cattle or horses. It is eaten by sheep and goats. Pigs root up and eat the tubers. The tubers are also eaten by many species of small mammals. It will withstand limited soil salinity. It is a valuable soil stabilizer, but it can become a serious weed. Indian rushpea is an attractive, drought-tolerant plant that is used in xeriscaping in the southwestern part of the region. ETHNOBOTANY: Its tuberous roots were eaten raw or roasted and eaten by members of many tribes of southwestern Great Plains Indians.

SYNONYMS: *Caesalpinia chicamana* Killip & J.F. Macbr., *C. glauca* (Ortega) Kuntze, *Hoffmannseggia densiflora* Benth., *H. falcaria* Cav., *H. stricta* Benth., *Larrea densiflora* (Benth.) Britton, *L. glauca* Ortega. OTHER COMMON NAMES: hog potato, mouse sweetpotato, pignut, rat sweetpotato, shoestring weed, sicklefruit hoffmannseggia.

SIMILAR SPECIES: *Hoffmannseggia drepanocarpa* A. Gray, sicklepod rushpea, is similar, except that its stems and inflorescences are not stipitate-glandular. It also grows in the southwestern Great Plains, and its range overlaps with that of *Hoffmannseggia glauca*.

Senna Mill.

senna (Ara.): brightness, from an Arabic name for this genus.

Perennial herbs (occasionally shrubs or trees in other regions), rarely annual herbs; leaves compound, leaflets 2 to many; petiole variously glandular; stipules small or obsolescent, not striate; inflorescences compound racemes, axillary or terminal; sepals small, connate basally; corollas symmetrically zygomorphic to irregular, generally yellow; stamens 10, functional stamens usually 7; fruits legumes, linear, terete or 4-angled, elongate, tardily dehiscent; seeds few to many. x=11, 12, 13, 14.

A genus of about 250 species predominantly in the New World. Some occur in Africa, Asia, and Pacific islands. Five grow in the Great Plains, and only one is common.

Wild senna
Senna marilandica (L.) Link

Handbuch zur Erkennung der nutzbarsten und am häufigsten vorkommenden Gewächse 2:140. 1829.

marilandica: of or from Maryland.

GROWTH FORM: forb, flowers July to August, reproduces from seeds. LIFE-SPAN: perennial. ORIGIN: native. HEIGHT: 0.5–2 m. STEMS: herbaceous, erect to ascending, glabrous or sparsely pubescent, from a woody caudex surmounting a taproot. LEAVES: alternate, even-pinnately compound (10–30 cm long), leaflets 8–24; leaflets elliptic to oblong (2–8 cm long, 5–30 mm wide); apex acute, mucronate; base slightly oblique to rounded; margins entire, ciliate; surfaces glabrous or lower surface glaucous; petioles 3–6 cm long; gland ovoid, near base of the petiole; stipules linear-lanceolate (7–10 mm long, about 1 mm wide), not striate, caducous. INFLORESCENCES: racemes terminal and upper axillary; flowers 6–9; peduncles 1–1.5 cm long. FLOWERS: perfect, irregular; sepals 5, oblong to ovoid (5–8 mm long, 3–4 mm wide), connate basally; petals 5; petals obovate to obtriangular (8–15 mm long), somewhat unequal; apex rounded or slightly emarginate; corolla yellow, becoming cream-colored with brown veins when dry; stamens 10, dark red; upper 3 erect, sterile; middle 4 projecting forward, fertile; lower 3 longer than the others, fertile; pistil silky; ovary appressed-pubescent; hypanthium short; pedicels 7–16 mm long. FRUITS: legumes linear (4–11 cm long, 8–12 mm wide), straight or slightly curved, tipped by a persistent style, terete to slightly compressed, glabrous (rarely sparsely pubescent), dark brown to black, separated between the seeds, tardily dehiscent; seeds usually 10–25. SEEDS: oblong or obovate (4.5–5.5 mm long, 2–2.5 mm wide), slightly compressed; central portion dull, gray; outer portion lustrous, darker gray. $2n=28$.

HABITAT: Wild senna is infrequent to common in prairies, roadsides, ravines, hillsides, and streambanks. It is most common on moist, sandy soils. USES AND VALUES: Cattle, horses, and sheep occasionally eat wild senna, but they usually avoid this cathartic plant. Ingesting large quantities may be toxic. Seeds are valuable food for many kinds of birds and small mammals. It is pollinated by bumblebees and various other bees. The petiolar gland provides nectar to ants. ETHNOBOTANY: Seeds have been used as coffee adulterants, but they can also be toxic to humans. Some Great Plains Indians prepared an infusion of roots for heart disorders and fevers. A poultice of roots was used to treat sores on humans and horses.

SYNONYMS: *Cassia acuminata* Moench, *C. marilandica* L., *C. medsgeri* Shafer, *C. reflexa* Salisb., *Ditremexa marilandica* (L.) Britton & Rose, *D. medsgeri* (Shafer) Britton & Rose, *D. nashii* Britton & Rose, *Senna riparia* Raf. OTHER COMMON NAMES: American senna, cassia, false acacia, Maryland senna, ûnnagéi (Cherokee).

SIMILAR SPECIES: *Senna obtusifolia* (L.) H.S. Irwin & Barneby, sicklepod, has a single gland between the two lowest leaf pairs; *Senna occidentalis* (L.) Link, septicweed, flowers have a conspicuous margin; *Senna pumilio* (A. Gray) H.S. Irwin & Barneby, dwarf senna, is diminutive (less than 10 cm tall); and *Senna roemeriana* (Scheele) H.S. Irwin & Barneby, twoleaved senna, leaves have only 2 leaflets. All four grow in the southern Great Plains.

II. *FABACEAE* LINDL.

Trees, shrubs, or herbs (annual or perennial); stems usually unarmed; leaves alternate or clustered from spurs, pinnately compound (even or odd) or palmately compound, rarely simple, not bipinnately compound, some with tendrils; commonly with stipules and stipels; inflorescences usually racemes, occasionally clusters, axillary or terminal; flowers typically perfect and zygomorphic; calyx gamosepalous; lobes typically 5 (rarely 1); lobes usually unequal; corollas variously colored; petals usually 5 (sometimes 1), imbricate, usually papilionaceous; banner uppermost and outermost; wings lateral; keel petals lowermost, inside; stamens usually 10, commonly diadelphous or monadelphous (sometimes distinct); ovary sessile or stipitate; fruit usually a legume; legumes straight or curved, may be winged, dehiscent or seldom indehiscent; fruit sometimes a loment, separating into 1-seeded joints at maturity.

This family contains about 400 genera and 10,000 species worldwide. It is most common in the tropics, but is well represented in temperate regions. Not including common garden plants, the family is represented in the Great Plains by 33 genera and 192 species.

Key to the Genera

1. Leaves simple, plants herbaceous
    2. Annuals; corollas yellow; stamens monadelphous. .... *Crotalaria*
    2. Perennials; corollas purple to white; stamens diadelphous. ....
    .... (in part) *Astragalus*
1. Leaves compound; plants herbaceous or woody
    3. Plants woody
        4. Vines, extensive, becoming woody, trailing or climbing; leaflets large; may not flower in the northern portion of its range. .... *Pueraria*
        4. Trees or shrubs
            5. Flowers not papilionaceous, petals 1; corolla purple to blue; leaflets glandular-punctate. .... *Amorpha*
            5. Flowers papilionaceous
                6. Leaves even-pinnately compound; corollas yellow. .... *Caragana*
                6. Leaves odd-pinnately compound; corollas white or rose to purple, sometimes marked with yellow
                    7. Shrubs; corollas rose to purple with some yellow; legumes less than 4 mm long. .... (in part) *Dalea*
                    7. Trees; flowers white; legumes 5 mm long or more. .... *Robinia*
    3. Herbaceous annuals and perennials
        8. Leaves with tendrils; even-pinnately compound

9. Stipules as large or larger than the leaflets. .... *Pisum*
9. Stipules smaller than the leaflets, often hastate to obliquely semisagittately lobed
   10. Wing petals coherent with the keel petals; style filiform, bearded with a tuft or ring of hairs at the apex. .... *Vicia*
   10. Wing petals free or nearly so; style compressed, bearded down the inner face. .... *Lathyrus*
8. Leaves without tendrils; odd-pinnately compound, palmate, or trifoliate
  11. Leaflets more than 3 on well-developed leaves
    12. Inflorescence an umbel or capitate cluster
      13. Inflorescence many-flowered (8 or more); corollas pink to white; fruit a loment. .... *Securigera*
      13. Inflorescence few-flowered (8 or fewer); corolla orangish-red to white; fruit a legume. .... (in part) *Lotus*
    12. Inflorescence a terminal or axillary raceme or spike
      14. Leaves palmately compound
        15. Leaflets 5–11; foliage not glandular-punctate; seeds 2 to many .. *Lupinus*
        15. Leaflets 3–5; foliage glandular-punctate; seeds 1
          16. Calyx campanulate, not enlarging in fruit; fruit indehiscent. .... *Psoralidium*
          16. Calyx becoming gibbous in flower, enlarging in fruit; fruit usually dehiscent. .... (in part) *Pediomelum*
      14. Leaves pinnately compound
        17. Leaves glandular-punctate
          18. Flowers in terminal spikes; stamens monadelphous. .... (in part) *Dalea*
          18. Flowers in axillary racemes or spikes; stamens diadelphous; legumes with hooked prickles; seeds 3 or more. .... *Glycyrrhiza*
        17. Leaves not glandular-punctate
          19. Plants twining; calyx lobes unequal, the lowest lobe surpassing the others. .... *Apios*
          19. Plants not twining; calyx lobes nearly equal in length
            20. Inflorescences usually terminal on leafy stems
              21. Calyx lobes shorter than the tube; stamens distinct; legumes constricted between the seeds, turgid. .... *Sophora*
              21. Calyx lobes equal to or longer than the tube; upper stamen free; legumes linear, laterally compressed. .... *Tephrosia*
            20. Inflorescences axillary or borne on scapes
              22. Legumes short (8 mm long or less), compressed, short, blunt spines on the margin. .... *Onobrychis*
              22. Legumes ovoid to subspherical or subglobose to linear, not spiny

23. Keel obtuse to acute. .... (in part) *Astragalus*

23. Keel abruptly narrowed to a distal beaklike appendage.
.... *Oxytropis*

11. Leaves trifoliate

  24. Leaves palmately compound

    25. Hairs dolabriform; plants prostrate, mat-forming. .... *Orophaca*

    25. Hairs basifixed or glabrous and glaucous; plants ascending to erect,
not mat-forming. .... (in part) *Baptisia*

  24. Leaves pinnately compound or appearing so

    26. Leaf margins serrulate (apparently entire in some *Kummerowia* and
Trifolium)

      27. Terminal leaflet sessile or nearly so. .... (in part) *Trifolium*

      27. Terminal leaflet with a petiolule (distinctly stalked)

        28. Inflorescences of globose heads, short spikes, or short racemes

          29. Flowers chasmogamous and cleistogamous intermixed
or in separate inflorescences. .... *Kummerowia*

          29. Flowers perfect, papilionaceous

            30. Corollas purple to violet, blue, or yellow; caducous
after withering; inflorescences axillary racemes;
legumes reniform, falcate, or coiled. ...... *Medicago*

            30. Corollas yellow, persistent after withering; inflores-
cences capitate, or short racemes; legumes short,
straight. .... (in part) *Trifolium*

        28. Inflorescences of elongate racemes; flowers yellow or white;
legumes obovate to rotund; seeds usually 1. ...... *Melilotus*

    26. Leaf margins entire

      31. Corollas usually yellow (sometimes white to light purple); flowers
in open, peduncled racemes

        32. Foliage blackening on drying; legume inflated. ....
.... (in part) *Baptisia*

        32. Foliage not blackening on drying; legume linear, laterally
compressed. .... *Thermopsis*

      31. Corollas white, cream, blue, purple, pink, or orangish-yellow in
globose heads, spikes, racemes, panicles, or solitary

        33. Corollas cream, usually with a purple spot on the banner. ..
.... (in part) *Lespedeza*

        33. Corollas white, blue, purple, pink, or orangish-yellow

          34. Plants with stems normally twining

            35. Inflorescences racemes; chasmogamous and cleistoga-
mous flowers in separate locations; keel petals straight.
.... *Amphicarpaea*

            35. Inflorescences subcapitate racemes; flowers all chas-
mogamous; keel petals strongly curved. *Strophostyles*

          34. Plants with stems erect to procumbent, not twining

36. Flowers in terminal racemes or spikes
    37. Flowers in dense cylindrical spikes. .... (in part) *Dalea*
    37. Flowers in racemes or panicles of racemes
        38. Fruit a loment; terminal leaflet with a petiolule (distinctly stalked). .... ***Desmodium***
        38. Fruit a legume; terminal leaflet sessile or with a petiolule no longer than the petiolules of the lateral leaflets. .... (in part) ***Baptisia***
36. Flowers axillary; solitary or clustered in umbels or racemes
    39. Flowers solitary (rarely in 2s)
        40. Corollas pink or white; legume glabrous. .... (in part) ***Lotus***
        40. Corollas orangish-yellow to white; loment segments 1 or 2. .... ***Stylosanthes***
    39. Flowers several to many in a cluster or raceme
        41. Foliage glandular-punctate. .... (in part) ***Pediomelum***
        41. Foliage not glandular-punctate. .... (in part) ***Lespedeza***

Amorpha L.

amorphous (Gk.): no form or deformed, in reference to the flowers having only one petal.

Erect shrubs or suffrutescent shrubs, sometimes with rhizomes; leaves alternate, odd-pinnately compound; leaflets numerous; apex commonly mucronate; margins entire (occasionally crenate); surfaces often glandular-punctate; inflorescences racemes, usually spikelike, terminal and from upper axils; flowers perfect, irregular, not papilionaceous, pedicellate, small; calyx obconic to turbinate, usually persistent, lobes 5; lobes acuminate to obtuse, equal or upper lobes shortest; corolla purple to blue (rarely white); petals 1 (banner), obovate, clawed, wrapped around stamens and style; stamens 10, diadelphous; ovary short; fruits legumes, usually longer than the calyx, often curved, glandular-punctate; seeds 1 or 2. $x=10$.

About 20 species in North America and most numerous in the southern United States. Only 3 species are commonly found in the Great Plains.

1. Tall shrubs, usually 1.5–4 m tall; petioles 2–5 cm long; leaflets 2–4.5 cm long. .... *Amorpha fruticosa*
1. Low, suffrutescent shrubs, rarely over 1 m tall; petioles mostly less than 1 cm long (sometimes longer); leaflets less than 2 cm long.
    2. Plants gray-canescent to tomentose throughout; calyx tube pubescent to canescent; racemes usually several in a cluster. .... *Amorpha canescens*
    2. Plants green, mostly glabrous, foliage green; calyx tube glabrate; racemes usually solitary. .... *Amorpha nana*

Leadplant
Amorpha canescens Pursh

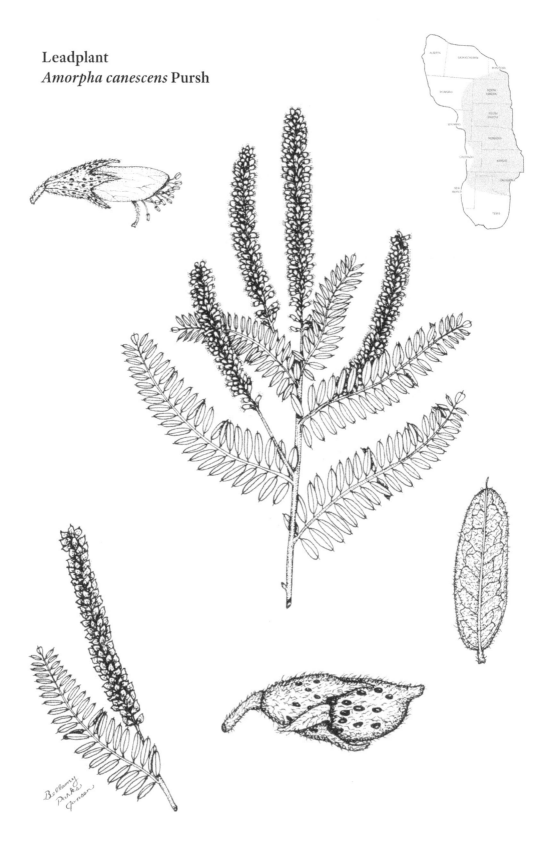

Flora Americae Septentrionalis 2:467. 1814.

canescens (Lat.): gray or pale, in reference to the dense covering of fine hairs.

GROWTH FORM: shrub, flowers June through August, reproduces from rhizomes and seeds. LIFE-SPAN: perennial. ORIGIN: native. HEIGHT: 0.3–1 m (may be taller in nongrazed and protected areas). STEMS: suffrutescent, erect or ascending, 1 to several, branching; twigs usually tomentose, often becoming glabrate with age. LEAVES: alternate, odd-pinnately compound (3.5–10 cm long); leaflets 11–51, crowded to imbricate, elliptic to oblong (7–18 mm long, 3–6 mm wide), sparingly glandular-punctate; apex acute to obtuse, usually mucronate; base rounded; margins entire; upper surface woolly, darker green; lower surface gray-canescent, lighter green; petiole 0.5–1 mm long (sometimes longer), tomentose; stipules subulate (1–3 mm long), caducous. INFLORESCENCES: racemes spikelike (5–15 cm long), terminal and axillary, 1 to several, often clustered in the upper leaf axils; central raceme longest and first to flower; flowers many, crowded, rachis densely villous. FLOWERS: perfect, irregular, not papilionaceous; calyx tube turbinate (3–4.5 mm long), lobes 5 (1–1.6 mm long), pubescent to canescent, glandular-punctate, resinous; corolla bright purple or rose-purple (occasionally light blue to violet blue); petals 1 (4–5 mm long, about 3 mm wide), broadly ovate, enclosing the pistil and stamens, claw slender; stamens 10, diadelphous, exserted; anthers yellow to yellowish-orange, conspicuous; ovary densely pilose. FRUITS: legumes slightly curved (3.5–4.5 mm long, 1.6–2 mm wide), slightly beaked, densely tomentose to canescent, glandular-punctate, resinous, calyx persistent; seeds 1. SEEDS: elliptical (2–2.8 mm long), beaked slightly, orangish-brown, smooth, indurate. $2n=20$.

HABITAT: Leadplant is infrequent to abundant on well-drained soils of prairies, rangelands, and open woodlands. On burned or mowed prairies, it has the appearance of a large herbaceous plant, rather than a shrub. USES AND VALUES: Leadplant is a desirable species that provides excellent forage for livestock and wildlife. It is highly palatable, and forage quality is high. Animals tend to select it before eating herbage of most other species. Leadplant decreases with heavy grazing and is rarely abundant on improperly grazed rangelands. It produces excellent forage for deer, pronghorn, elk, and bighorn sheep. Its seeds are important food for ground-foraging birds and small mammals. Leadplant attracts butterflies and other beneficial insects. It can be easily transplanted and is becoming increasingly popular as a landscape plant. It is an important component of many prairie restorations. ETHNOBOTANY: Some Plains Indians dried the leaves for tea and to be mixed with bison fat and smoked. It was used to treat neuralgia and rheumatism by cutting stems into small pieces, attaching one end of each piece to the skin over the affected area by first moistening it, then lighting it, and allowing the fire to burn down to the skin as a counterirritant.

SYNONYM: *Amorpha brachycarpa* E.J. Palmer. OTHER COMMON NAMES: buffalo bellows, downy amorpha, downy indigobush, leadplant amorpha, prairie shoestring, te huto hi (Omaha-Ponca), zinŋkátȟačhaŋ (Lakota). It was called buffalo bellows because it flowered during the bison breeding season.

False indigo
Amorpha fruticosa L.

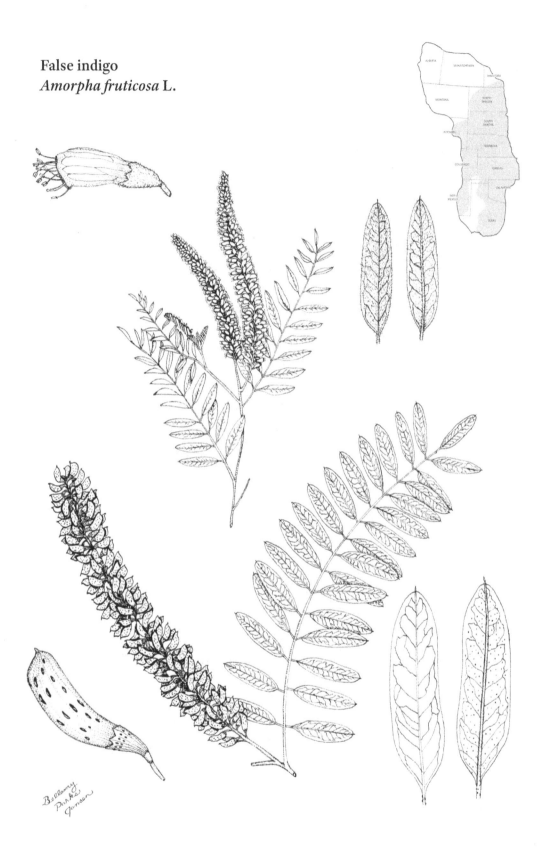

Species Plantarum 2:713. 1753.
fruticosus (Lat.): shrubby, for its woody twigs and general appearance.

GROWTH FORM: shrub, flowers May and June (as early as April in the southern portion of the region), reproduces from rhizomes and seeds. LIFE-SPAN: perennial. ORIGIN: native. HEIGHT: usually 1.5–4 m (rarely to 5 m). STEMS: erect to ascending with a single stem or a few clustered stems, usually from rhizomes, sometimes forming dense stands, freely branching near the top; glabrous; twigs finely ribbed; bark of juvenile trunks smooth, brownish-gray; lenticels transverse, prominent; bark of mature trunks gray, slightly fissured. LEAVES: alternate, odd-pinnately compound (12–22 cm long); leaflets 9–31, elliptic to oblong or lanceolate (2–4.5 cm long, 8–18 mm wide); apex acute or obtuse, generally mucronate; base acute or obtuse; margins entire; upper surface glabrous to appressed-pubescent (when mature), dull grayish-green to dark green; lower surface glabrous, without glands or usually sparingly glandular-punctate; petioles 2–5 cm long, pubescent; stipules linear to subulate (2–4 mm long), caducous. INFLORESCENCES: racemes (2–15 cm long) terminal, 1–5 (usually 2 or 3), densely flowered. FLOWERS: perfect, irregular, not papilionaceous, sweetly fragrant; calyx tube obconic (2–4 mm long), glabrate; lobes 5, broadly rounded to acute (0.5–1.2 mm long), upper lobes shortest, pubescent, sometimes glandular-punctate; corolla violet-purple (rarely white or blue), petals 1 (5–6 mm long), broadly ovate, folded to enclose the stamens and pistil, claw indistinct; stamens 10 (6–8 mm long), exserted, diadelphous; anthers yellow to yellowish-orange, conspicuous. FRUITS: legumes straight or curved (6–8 cm long, 1.5–3 mm wide), brown, glandular-punctate; seeds usually 1 or 2. SEEDS: oblong to oval or cordiform (3–4.5 mm long, 1.2–1.5 mm wide), beaked slightly, tan to brown or olive, smooth, glossy, indurate. $2n=20, 40$.

HABITAT: False indigo grows on moist stream banks, seeps, open woodlands, prairie gullies, and along lake and pond shores. USES AND VALUES: It is eaten by livestock, though it is rarely present in large enough quantities to be important in their diets. Browsing wildlife utilize the foliage, and small mammals and birds eat the seeds. It is sometimes used for commercial and residential landscaping, roadside plantings, and wildlife plantings. Many species of bees, other pollinators, and butterflies visit the flowers. ETHNOBOTANY: Pawnees gathered the leafy branches and placed them on the ground near the place where animals were butchered. Fresh meat was placed on the branches to keep the meat clean. Lakotas cut false indigo to feed to horses and made arrow shafts from the stems.

SYNONYMS: *Amorpha angustifolia* (Pursh) F.E. Boynton, *A. arizonica* Rydb., *A. bushii* Rydb., *A. croceolanata* P. Watson, *A. curtissii* Rydb., *A. dewinkeleri* Small, *A. emarginata* Sweet, *A. humilis* Tausch, *A. occidentalis* Abrams, *A. pendula* Carrière, *A. tennesseensis* Shuttlew. *ex* Kuntze, *A. virgata* Small. OTHER COMMON NAMES: desert falseindigo, false indigobush, fragrant amorpha, indigobush amorpha, kitsuhast (Pawnee), streambank amorpha, śunktáwote (Lakota), te-hunton-hi (Omaha-Ponca), zinjkala tačaŋ (Lakota).

Dwarf wildindigo
Amorpha nana Nutt.

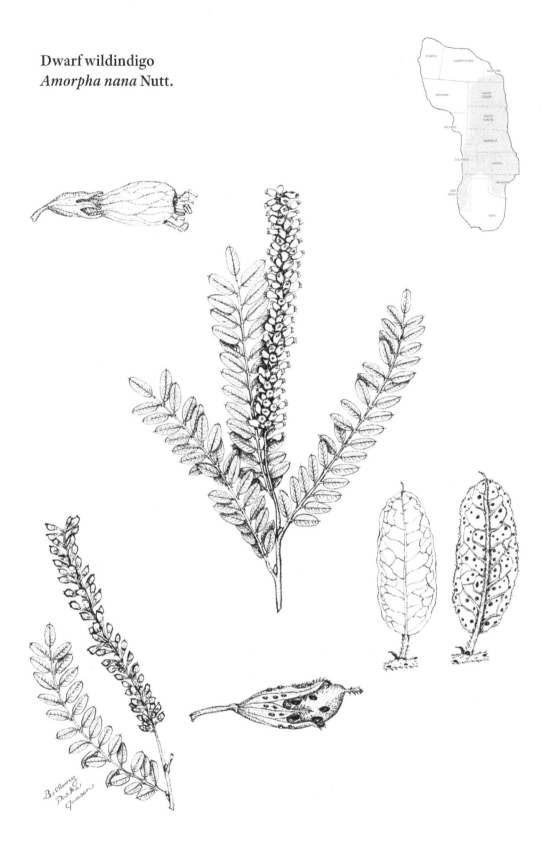

Catalogue of New and Interesting Plants Collected in Upper Louisiana No. 5. 1813.
nanos (Gk.): dwarf or small, describing the stature of the plant.

GROWTH FORM: shrub, flowers May to July, reproduces from rhizomes and seeds. LIFE-SPAN: perennial. ORIGIN: native. HEIGHT: 30–60 cm, rarely to 1 m. STEMS: suffrutescent, ascending to erect, usually single from rhizomes, with numerous distal branches, mostly glabrous, sometimes strigulose to glabrescent, hairs appressed; twig bark brown when immature, becoming gray with maturity; splitting longitudinally, slightly rough. LEAVES: alternate, odd-pinnately compound (3–10 cm long); leaflets 9–41, elliptic to broadly oblong (3–15 mm long, 2–6 mm wide); apex obtuse to truncate or acute, generally mucronate; base rounded to cuneate; margins shallowly crenate and often ciliate; upper surface mostly glabrous, sometimes puberulent or glabrate, dark yellowish-green; lower surface lighter, glandular-punctate, sometimes sparingly pubescent; petiole pubescent (3–10 mm long); stipules linear to subulate (3–5 mm long), caducous; reddish-brown gland on the side of petiole opposite of the stipule. INFLORESCENCES: racemes (3–9 cm long, 1–2 cm wide) terminal, usually solitary, densely flowered. FLOWERS: perfect, irregular, not papilionaceous, sweetly fragrant; calyx tube obconic (1.5–3 mm long), glabrate, lobes 5; lobes acuminate (0.9–1.4 mm long), upper lobes shortest, ciliate; petals reddish-purple with darker purple veins; petals 1, obcordate (3.5–5 mm long, 2.5–4.5 mm wide); apex obtuse and usually emarginate, enclosing the stamens and pistil, claw slender; stamens 10, exserted, diadelphous, bright red, filaments united below; ovary glabrous. FRUITS: legumes straight or curved (4–6 mm long, 2–5 mm wide), slightly beaked, initially green changing to reddish-brown, glabrous except pubescent at the apex, glandular-punctate, calyx persistent; seeds usually 1. SEEDS: narrowly oval (about 2.5 mm long, 1.5 mm wide), flattened, smooth, olive-brown, indurate. $2n=20$.

HABITAT: Dwarf wildindigo is scattered but may be locally common on dry prairies and rocky or sandy hillsides. It grows best in well-drained soils but does not grow well in extremely sandy soils. It is often overlooked on prairies because it is frequently browsed down to less than 5 cm in height. USES AND VALUES: It provides excellent browse for livestock and wildlife. This desirable species has high forage quality and is highly palatable. Animals frequently select it before other species. It decreases on heavily grazed rangelands. It is considered to be excellent forage for elk, deer, pronghorn, and bighorn sheep. The legumes are collected and eaten by small mammals, and the seeds are eaten by small mammals and ground-foraging birds. The bright red pollen and honeylike fragrance attracts butterflies, bees, and other insects. It is not used for erosion control. This drought-tolerant shrub can be used in border plantings or as a specimen plant in rock gardens. Transplanting and propagation from seeds and softwood cuttings have been successful. ETHNOBOTANY: Some Plains Indians dried the leaves for smoking and for tea.

SYNONYMS: *Amorpha microphylla* Pursh, *A. punctata* Raf. OTHER COMMON NAMES: dwarf falseindigo, dwarf indigo, dwarf leadplant, smooth leadplant.

Amphicarpaea Elliott *ex* Nutt.

amphi (Gk.): of both kinds; + *karpos* (Gk.): fruit, in reference to two kinds of fruit.

Low perennial herbs, twining; leaves pinnately trifoliate; leaflets well-petioled, sub-tended by stipules; leaflets elliptic to ovate or rhombic, margins entire; flowers of 2 kinds, those of the upper branches chasmogamous in axillary racemes, papilionaceous, calyx slightly irregular, corollas purple to white (sometimes yellow), keel petals straight; each pedicel subtended by a striate-veined bractlet, stamens diadelphous, style elongate; fruits legumes, oblong, straight, or oblique; laterally compressed, valves papery, coiled after dehiscence; flowers cleistogamous at the base of the plant, borne on threadlike branches, calyx reduced, corollas rudimentary or wanting, stamens few; legumes obovoid to ellipsoid, may be subterranean. x=10.

A genus with only 7 species. Six species are found in East Asia, and 1 grows in the eastern Great Plains.

Hog peanut
Amphicarpaea bracteata (L.) Fernald

Rhodora 35 (416):276. 1933.

bracta (Lat.): thin, referring to the scalelike leaves subtending the flowers.

GROWTH FORM: forb, flowers August to September, reproduces from seeds. LIFE-SPAN: perennial. ORIGIN: native. LENGTH: 0.3–2 m. STEMS: herbaceous, twining vine, densely hairy, from taproots. LEAVES: alternate, pinnately trifoliate; leaflets ovate to broadly ovate or rhombic (2–10 cm long, 1.5–7 cm wide); apex acute or acuminate; base of center leaflet obtuse, bases of lateral leaflets oblique; margins entire; both surfaces pubescent, veins prominent; petioles 2–7 cm long; terminal petiolules 5–40 mm long; lateral petiolules 1–2.5 mm long; stipules membranous (3–8 mm long), persistent, pubescent. INFLORESCENCES: two types; chasmogamous inflorescences racemes, axillary; flowers 5–20, individual or pairs of flowers subtended by an obtriangular bract; pedicels 1–6 cm long; cleistogamous flowers borne on filiform runners from the lower stem nodes, may be subterranean. FLOWERS: chasmogamous flowers perfect, papilionaceous; calyx tube gibbous (5–7 mm long), lobes 4 (2 upper sepals normally forming 1 lobe); lobes lanceolate to deltoid (0.5–2 mm long); corollas purple to white or rarely red; banner 1.2–1.4 cm long, margins recurved; wing and keel petals nearly straight; stamens diadelphous; cleistogamous flowers reduced, lacking well-developed petals. FRUITS: aerial legumes elliptic (1.5–4 cm long, to 7 mm wide), laterally compressed; valves pubescent, reticulate, glabrous to glabrate, coiled after dehiscence; seeds usually 3 or 4; subterranean legumes fleshy (6–12 mm in diameter); seeds 1. SEEDS: round to reniform (3–6 mm long), somewhat compressed, reddish-brown, mottled with black. $2n=20, 40$.

HABITAT: Hog peanut grows in brushy ravines, open woodlands, and thickets. USES AND VALUES: The herbage is eaten by cattle, and the underground legumes are rooted up and consumed by hogs. Seeds and subterranean legumes are eaten by deer, small mammals, song birds, and upland gamebirds. ETHNOBOTANY: The subterranean legumes were an important source of food for some Plains Indians. They searched for rodent caches of the legumes and dug them out. Up to 2 liters of legumes could be obtained from one woodrat cache. They often refilled the cache with an equal amount of corn.

VARIETIES: Var. *comosa* (L.) Fernald is most common in the northern Great Plains. Var. *bracteata* is a smaller and more slender plant growing as far north as southeastern Kansas. SYNONYMS: *Amphicarpaea chamaecaulis* B. Boivin & Raymond, *A. comosa* (L.) G. Don *ex* Loudon, *A. elliottii* Raf., *A. heterophyla* Raf., *A. monoica* Nutt., *A. pitcheri* Torr. & A. Gray, *A. sarmentosa* Elliott *ex* Nutt., *Falcata comosa* (L.) Kuntze, *F. pitcheri* (Torr. & A. Gray) Kuntze, *Glycine comosa* L., *Lobomon sarmentosum* (Elliott *ex* Nutt.) Raf., and about 15 additional synonyms. OTHER COMMON NAMES: American hogpeanut, hoŋbŏineu (Osage), makatominiča (Lakota), Pitcher hogpeanut, southern hogpeanut.

SIMILAR SPECIES: *Arachis hypogaea* L., common peanut, a crop in the southern Great Plains, also has subterranean fruits, but it has 4 leaflets. *Glycine max* (L.) Merr., soybean, an important crop in the Great Plains, has trifoliate leaves and is an erect annual.

Apios Fabr.

apios (Gk.): a pear, in reference to the occasional shape of the tubers.

Twining, perennial herbs from rhizomes with fleshy tubers; leaves alternate, odd-pinnately compound, well-petioled; leaflets 5–9 on petiolules (3–5 mm long), not glandular; stipels minute; stipules setaceous, caducous; inflorescences racemes, axillary; each flower subtended by a pair of linear bractlets; bractlets caducous, veins 1; calyx campanulate, lobes 5; lobes unequal, 4 shorter, lowest slightly longer; corollas papilionaceous; banner round or obovate, claws short; wings oblong or obovate, deflexed below the keel; keel longer than the banner and wings, strongly curved, tips blunt; stamens 10, diadelphous; style curved with the keel; fruits legumes, linear, straight or curved, laterally compressed, seeds several, splitting into 2 valves; valves coiled after dehiscence. x=11.

A genus consisting of about 10 species, 6 of which occur in eastern Asia. Two species grow in North America, and 1 is found in the Great Plains.

Groundnut
Apios americana Medik.

Vorlesungen der Churpfälzischen physikalisch-ökonomischen Gesellschaft 2:355. 1787.
americana: of or from America.

GROWTH FORM: twining vine, flowers August to September, reproduces from slender
rhizomes and seeds. LIFE-SPAN: perennial. ORIGIN: native. LENGTH: usually 1–3 m
(rarely to 5 m). STEMS: herbaceous, twining, smooth, glabrous or sparingly pubescent
(pubescence usually not visible without magnification); rhizomes bearing a chain
of 2–40 fleshy tubers; tubers oblong to globose (1–2 cm thick). LEAVES: alternate,
odd-pinnately compound, leaflets 3–9 (usually 5–7); leaflets lanceolate to ovate (3–9
cm long, 1–4 cm wide); apex acute to acuminate, usually mucronate; base rounded;
margins entire; surfaces glabrous to sparsely pubescent; petioles 1.5–8 cm long; petiolules 3–5 mm long; stipules subulate to setaceous (4–7 mm long), caducous. INFLO-
RESCENCES: racemes 3–6 cm long, axillary, dense; flowers 5 to many, often flowering
for half or more of the axis length at a time, each node of the inflorescence subtended
by 2 bractlets; bractlets minute, caducous; pedicels 1–4 mm long. FLOWERS: perfect,
papilionaceous; calyx tube 4–6 mm long, glabrous to sparsely pubescent, lobes 5;
upper 4 lobes small, lower lobe prominent; corolla not uniformly colored; banner
petals broader than long (9–13 mm long), apically notched, usually retuse, white
outside and red to brown inside; wing petals curved downward, auricles near base,
brownish-purple; keel petals falcate or spirally incurved, purple or green. FRUITS:
legumes linear (5–10 cm long, 5–6 mm wide), valves slightly compressed between
the seeds, coiling after dehiscence; seeds few. SEEDS: oblong (4–5 mm long), plump
but slightly compressed, surfaces wrinkled, dark brown. $2n=22$, $3n=33$.

HABITAT: Groundnut is occasionally found to locally common in rich, moist soils of prairie
ravines, pond and stream banks, and thickets. USES AND VALUES: The herbage is eaten
by livestock, especially horses. Seeds are eaten by upland gamebirds and songbirds.
Tubers are eaten by hogs, rabbits, and small mammals. The plants have little potential
for erosion control or landscaping. ETHNOBOTANY: The starchy tubers furnished food
for some Plains Indians. They gathered tubers in the autumn, often taking them from
rodent caches. Members of a few tribes cultivated groundnut. Seeds were prepared
for consumption as peas are prepared today. Tubers were roasted or boiled and are
said to taste somewhat like roasted sweet potatoes. They were sometimes dried,
ground, and used to thicken soups. Tubers are rich in antioxidants and contain about
17 percent crude protein, or about 2–3 times the crude protein in potatoes.

SYNONYMS: *Apios tuberosa* Moench, *Glycine apios* L. OTHER COMMON NAMES: American
potatobean, blo (Lakota, Dakota), do (Osage), its (Pawnee), groundpea, hopniss,
Indian potato, mdo (Dakota), modo (Lakota), noa (Missouri tribes), nu (Omaha-
Ponca), pig potato, pomme de terre (French), wild potato.

SIMILAR SPECIES: *Galactia volubilis* (L.) Britton, hairy milkpea, is another prostrate
twining and trailing vine. It grows in eastern Oklahoma and northeastern Texas. Its
pinkish-purple flowers are few and occur in the axils. Its keels are carinate or moderately incurved, and its leaflets are 2–5 cm long.

Astragalus L.

astragalos (Gk.): ancient name for a leguminous plant.

Annual or perennial; stems herbaceous, caulescent or not, simple or branching; perennials from a rhizome or caudex surmounting a taproot; leaves odd-pinnately compound (occasionally palmately trifoliate or rarely reduced to one leaflet); leaflets several to numerous; petioles present or absent; stipules connate or distinct; inflorescences racemes, axillary or borne on scapes, loose or dense, elongated or contracted; calyx tube campanulate to cylindric, often oblique, lobes 5; lobes short, triangular or subulate; corolla white, ochroleucous, yellow, pink, or purple; banner obovate-oblong to rotund, clawed, usually reflexed from the wings, usually exceeding the wings; wings clawed, apex usually notched or rounded; keel shorter than the wings, obtuse to acute, clawed; stamens 10, diadelphous; fruits legumes, short or long, sessile to conspicuously stipitate, inflated or compressed, thick- or thin-walled, mostly fleshy in texture, sometimes the lower margin or both margins depressed or intruded as a false partition; the fruit, thus, is bilocular or unilocular; dehiscent or rarely indehiscent, subtended by the remnants of the stamen tube and ruptured calyx; seeds 1 to many. x=8, 11, 12, 13, 14, 15, 16.

A widespread genus containing about 1,500 species, which primarily grow in the northern temperate and arctic zones. It is a complex, highly variable, and difficult genus. Mature legumes are often necessary for identification of the species. Some species, often called locoweeds or poisonvetches, contain various alkaloids or accumulate selenium from the soil and are toxic to livestock. Thirty-five Great Plains species are included, and the 20 most common are described in detail.

1. Leaves primarily simple; some lower leaves may be compound
    2. Leaflets spatulate, mostly basal; subcaulescent; legumes narrowly oblong and solid in color. .... *Astragalus spatulatus*
    2. Leaflets filiform; mostly caulescent; legumes inflated and mottled. .... *Astragalus ceramicus*
1. Leaves all compound
    3. Leaflets spine-tipped. .... *Astragalus kentrophyta*
    3. Leaflets not spine-tipped
        4. Pubescence of the leaves mostly dolabriform
            5. Plants obviously caulescent
                6. Legumes glabrous; inflorescences not extending above the subtending leaves. .... *Astragalus canadensis*
                6. Legumes strigulose; inflorescences surpassing the subtending leaves. .... *Astragalus laxmannii*

5. Plants cespitose to short-caulescent
   7. Flowers of 2 kinds (chasmogamous and cleistogamous); banner 1.4 cm long or less; calyx tube 5 mm long or less. .... *Astragalus lotiflorus*
   7. Flowers of 1 kind (chasmogamous); banner 1.5 cm long or more; calyx tube 6 mm long or more. .... *Astragalus missouriensis*
4. Pubescence of the leaves basifixed
   8. Stems and leaves villous with hairs 1–2 mm long from minutely bulbous bases. .... *Astragalus drummondii*
   8. Stems and leaves variously pubescent with shorter hairs without bulbous bases
     9. Stipules distinct
       10. Banner less than 1 cm long; legumes glabrous to strigulose. .... *Astragalus nuttallianus*
       10. Banner 1.5 cm long or more; legumes glabrous to puberulent
         11. Plants caulescent; inflorescences not extending above the subtending leaves; foliage strigulose. .... *Astragalus crassicarpus*
         11. Plants subcaulescent to short-caulescent; inflorescences about equaling the height of the subtending leaves; foliage villous-tomentose. .... *Astragalus mollissimus*
     9. Stipules connate at least at the lower nodes
       12. Banner 1 cm long or less
         13. Legumes laterally compressed; calyx tube with at least some black hairs. .... *Astragalus multiflorus*
         13. Legumes terete or angular, not compressed; calyx tube without black hairs
           14. Leaflets linear, 2 mm wide or less; legumes less than 1 cm long. .... *Astragalus gracilis*
           14. Leaflets oblanceolate, 2 mm wide or more; legumes more than 1 cm long. .... *Astragalus flexuosus*
       12. Banner more than 1 cm long
         15. Leaflets sessile or nearly so
           16. Legumes bilocular and inflated; banner 1.5 cm long or less. .... *Astragalus cicer*
           16. Legumes unilocular and slightly compressed; banner 1.5 cm long or more. .... *Astragalus pectinatus*
         15. Leaflets with petiolules
           17. Legumes sessile or with a stipe 1.5 mm long or less, usually bilocular
             18. Racemes 4 cm long or less; pedicels less than 2 mm long; legumes 1 cm long or less. .... *Astragalus agrestis*
             18. Racemes 4 cm long or more; pedicels 2 mm long or more; legumes 1–2 cm long. .... *Astragalus plattensis*

17. Legumes with a stipe longer than 1.5 mm long, unilocular
    19. Legume strongly 2-grooved on the ventral face; calyx lobes often red. .... *Astragalus bisulcatus*
    19. Legume not grooved ventrally; calyx lobes green. ....
        .... *Astragalus racemosus*

Field milkvetch
Astragalus agrestis G. Don

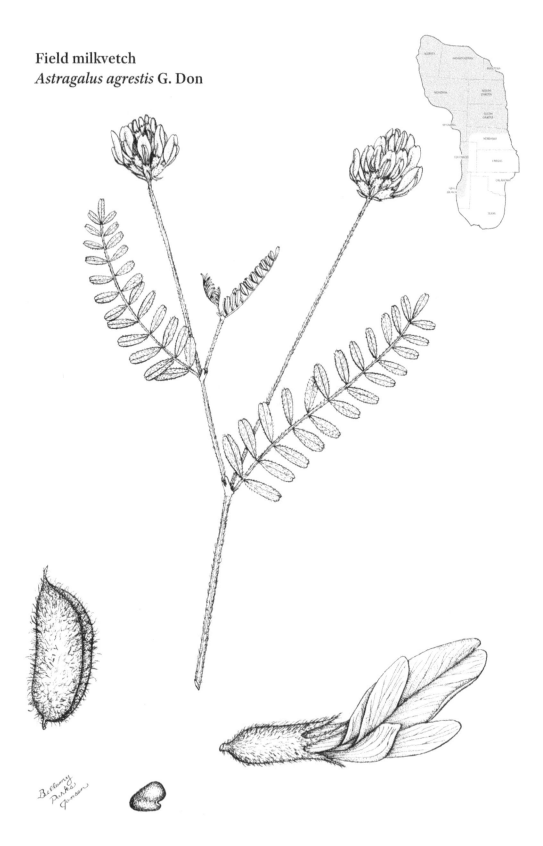

A General History of the Dichlamydeous Plants 2:258. 1832.
agrestis (Lat.): pertaining to fields or land, in reference to the habitat.

GROWTH FORM: forb, flowers May to August, reproduces from rhizomes and seeds. LIFE-SPAN: perennial. ORIGIN: native. HEIGHT: usually 10–30 cm (rarely to 40 cm). STEMS: herbaceous, decumbent to ascending, 1 to several, from shallow rhizomes or a buried crown, sometimes branching basally, diffuse, loosely tufted, patch-forming, slender, glabrous or sparsely strigose, hairs basifixed. LEAVES: alternate, odd-pinnately compound (3–12 cm long), leaflets 9–27; leaflets linear-oblong or elliptic-oblong to lanceolate (4–20 mm long, 2–4 mm wide); apex obtuse to retuse; base obtuse to rounded: margins entire; upper and lower surfaces sparingly pilose, hairs basifixed; sessile or nearly so; petioles 1–7 mm long; petiolules about 1 mm long; stipules linear to ovate (2–9 mm long), connate, sometimes split by stem expansion; upper stipules fully or distally herbaceous; lower stipules papery, scarious. INFLORESCENCES: racemes, ovoid to subcapitate (2–4 cm long, scarcely elongating during anthesis), headlike, axillary; flowers 5–20, narrowly ascending; peduncles erect to incurved (1.5–10 cm long). FLOWERS: perfect, papilionaceous; calyx tube short-cylindric (4–8 mm long), villous with mixed black and white hairs, lobes 5; lobes subulate (3–5 mm long); corolla colors variable, pinkish-purple, blue, lavender (sometimes white, ochroleucous, or yellow), drying violet or stramineous; banner oblanceolate to elliptic (1.6–2.3 cm long), slightly reflexed, notched; wings 1.4–1.8 cm long, claws 6–8 mm long; keels 1–1.4 cm long, claws 6–9 mm long; stamens 10, diadelphous; pedicels 1–1.5 mm long. FRUITS: legumes clustered, erect, obliquely ovoid (7–10 mm long, 3–5 mm wide), obcompressed to somewhat triquetrous, bases rounded or subcordate; shortly stipitate (stipe about 1 mm long); slightly beaked; bilocular; valves subcoriaceous, first green and then turning black, densely hirsute with white hairs, deeply sulcate on lower margins; seeds few to several. SEEDS: reniform (1.5–2 mm long), brown to black, smooth, indurate. $2n=16$.

HABITAT: Field milkvetch is infrequent to locally common in moist prairies and meadows, along margins of lakes and streams, open woodlands, thickets, and roadsides. USES AND VALUES: Palatability of field milkvetch is low, and it is rarely eaten by livestock or wildlife. It may be toxic because of accumulated selenium, but there are no substantiated reports of it causing poisoning. The seeds can be important for ground-foraging birds, and small mammals eat the legumes and seeds. It is visited by butterflies and is usually pollinated by bees. It is not used for erosion control. This attractive low-growing plant is sometimes planted in rock gardens. Care must be taken when using field milkvetch in landscaping because it can rapidly spread from rhizomes.

VARIETIES: This variable species has been divided into var. *bracteatus* (Osterh.) M.E. Jones and the several-seeded var. *polyspermus* (Torr. & A. Gray) M.E. Jones. SYNONYMS: *Astragalus dasyglottis* Fisch. *ex* DC., *A. goniatus* Nutt., *A. tarletonis* Rydb., *A. virgultulus* E. Sheld. OTHER COMMON NAMES: Don meadow-milkvetch, nickleleaf vetch, purple loco, purple milkvetch.

Twogrooved poisonvetch
Astragalus bisulcatus (Hook.) A. Gray

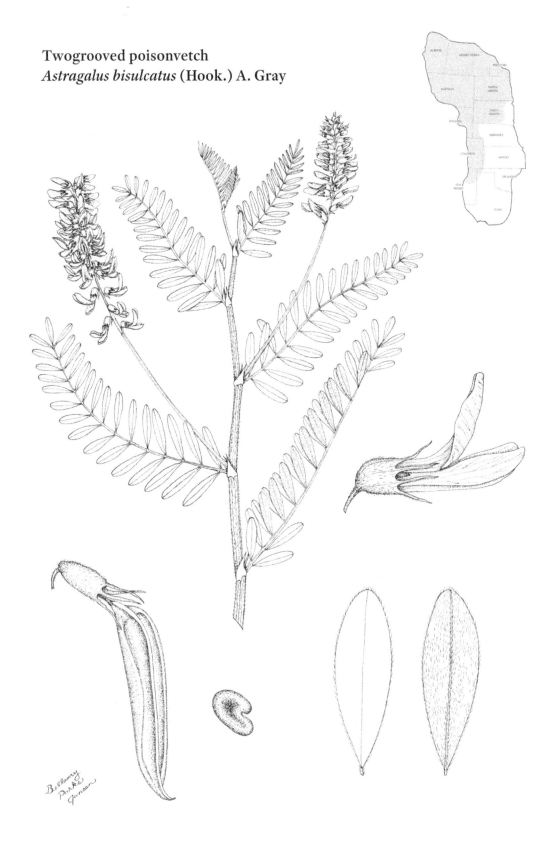

Pacific Railroad Report 12(2):42. 1860.

bi (Lat.): two; + *sulcus* (Lat.): furrow, referring to the two furrows, or grooves, in the legume.

GROWTH FORM: forb, flowers May to August, reproduces from seeds. LIFE-SPAN: perennial. ORIGIN: native. HEIGHT: 20–80 cm. STEMS: herbaceous, ascending to erect (sometimes decumbent), few to many from a caudex and taproot, forming a clump, simple or few-branched, glabrate or with basifixed hairs. LEAVES: alternate, odd-pinnately compound (4–13 cm long), leaflets 13–35; lower leaflets ovate to elliptic (9–35 mm long), upper leaflets elliptic to linear (5–25 mm long); apex acute to obtuse (occasionally mucronate); base cuneate to attenuate; margins entire; upper surfaces glabrate; lower surfaces with basifixed hairs; petioled below to sessile above; petiolules about 1 mm long; upper stipules distinct (3–10 mm long); lower stipules connate. INFLORESCENCES: racemes, ovoid to cylindric (3–12 cm long), axillary; flowers many, reflexed; peduncles 3–12 cm long. FLOWERS: perfect, papilionaceous; calyx campanulate to subcylindric (3.5–6 mm long), gibbous, strigulose, lobes 5; lobes subulate to setaceous (1.5–5 mm long), often reddish; corollas purple, white with a purple maculate keel, or entirely white to lavender or pink; banner obovate to oblanceolate (1–1.7 cm long), recurved; keel obtuse (6–14 mm long); wings 8–16 mm long; stamens 10, diadelphous; pedicels 1–3 mm long. FRUITS: legumes pendulous, oblong or shortly ellipsoid (7–20 mm long, 2–5 mm wide), obcompressed, stipitate or exserted-stipitate (stipe 1.5–5 mm long), unilocular, strongly 2-grooved on the ventral face, valves thick-papery, smooth or transversely rugose, strigulose to glabrate; seeds many. SEEDS: reniform (3–3.5 mm long), flattened, yellow or brown to black, indurate. $2n=24$.

HABITAT: Twogrooved poisonvetch grows in dry, alkaline soils of rangelands, prairies, and badlands. It grows only in soils containing selenium. USES AND VALUES: Forage quality is poor, and it is usually not grazed when other forage is available. Selenium from the soil accumulates in plant tissues, making it a toxic plant. Acute selenium poisoning is not common, but it may cause death. These plants also contain dangerous indolizidine alkaloids (swainsonine). Animals eating these plants are said to acquire "alkali disease" and may exhibit blindness, wandering, excitement, and depression. Respiratory failure and death may follow. It is not used in landscaping because the plants emit a urinelike odor. The flowers attract honeybees and bumblebees.

SYNONYMS: *Astragalus demissus* Greene, *A. diholcos* Tidestr., *A. grallator* S. Watson, *A. haydenianus* A. Gray, *A. scobinatulus* E. Sheld., *Diholcos bisulcatus* (Hook.) Rydb., *D. micranthus* Rydb., *Phaca bisulcata* Hook., *Tragacantha bisulcata* (Hook.) Kuntze. OTHER COMMON NAMES: grooved milkvetch, skunkweed, twogrooved locoweed, twogrooved milkvetch, twotoothed milkvetch.

SIMILAR SPECIES: *Astragalus alpinus* L., alpine milkvetch, is similar. However, the stems arise from rhizomes, and the plants can produce stolons. It grows in the Black Hills and the far western Great Plains. *Astragalus bodinii* E. Sheld., Bodin milkvetch, is similar, but all of its stipules are connate, and its banner is smaller. It also grows in the western Great Plains.

Canada milkvetch
Astragalus canadensis L.

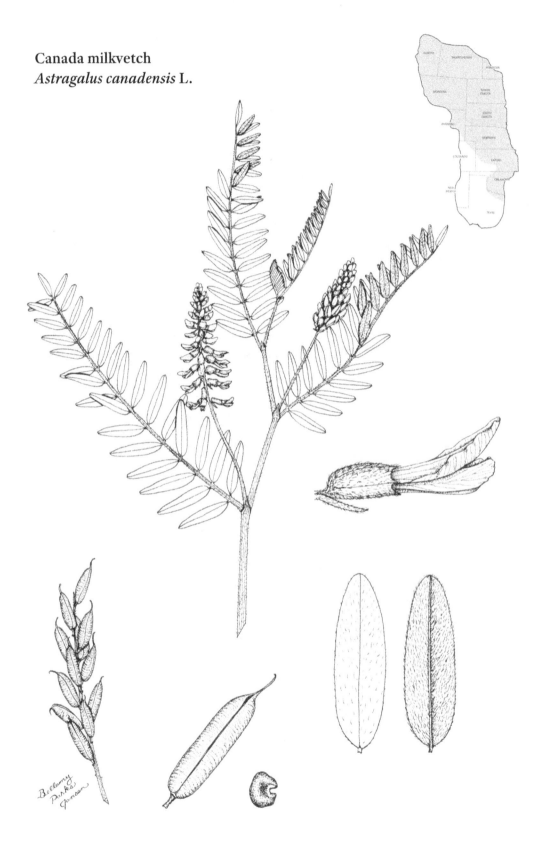

Species Plantarum 2:757. 1753.
canadensis: of or from Canada.

GROWTH FORM: forb, flowers May to August, reproduces from short rhizomes and seeds.
LIFE-SPAN: perennial. ORIGIN: native. HEIGHT: 0.2–1.6 m. STEMS: herbaceous, caulescent,
erect to ascending, robust, few-branched, solid or hollow, glabrous to thinly strigulose,
hairs dolabriform. LEAVES: alternate, odd-pinnately compound (5–35 cm long), leaflets
13–35; leaflets oblong to elliptic to narrowly lanceolate (1–4 cm long, 5–15 mm wide);
apex obtuse; base obtuse to rounded; margins entire; upper surfaces glabrous or rarely
strigulose; lower surfaces strigulose; hairs dolabriform, one arm longer than the other;
sessile to short-petioled; stipules connate, lanceolate to deltoid (3–18 mm long), membranous. INFLORESCENCES: racemes shorter than the subtending leaves (3–20 cm long),
axillary, dense; flowers many, imbricate, spreading or declining; peduncles 4–10 cm
long. FLOWERS: perfect, papilionaceous; calyx tube campanulate to broadly cylindric
(4–7 mm long), strigose to glabrate, lobes 5; lobes acuminate (1–5 mm long, usually
1–2 mm long); corollas greenish-white or ochroleucous, sometimes tinged with purple,
often maculate on the keel; banner 1.1–1.7 cm long; wings 1–1.5 cm long, clawed; keel
9–13 mm long, clawed; stamens 10, diadelphous. FRUITS: legumes, oblong to elliptic
(1–1.8 cm long, 5–6 mm wide), beaks filiform (to 7 mm long); subterete in cross section, bilocular, valves coriaceous, both surfaces convex or only the dorsal surface convex, rugose, glabrous (rarely strigulose), tardily dehiscent; seeds usually 3–5. SEEDS:
obliquely cordiform to reniform (2–2.5 mm long), brownish-yellow, smooth, indurate.
$2n=16$.

HABITAT: Canada milkvetch is locally common in moist rangelands, prairies, open woodlands, roadsides, and stream banks in all types of soil. USES AND VALUES: Canada
milkvetch is palatable to livestock and furnishes good to excellent forage. Deer, elk,
bighorn sheep, pronghorn, and rabbits eat the herbage. The seeds are eaten by gamebirds, songbirds, and small mammals. It can become rather large for landscaping, but
it is an excellent specimen plant. Its creamy flowers in summer and groups of legumes
in late summer and early fall are attractive. ETHNOBOTANY: Some Plains Indians pulverized and chewed the roots for back and chest pains. Tea was made from the roots
to control coughing, and a poultice of chewed roots was applied to cuts. Roots were
gathered in the spring and eaten raw or after boiling.

SYNONYMS: *Astragalus brevidens* Rydb., *A. carolinianus* L., *A. halei* Rydb., *A. oreophilus*
Rydb., *A. torreyi* Rydb. OTHER COMMON NAMES: cow vetch, little rattlepod, ločipišni
pežixota (Lakota), pejúta ska hu (Lakota), rattlepod.

SIMILAR SPECIES: *Astragalus miser* var. *hylophilus* (Rydb.) Barneby, woodland milkvetch,
is similar in appearance with white or ochroleucous corollas. However, it has unilocular fruits. It is infrequently found in the Black Hills and the far western Great
Plains. *Astragalus soxmaniorum* Lundell, Soxman milkvetch, also has greenish-white or
ochroleucous corollas and tardily dehiscent legumes. It grows in sterile, sandy post oak
woodlands and abandoned fields in Texas at the southeastern edge of the Great Plains.

Painted milkvetch
Astragalus ceramicus E. Sheld.

Minnesota Botanical Studies 1(1):19. 1894.
keramikos (Gk.): a vessel, pot, or other articles made from clay and hardened by heat, referring to the shape and coloration of the legume.

GROWTH FORM: forb, flowers May to July, reproduces from rhizomes and seeds. LIFE-SPAN: perennial. ORIGIN: native. LENGTH: 5–40 cm, mostly caulescent. STEMS: herbaceous, decumbent to ascending, simple or few-branched from a caudex or extensive rhizomes, flexuous, strigulose; hairs basifixed (rarely dolabriform), straight. LEAVES: alternate, simple or reduced to the filiform rachises or phyllodes, lowest chartaceous, connate, curved or somewhat hooked at the apex, leaflets generally none (lowest leaves may have 5–7 leaflets); upper leaves connate only at the base or halfway; phyllodes 3–9, simple or odd-pinnately compound (4–15 cm long), involute, filiform (5–30 mm long), surfaces pubescent; hairs short, white, unbranched; petioles of compound leaves 2–9 cm long, lower stipules connate. INFLORESCENCES: racemes 2–15 cm long, loose; ascending in flower, spreading, deflexed in fruit; flowers 2–15; peduncles 1–7 cm long. FLOWERS: perfect, papilionaceous; calyx tube campanulate (2–4 mm long), white or black strigose, lobes 5; lobes 1–3 mm long, strigose; corollas white to pink (rarely purple), sometimes purple-veined; banner recurved (6–11 mm long); wings 7–9 mm long; keel 6–8 mm long, usually maculate; stamens 10, diadelphous. FRUITS: legumes pendulous, ellipsoid to broadly ovoid (1.5–15 cm long, 1–2.5 cm wide), inflated, bladderlike; stipitate (stipe 1–3 mm long) or sessile; unilocular; glabrous, mottled with red or purple or red to purple with white mottling; seeds several. SEEDS: reniform (2–3 mm long), brown, smooth, compressed, indurate. $2n$=22.

HABITAT: Painted milkvetch is found growing almost exclusively in sand. It is rare to locally common in prairies, hillsides, dunes, blowouts, and roadsides. USES AND VALUES: Livestock occasionally eat painted milkvetch, but it is unimportant in their diets because of its small stature, and it is usually scattered. Ground-foraging birds, songbirds, and small mammals eat the seeds. It has little value for erosion control. The inflated legumes make it a horticultural curiosity. It is sometimes added into landscapes as a specimen plant. ETHNOBOTANY: The sweet-tasting roots were a favorite food of Great Plains Indians.

VARIETIES: The most common variety in the Great Plains is var. *filifolius* (A. Gray) F.J. Herm. The others grow outside of the region. Var. *apus* Barneby grows in southeastern Idaho and adjacent Montana, and var. *ceramicus* grows west of the Great Plains. SYNONYMS: *Astragalus angustus* var. *ceramicus* (E. Sheld.) M.E. Jones, *A. pictus* (A. Gray) A. Gray, *Phaca picta* A. Gray, *Psoralea longifolia* Pursh. OTHER COMMON NAMES: birdegg milkvetch, longleaf milkvetch, tasúsu canhologan (Lakota).

SIMILAR SPECIES: *Astragalus lonchocarpus* Torr., rushy milkvetch, is another species with phyllodes but it should not be confused with painted milkvetch. Rushy milkvetch plants are shrubby in appearance, large (to 90 cm), and have large (1.2–2.3 cm long) white corollas. It grows in the Great Plains in southeastern Colorado and northeastern New Mexico.

Cicer milkvetch
Astragalus cicer L.

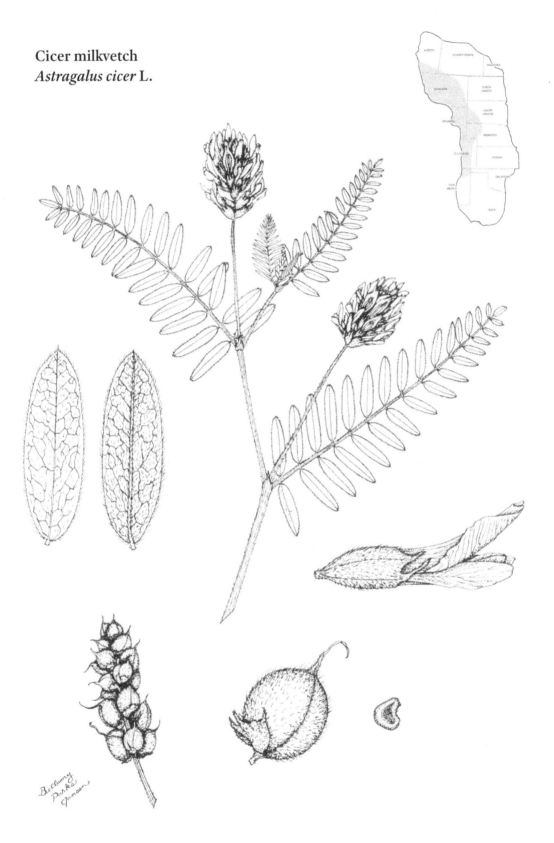

Species Plantarum 2:757. 1753.

cicero (Lat.): chickpea, in reference to its papilionaceous flowers.

GROWTH FORM: forb, flowers June to July, reproduces from stout rhizomes and seeds. LIFE-SPAN: perennial. ORIGIN: introduced (from Europe). LENGTH: 0.3–1 m (sometimes to 3 m). STEMS: herbaceous, initially ascending to erect, then becoming decumbent, tangle-forming, branching, hollow, surfaces glabrate to strigulose, hairs basifixed. LEAVES: alternate, odd-pinnately compound (5–20 cm long), leaflets 17–35; leaflets elliptic to oblong (5–30 mm long, 2–10 mm wide); apex acute to retuse but usually obtuse, often mucronate; base rounded to obtuse; margins entire; upper surface glabrate; lower surface sparsely pubescent, hairs basifixed; sessile or nearly so; lower stipules connate, chartaceous; upper stipules herbaceous, connate at the base to half their length. INFLORESCENCES: racemes shorter than or slightly surpassing the sub-tending leaves, compactly ovoid to short-spicate (10–20 cm long), dense, axillary; flowers 10–60, ascending; peduncle 3–15 cm long. FLOWERS: perfect, papilionaceous; calyx tube cylindric to campanulate (3–5 mm long), lobes 5; lobes 1.5–2.5 mm long, sometimes irregular, strigulose, hairs black or brown or white; corollas pale yellow to white or ochroleucous (1.2–1.5 cm long); banner oblanceolate (1.2–1.5 cm long), recurved; wings rounded to emarginate (1.1–1.4 cm long); keel incurved (1–1.3 cm long); stamens up to 10, diadelphous; pedicels in flower 0.3–0.8 mm long, in fruit 0.5–1.5 mm long. FRUITS: legumes clustered, spreading to ascending, ovoid to ellipsoid (1–1.5 cm long), remains of the style forming a slight beak, inflated, subterete in cross section, stipitate or sessile; bilocular, ventrally sulcate; valves thickly papery, coriaceous, pustulate-hirsute, initially stramineous or reddish-green, turning black, tardily dehiscing; seeds 6–10. SEEDS: reniform to ovate (4–6 mm long), bright yellow to pale green, lustrous, compressed, indurate. $2n=64$.

HABITAT: Cicer milkvetch was introduced into the United States from Sweden in 1926. It is occasionally planted under dryland and irrigated conditions in the western Great Plains. It has rarely escaped to fields, waste places, and roadsides. It is best adapted to cool, moist sites with moderately coarse-textured soils. However, it will grow in clayey soils. It withstands slightly acidic to moderately alkaline conditions and grows well in the Great Plains under irrigation. USES AND VALUES: Forage value of cicer milkvetch is excellent and comparable to alfalfa. It is eaten by all classes of livestock, either as hay or grazed in pastures. It resists heavy grazing because of its vigorous rhizomes, and recovery following grazing is rapid. It is grazed by deer, elk, bighorn sheep, and pronghorn. No cases of bloat have been noted. Cicer milkvetch contains no harmful alkaloids, nor does it accumulate toxic amounts of selenium. It is valuable for critical area plantings, soil conservation, and wildlife habitat. It is grown alone or mixed with grasses. It attracts many species of insects and is pollinated by bumblebees and smaller bees. Cicer milkvetch is occasionally grown as an ornamental novelty.

OTHER COMMON NAME: chickpea milkvetch.

Groundplum milkvetch
Astragalus crassicarpus Nutt.

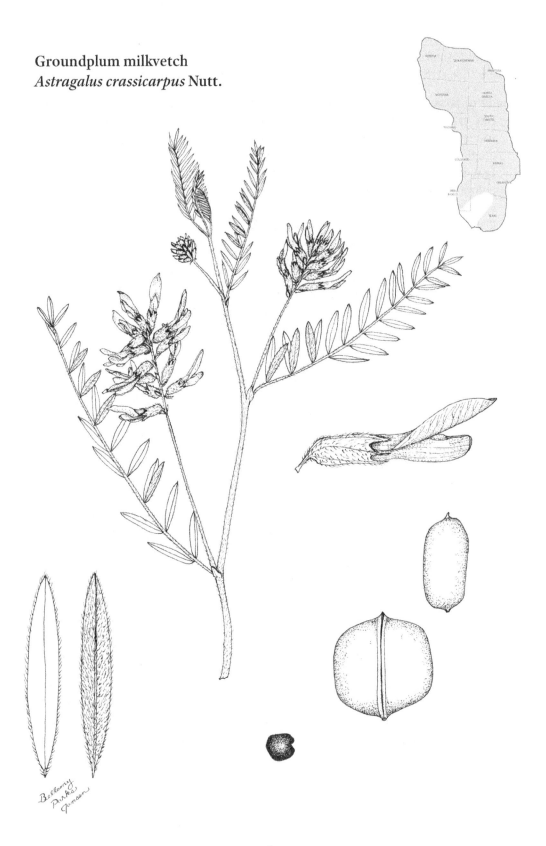

Catalogue of New and Interesting Plants Collected in Upper Louisiana No. 6. 1813. *crassus* (Lat.): thick; + *karpos* (Gk.): fruit, in reference to its plum-shaped fruit.

GROWTH FORM: forb, flowers March to May, reproduces from seeds. LIFE-SPAN: perennial. ORIGIN: native. LENGTH: 10–60 cm, caulescent. STEMS: herbaceous, ascending to prostrate with ascending tips, robust, few-branched, glabrate to strigulose with basifixed hairs, from a strong caudex and taproot. LEAVES: alternate, odd-pinnately compound (2–14 cm long), mostly subsessile, leaflets 11–33; leaflets oblanceolate to elliptic (3–17 mm long, 2–6 mm wide); apex obtuse to acute, often mucronate; base cuneate; margins entire; upper surface usually glabrate; lower surface strigulose, hairs basifixed; petiole present or absent; stipules lanceolate (3–9 mm long), membranous, distinct. INFLORESCENCES: racemes not extending above the subtending leaves, 1–8 cm long, axillary, dense, flowers many; flowers spreading or declining; peduncles initially short (2–3 cm long) remaining so or elongating (to 15 cm). FLOWERS: perfect, papilionaceous; calyx tube subcylindric to campanulate (6–13 mm long), strigose to tomentose with black or white hairs, lobes 5; lobes acuminate or long-acuminate (1–4 mm long); corolla purple to blue to pinkish-white or greenish-white to ochroleucous; banner obovate to lanceolate (1.5–2.6 cm long), moderately reflexed, notched; wings 1.4–1.8 cm long, clawed; keel 9–15 mm long, apex obtuse, clawed; stamens up to 10, diadelphous; pedicels 2–7 mm long. FRUITS: legumes subglobose (1.5–4 cm long, about 2 cm in diameter), round in cross section, initially succulent, sessile; bilocular, valves thick-walled, glabrous, tardily dehiscent to indehiscent; seeds usually 25–75. SEEDS: cordiform to reniform (2–4 mm long), black, smooth or pitted, indurate. $2n=22$.

HABITAT: Groundplum milkvetch is scattered to locally common on rocky or sandy prairies. USES AND VALUES: It grows low to the ground and is not readily grazed by livestock. It is grazed by deer and pronghorn. Bees and butterflies visit the flowers. Small mammals cache and eat the fruits. ETHNOBOTANY: Lakotas ate the fruits. Chippewas used the fruits in compounding medicines for convulsions and hemorrhages from wounds. For reasons unknown, Pawnees customarily placed the fruits in the water in which they soaked their corn seeds before planting.

VARIETIES: Three varieties grow in the Great Plains. Var. *crassicarpus* has blue or purple corollas, var. *paysonii* (E.H. Kelso) Barneby corollas are white, and var. *trichocalyx* (Nutt. *ex* Torr. & A. Gray) Barneby *ex* Gleason has greenish-white to ochroleucous corollas. SYNONYMS: *Astragalus carnosus* Nutt., *A. caryocarpus* Ker Gawl., *A. mexicanus* A. DC., *A. succulentus* Richardson, *Geoprumnon crassicarpum* (Nutt.) Rydb., *G. succulentum* (Richardson) Rydb. *Tragacantha caryocarpa* (Ker Gawl.) Kuntze. OTHER COMMON NAMES: bi'jikwi'bugesan (Chippewa), buffalo apple, buffalo bean, buffalo pea, ground plum, pe ta wote (Lakota), prairie apple, prairie plum, tdika shande (Omaha-Ponca).

SIMILAR SPECIES: *Astragalus distortus* Torr. & A. Gray, Ozark milkvetch, also has decumbent stems and purple corollas. Its stems are glabrous and the banner is 1.5 cm or less in length. It grows in the eastern Great Plains from Texas to Kansas.

Drummond milkvetch
Astragalus drummondii Douglas

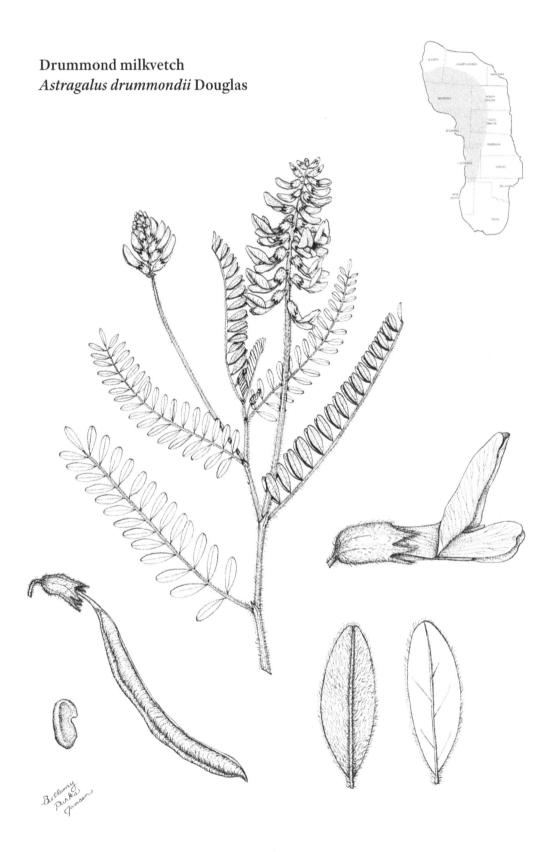

Flora Boreali—Americana 1(3):153–54. 1831.

drummondii: named for Thomas Drummond (1793–1835), Scottish botanical collector who collected plants from 1825 to 1835 in North America, including plants from the southern Great Plains.

GROWTH FORM: forb, flowers June to July, reproduces from short rhizomes and seeds. LIFE-SPAN: perennial. ORIGIN: native. HEIGHT: 20–70 cm. STEMS: herbaceous, erect to ascending, few to several from a caudex and rhizomes, usually clustered, robust, branching or spurred upward, naked at the base, surfaces villous, hairs basifixed, spreading (1–2 mm long), from minutely bulbous bases. LEAVES: alternate, odd-pinnately compound (5–14 cm long), leaflets 11–35, sometimes not paired; leaflets elliptic-lanceolate to oblanceolate (5–25 mm long, 2–10 mm wide); apex obtuse to truncate or rounded; base obtuse to truncate; margins entire; upper surface bright green, glabrate to glabrous; lower surface pallid green, villous to glabrate, hairs basi-fixed (1–2 mm long), from a bulbous base; petioles 2–5 mm long or subsessile; upper stipules lanceolate (3–12 mm long), distinct; lower stipules submembranous, connate. INFLORESCENCES: racemes extending above the subtending leaves, ovoid to cylindric (2–10 cm long), elongating in fruit (to 22 cm long), axillary; flowers 12–50, spreading to declining; peduncles erect (3–13 cm long), stout. FLOWERS: perfect, papilionaceous; calyx tube campanulate to subcylindric (4–8 mm long), pubescence white or black or mixed, lobes 5; lobes subulate to setaceous (2–4 mm long); corollas white to pale yellow sometimes with a purple-tipped or maculate keel; banner moderately recurved (1.5–2.5 cm long, 7–10 mm wide); wings 1.5–2 cm long, claws 4–6 mm long; keel 1.2–1.5 mm long, bluntly deltate at the apex, claws 6–8 mm long; stamens 10, diadelphous; pedicels 1–3 mm long. FRUITS: legumes pendulous, stipitate, oblong to oblanceolate (1–4 cm long, 3–5 mm wide), straight or slightly curved, obcordate or subtriquetrous in cross section, stipitate (stipe 3–12 mm long); bilocular or nearly so, may be unilocular at the apices, dorsally sulcate, suture prominent; valves thick-papery, glabrous, first green, then becoming stramineous and subcoriaceous; tardily dehiscent; seeds many. SEEDS: reniform (2–3 mm long), lustrous to dull, indurate. $2n=22$.

HABITAT: Drummond milkvetch is frequent to locally common in prairies, rangelands, foothills, stream terraces, and roadsides. It grows in a wide range of soil textures, as well as in light and dry to heavy and moderately wet soils. USES AND VALUES: Drummond milkvetch produces coarse, low-quality forage. Cattle will eat it if other forage is not available or sufficient. Pronghorn, deer, and elk occasionally eat the herbage. It increases with improper grazing. Reports on the toxicity of the plant vary. Selenium from the soil may accumulate in plant tissues. These plants contain dangerous indolizidine alkaloids (swainsonine). However, selenium or swainsonine poisoning of livestock by this species has not been substantiated. It attracts bumblebees, honey bees, and other nectar-seeking insects. It is not used in landscaping or for erosion control.

SYNONYMS: *Tium drummondii* (Douglas) Rydb., *Tragacantha drummondii* (Douglas) Kuntze. OTHER COMMON NAME: pale-yellow milkvetch.

Pliant milkvetch
Astragalus flexuosus (Hook.) Douglas *ex* G. Don

A General History of the Dichlamydeous Plants 2:256. 1832.
flexuosus (Lat.): bending, winding, in reference to the pliant stems.

GROWTH FORM: forb, flowers May to July, reproduces from short rhizomes and seeds. LIFE-SPAN: perennial. ORIGIN: native. HEIGHT: 15–60 cm. STEMS: herbaceous, decumbent to ascending, from a caudex and rhizomes, branched or spurred from early emerged nodes; internodes often appearing in a zigzag pattern, thinly canescent; pubescence straight or incurved, hairs basifixed. LEAVES: alternate, odd-pinnately compound (3–8 cm long), leaflets 11–27; leaflets oblanceolate or oblong to linear (5–15 mm long, 2.5–4.5 mm wide), often folded along the midrib; apex truncate to retuse; base obtuse to acute; margins entire; upper surface glabrous (rarely pubescent); lower surface strigose; petiolate below to subsessile above; upper stipules distinct (1.5–7 mm long) or connate at the base or nearly free; lower stipules connate in a campanulate sheath, papery, membranous (3–12 mm long). INFLORESCENCES: racemes, 2–3 cm long at anthesis, much longer at maturity, axillary, loose; flowers 11–49, first spreading, then declining and nodding with age; peduncles erect (2–13 cm long), stout. FLOWERS: perfect, papilionaceous; calyx tube campanulate (2.5–4 mm long), strigulose with black or white hairs, lobes 5; lobes triangular (0.5–2 mm long); corollas white to pale purple or pinkish-purple; banner moderately recurved (7–10 mm long); wings 7–9 mm long; keel recurved (5–6 mm long), acute at the apex; stamens 10, diadelphous. FRUITS: legumes stipitate to subsessile (stipe 0.5–4 mm long), oblanceolate to ellipsoid (1.1–4 cm long, 3–5 mm wide), straight or slightly curved, slightly flattened dorsally but nearly terete, turgid or inflated, not sulcate; unilocular; valves papery, pubescent, reticulate to cross pendulous, rugose with reddish streaks or specks; seeds 12–25. SEEDS: reniform (2–2.5 mm long), pale brown, smooth, indurate. 2*n*=22.

HABITAT: Pliant milkvetch is infrequent on prairies, rangelands, roadsides, rocky hilltops, and bluffs. It is most abundant on shale to sandy clay soils. USES AND VALUES: Pliant milkvetch is grazed by all classes of livestock, but palatability and forage quality are rated as low. Cattle will eat it if other forage is not available. Pronghorn, deer, bighorn sheep, and elk occasionally eat the herbage. It increases with improper grazing. There are no reports of toxicity. It is visited by many types of nectar-seeking insects. The legumes are cached by small mammals, and ground-foraging birds eat the seeds. It is not used for erosion control or landscaping.

VARIETIES: This species has been divided into several varieties. Var. *flexuosus* is most common in the Great Plains. Var. *diehlii* (M.E. Jones) Barneby and var. *greenei* (A. Gray) Barneby are both found south and west of the region. SYNONYMS: *Astragalus fendleri* (A. Gray) A. Gray, *Homalobus fendleri* (A. Gray) Rydb., *H. flexuosus* (Hook.) Rydb., *Phaca elongata* Hook., *P. fendleri* A. Gray, *P. flexuosa* Hook., *Pisophaca elongata* (Hook.) Rydb., *P. flexuosa* (Hook.) Rydb., *P. ratonensis* Rydb., *P. saundersii* Rydb., *P. sierrae-blancae* Rydb., *Tragacantha fendleri* (A. Gray) Kuntze, *T. flexuosa* (Hook.) Kuntze. OTHER COMMON NAMES: flexile milkvetch, slender milkvetch.

Slender milkvetch
Astragalus gracilis Nutt.

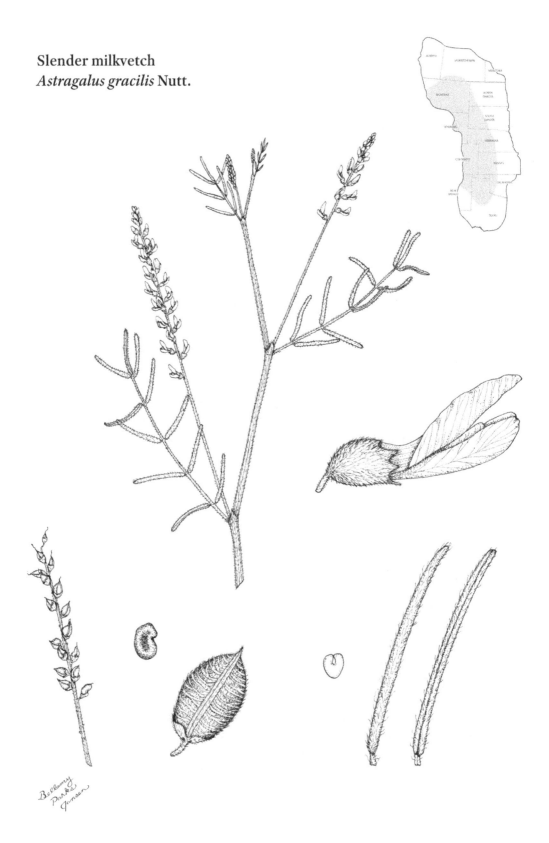

The Genera of North American Plants 2:100. 1818.
gracilis (Lat.): slender, referring to the stems and leaves.

GROWTH FORM: forb, flowers May to July, reproduces from rhizomes and seeds. LIFE-SPAN: perennial. ORIGIN: native. HEIGHT: 15–40 cm (rarely to 80 cm). STEMS: herbaceous, erect to nearly prostrate, from a caudex and short rhizomes, 1 to several, slender, much-branched from the base, branches divergent, strigulose with appressed basifixed hairs, rarely glabrous. LEAVES: alternate, odd-pinnately compound (2–9 cm long), leaflets 9–17; leaflets linear to narrowly oblong (4–25 mm long, 1–2 mm wide), often involute, nearly cylindric in cross section; apex rounded to truncate or emarginate, slightly notched; base rounded; margins entire; upper surface pubescent to glabrate and brighter green; lower surface strigulose; hairs basifixed; sessile or petioles to 3 cm; stipules of upper leaves connate (1–4 mm long), stipules of lower leaves membranous-papery. INFLORESCENCES: racemes longer than the subtending leaves, lax, loose; terminal and axillary; flowers 5–50; peduncles ascending (4–22 cm long), incurved or divergent, naked. FLOWERS: perfect, papilionaceous; calyx tube campanulate (1.5–2.6 mm long), lobes 5; lobes deltoid to triangular-subulate or short-acuminate (0.5–1 mm long), sometimes broader than long, slightly irregular, strigulose with white or rarely a few black hairs; corollas light purple or lavender to dark purple (rarely white), keel purple-tipped, fading to yellowish; banner moderately to strongly reflexed (6–9 mm long), veins often purple; wings 5–8 mm long, claws 1.5–3 mm long; keel incurved (4–6 mm long), claws 1.5–3 mm long; stamens 10, diadelphous, 9 united, upper 1 nearly free; pedicels ascending in flower (0.5–1 mm long); pedicels in fruit arching outward or downward (1–2 mm long). FRUITS: legumes nearly sessile, declined or deflexed or from decumbent branches, horizontally spreading, ellipsoid to broadly ovoid (4–9 mm long, 2.5–4 mm in diameter), lower face flattened or openly to shallowly sulcate, upper face convex, obtuse at the base, beaked slightly, inflated, unilocular; valves strigose, coriaceous, cross rugose, tardily dehiscent; seeds usually 5–9. SEEDS: reniform (2.2–3.5 mm long), brown or greenish-brown, smooth to slightly pitted, compressed, indurate. $2n=22$.

HABITAT: Slender milkvetch is locally common on sandy prairies, but it also may be common on dry, rocky, or prairie hillsides, rangelands, uplands, stream valleys, and bluffs. USES AND VALUES: All classes of livestock, deer, pronghorn, elk, and bighorn sheep graze slender milkvetch, especially in early spring before many other plants have begun growth. It is not toxic to livestock. Ground-foraging birds eat the seeds, and small mammals eat the seeds and legumes. It is not used for erosion control or landscaping. ETHNOBOTANY: Some Lakota mothers chewed the roots to improve lactation.

SYNONYMS: *Astragalus microlobus* A. Gray, *A. microphacos* Cory, *A. parviflorus* (Pursh) MacMill., *Dalea parviflora* Pursh, *Microphacos gracilis* (Nutt.) Rydb., *M. microlobus* (A. Gray) Rydb., *M. parviflorus* (Pursh) Rydb., *Tragacantha microloba* (A. Gray) Kuntze, *T. parviflora* (Pursh) Kuntze. OTHER COMMON NAMES: Ender milkvetch, notched milkvetch, pejúťa skúya (Lakota).

Spiny milkvetch
Astragalus kentrophyta A. Gray

Proceedings of the Academy of Natural Sciences of Philadelphia 15(3):60. 1863.
kentron (Gk.): a point or spine, + *phyton* (Gk.): plant, in reference to the spiny leaflets.

GROWTH FORM: forb, flowers June to September, reproduces from seeds. LIFE-SPAN: perennial (occasionally annual). ORIGIN: native. LENGTH: 10–45 cm. STEMS: herbaceous (woody below), prostrate to ascending, pulvinate to mat-forming (mats to 40 cm in diameter), much-branched from a caudex and taproot or superficial root crown, strigulose or villosulous; hairs dolabriform or basifixed, appressed or spreading, straight or curved. LEAVES: alternate, pinnately to subpalmately compound, leaflets 3–9 (commonly 5); leaflets elliptic-lanceolate to subulate or oblanceolate (3–24 mm long, 1–2 mm wide); apex with a soft to stiff spine (to 1.2 mm long), becoming rigid as it matures; margins entire; surfaces pubescent; petiole 0–4 mm long; lowest stipules connate in a sheath (1.5–11 mm long); upper stipules more shortly connate, becoming sharply pointed. INFLORESCENCES: racemes, shorter than the subtending leaves, elongating in fruit, axillary; flowers usually 1–3, these ascending, then declining; peduncles obsolete or short (to 1.5 cm long), concealed by the stipules. FLOWERS: perfect, papilionaceous; calyx tubes campanulate to short-cylindric (1.5–3 mm long), strigulose with white or black hairs, lobes 5; lobes subulate or linear-bristle-like (0.5–2.5 mm long), irregular; corollas white to pinkish-purple; banner narrow or oblanceolate (4–6 mm long), recurved; wings narrow (4–6 mm long), purple-tipped; keel arcuate (3–4 mm long), spurred or gibbous; stamens 10, diadelphous; pedicels 0.5–8 mm long. FRUITS: legumes spreading or deflexed, lenticular-ovoid to ellipsoid or lance-acuminate (3.5–8 mm long, 1.5–4 mm wide), with or without a beak; subterete to laterally compressed; valves thick, papery, first green, then becoming stramineous, villous or strigose, freely dehiscing, seeds 1–10. SEEDS: cordiform, notched at one end, compressed, brown or olive-brown or black, smooth, often pitted, indurate. $2n=24$.

HABITAT: Spiny milkvetch is infrequent but widely distributed in barren and exposed habitats on ridges, hilltops, and rocky bluffs. USES AND VALUES: When immature, it is relatively palatable to sheep, elk, goats, and pronghorn. It becomes worthless to livestock when the spines on the leaves and stipules stiffen as the plants mature. It is visited by many species of small, nectar-seeking insects. The mat-forming growth of spiny milkvetch provides some protection to the soil from wind and water erosion. It is not used in landscaping. ETHNOBOTANY: It was a ceremonial plant and life medicine for some Navajos.

VARIETIES: This is a highly variable species with at least nine varieties. Var. *kentrophyta* is most common in the Great Plains, and most of the rest grow west of the region. SYNONYMS: *Astragalus aculeatus* A. Nelson, *A. jessiae* M. Peck, *A. montanus* (Nutt.) M.E. Jones, *A. tegetarius* S. Watson, *Homalobus wolfii* Rydb., *Kentrophyta minima* Rydb., *K. montana* Nutt., *K. viridis* Nutt. *ex* Torr. & A. Gray, *Phaca viridis* (Nutt. *ex* Torr. & A. Gray) Piper, *Tragacantha montana* (Nutt.) Kuntze. OTHER COMMON NAMES: kentrophyta, Nuttall kentrophyta, prickly milkvetch, thistle milkvetch.

Prairie milkvetch
Astragalus laxmannii Jacq.

Bellamy
Parks
Jonson

Hortus Botanicus Vindobonensis 3:22, pl. 37. 1776.

laxmannii: named after Finnish-born Erich Gustav Laxmann (1737–96), plant collector, member of the clergy, and glass technologist who worked in Russia and sent many plant specimens to Carolus Linnaeus.

GROWTH FORM: forb, flowers June to August, reproduces from seeds. LIFE-SPAN: perennial. ORIGIN: native. HEIGHT: 10–30 cm, rarely to 50 cm. STEMS: erect to decumbent, several, clustered, cespitose, simple from a caudex and a woody taproot, densely or thinly strigulose, hairs dolabriform. LEAVES: alternate, odd-pinnately compound (4–17 cm long), leaflets 9–25; leaflets elliptic-lanceolate to oblong-obovate (8–30 mm long, 3–9 mm wide); apex obtuse, subacute, or emarginate; base obtuse; margins entire; surfaces glabrate or strigose above, hairs dolabriform; petioles to 2.5 mm long, sparsely to loosely hairy; lower stipules connate for at least one-third of their length (5–15 mm long), papery, membranous. INFLORESCENCES: racemes longer than the subtending leaves, subcapitate to densely spicate (2–10 cm long); flowers 12–50; peduncles erect to ascending or incurved (4–14 cm long). FLOWERS: perfect, papilionaceous; calyx tubes campanulate or cylindric (5–11 mm long), strigulose with white and black hairs or only black hairs, lobes 5; lobes linear, subulate, or setaceous (1.5–5 mm long); corolla purple, lavender, dull blue, reddish-lilac, or white; banner emarginate, scarcely recurved (1.2–2 cm long, 4–8 mm wide); wings 9–17 mm long, claws 4–8 mm long; keel obtuse (8–15 mm long), claws 4–8 mm long; stamens 10, diadelphous; pedicels less than 1 mm long. FRUITS: legumes substipitate (stipe less than 2 mm long) to sessile, erect or ascending, ellipsoid to ellipsoid-lanceolate (7–12 mm long, 2.5–4 mm wide), straight or curved, beaked to 2.5 mm long, triquetrous to subterete in cross section, dorsally sulcate; bilocular or nearly so, valves thick-papery; surfaces closely strigulose, hairs basifixed; dehiscence along both sutures, valves tending to twist outward; seeds many. SEEDS: reniform to cordiform (2–2.8 mm long), notched, compressed, brown or nearly black, indurate. $2n$=16, 32.

HABITAT: Prairie milkvetch is locally common on prairies, rangelands, roadsides, and sand hills; often in gravelly and rocky places. USES AND VALUES: It is not palatable to livestock, and it is seldom eaten by wildlife. Selenium from the soil may accumulate in its tissues. However, acute selenium poisoning of livestock from this species is uncommon because it is rarely consumed. Birds and small mammals eat the seeds without poisoning. The flowers are visited by nectar-seeking insects. It is not used for erosion control, and it is infrequently used in landscaping. ETHNOBOTANY: Some Cheyennes ground leaves and stems and sprinkled the material on watery poison ivy rash.

VARIETIES: Var. *robustior* (Hook.) Barneby & S.L. Welsh grows on the Great Plains. SYNONYMS: *Astragalus adsurgens* Pall., *A. albus* Širj., *A. austrosibiricus* Schischk., *A. fujisanensis* Miyabe & Tatew., *A. inopinatus* Boriss., *A. longispicatus* Ulbr. OTHER COMMON NAMES: ascending milkvetch, ascending purple-milkvetch, Laxmann milkvetch, mahkhá nowas (Cheyenne), standing milkvetch, Tanana milkvetch.

Lotus milkvetch
Astragalus lotiflorus Hook.

Flora Boreali-Americana (Hooker) 1(3):152–53. 1831.

lotus (Lat.) from *lōtos* (Gk.): name of a legume; +*florus* (Lat.): flowers, in reference to the flowers resembling lotus flowers.

GROWTH FORM: forb, flowers March to July, reproduces from seeds. LIFE-SPAN: annual or short-lived perennial. ORIGIN: native. HEIGHT: usually 3–15 cm tall. STEMS: 1 to several, outer stems ascending to prostrate, inner stems nearly erect, cespitose from a caudex surmounting a taproot; pubescence appressed, straight or spreading and sinuous, hairs dolabriform. LEAVES: alternate, odd-pinnately compound (2–14 cm long), leaflets 3–17; leaflets oblong to elliptic or oblanceolate (5–20 mm long, 1.5–7 mm wide); apex and base obtuse to acute; margins entire; upper surface nearly glabrous to thinly canescent; lower surface canescent, hairs dolabriform; petioles 0–2 cm long; stipules lanceolate (2–8 mm long), distinct, somewhat membranous. INFLORESCENCES: racemes of chasmogamous flowers seldom surpassing the leaves, subcapitate to ovoid, terminal; flowers 3–17; peduncles 3–11 cm long; racemes of cleistogamous flowers inconspicuous, on short peduncles near the base of the plants, flowers 2–8. FLOWERS: chasmogamous flowers perfect, papilionaceous, calyx tube campanulate (3–5 mm long), densely canescent, lobes 5; lobes long-acuminate (2–4 mm long); corolla greenish-white to yellowish-white, sometimes suffused with lavender or bicolored; banners 8–14 mm long, strongly to moderately reflexed; wings 7–12 mm long, claws 3–4.5 mm long; keels 6–10 mm long, claws 3.5–4.5 mm long; stamens 10, diadelphous; cleistogamous flowers 4–7 mm long, often shorter than the calyx teeth. FRUITS: legumes sessile, spreading, oblong-lanceolate to ellipsoid or ovoid-acuminate (1.5–4 cm long, 5–8 mm in diameter), inflated, straight or curved, obcompressed to subterete, beak conspicuous; unilocular; upper margin nearly straight; lower margin convex, sometimes dorsally sulcate; valves fleshy and becoming coriaceous, strigulose, darkening to reddish-brown to tan at maturity; seeds several. SEEDS: cordiform to nearly oval (1.5–2.5 mm in diameter), compressed, yellow to brown, sometimes with purple spots, indurate. $2n=26$.

HABITAT: Lotus milkvetch is infrequent to common on sandy, gravelly, or rocky prairie uplands, bluffs, canyons, and roadsides. USES AND VALUES: It is rarely grazed by livestock. The seeds are eaten by ground-foraging birds, and the legumes and seeds are eaten by small mammals. Nectar-seeking insects visit the flowers. Lotus milkvetch is planted in rock gardens. It has little value for erosion control. ETHNOBOTANY: Some Plains Indians collected immature legumes and ate them raw or cooked.

VARIETIES: This highly variable species has been divided into about ten varieties. SYNONYMS: *Astragalus ammolotus* Greene, *A. batesii* A. Nelson, *A. elatiocarpus* E. Sheld., *A. nebraskensis* (Bates) Bates, *A. reverchonii* A. Gray, *Batidophaca cretacea* (Buckley) Rydb., *B. lotiflora* (Hook.) Rydb., *B. nebraskensis* (Bates) Rydb., *Cystopora elatiocarpa* (E. Sheld.) Lunell, *C. lotiflora* (Hook.) Lunell, *Phaca cretacea* Buckley, *P. elatiocarpa* (E. Sheld.) Rydb., *P. lotiflora* (Hook.) Torr. & A. Gray, *P. reverchonii* (A. Gray) Rydb., *Tragacantha lotiflora* (Hook.) Kuntze. OTHER COMMON NAMES: low milkvetch, Reverchon astragalus.

Missouri milkvetch
Astragalus missouriensis Nutt.

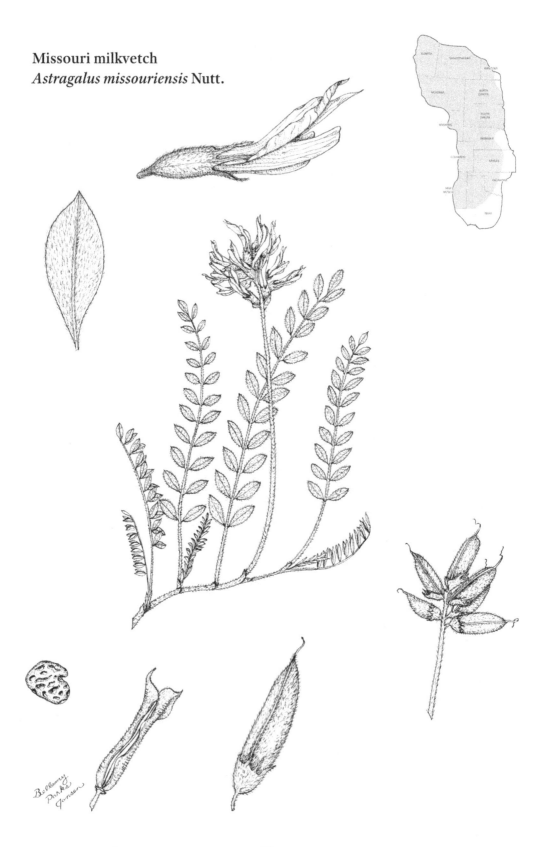

The Genera of North American Plants 2:99. 1818.
missouriensis: of or from Missouri.

GROWTH FORM: forb, flowers March to August, reproduces from seeds. LIFE-SPAN: perennial. ORIGIN: native. LENGTH: 6–20 cm. STEMS: several to many from a caudex and taproot, short-caulescent, usually cespitose; outer stems spreading, prostrate or ascending; inner stems erect; densely strigose-canescent, hairs dolabriform. LEAVES: mostly basal, sometimes alternate, odd-pinnately compound (4–14 cm long), leaflets 7–25; leaflets oblanceolate to elliptic (5–17 mm long, 2–6 mm wide), rather thick; apex acute to obtuse or mucronate; base obtuse; margins entire; both surfaces densely canescent giving a silvery to grayish-green appearance, hairs dolabriform; stipules lanceolate to lance-oval (2–9 mm long), distinct. INFLORESCENCES: racemes elevated above the leaves, short (5–20 mm long), terminal, scarcely elongating in fruit; flowers 3–15; peduncles 10–15 cm long, becoming prostrate in fruit. FLOWERS: perfect, papilionaceous; calyx tubes cylindric to campanulate (6–14 mm long), oblique, often dark-pigmented, strigulose with appressed hairs, lobes 5; lobes subulate to long-acuminate (1–4 mm long), strigulose with appressed hairs; corolla purplish-pink (rarely white or deep purple), drying blue, white patch on banner; banner moderately to strongly reflexed (1.5–2.5 cm long), notched; wings 1.3–2 cm long, claws 7–11 mm long; keel 1.2–1.9 cm long, claws 7–11 mm long; stamens 10, diadelphous, 9 united, 1 free; pedicels 1–3.5 mm long. FRUITS: legumes spreading, ascending or descending; oblong-cylindric or ellipsoid (1.5–2.5 cm long, 4–10 mm wide), slightly curved, slightly compressed in cross section; unilocular or subunilocular; glabrate to strigulose; hairs appressed; dull, reddish when immature, becoming coriaceous and black with age; dehiscent through the gaping beak (beak about 4 mm long); seeds many. SEEDS: cordiform to reniform (2–3 mm in diameter), notched, compressed, surface pitted, indurate. $2n=22$.

HABITAT: Missouri milkvetch is common in dry sandy to rocky soils of prairies, rangelands, hillsides, bluffs, washes, badlands, and stream banks. USES AND VALUES: Missouri milkvetch produces good forage for livestock, and it decreases with improper grazing. It does not accumulate large amounts of selenium and has not been shown to contain dangerous amounts of swainsonine. Deer, pronghorn, elk, and bighorn sheep graze Missouri milkvetch, and its seeds are eaten by ground-foraging birds and small mammals. Its purplish-pink flowers and drought tolerance make it an attractive addition to rock gardens. ETHNOBOTANY: Some Plains Indians and pioneers gathered the immature legumes for food.

VARIETIES: This species has been divided into four varieties. Var. *missouriensis* is most common in the Great Plains. Vars. *accumbens* (E. Sheld.) Isely, *humistratus* Isely, and *mimetes* Barneby grow west of the region. SYNONYMS: *Astragalus melanocarpus* Nutt., *Tragacantha missouriensis* (Nutt.) Kuntze, *Xylophacos missouriensis* (Nutt.) Rydb.

SIMILAR SPECIES: *Astragalus humistratus* A. Gray, groundcover milkvetch, is found in the western Great Plains in Texas and New Mexico. It is prostrate, with pale or dully varicolored corollas, and often elongate calyx lobes.

Woolly locoweed
Astragalus mollissimus Torr.

Annals of the Lyceum of Natural History of New York 2:178–79. 1827.
mollissimus (Lat.): softest, referring to the covering of silky hairs.

GROWTH FORM: forb, flowers April to June, reproduces from seeds. LIFE-SPAN: perennial. ORIGIN: native. HEIGHT: 2–15 cm, rarely to 25 cm. STEMS: herbaceous, subcaulescent to short-caulescent, outer stems prostrate, inner stems ascending, 1 to several from a woody taproot, densely to loosely tufted, usually simple, pubescent; pubescence villous-tomentose, hairs basifixed, without bulbous bases. LEAVES: alternate, odd-pinnately compound (5–26 cm long), ascending or arching, leaflets 15–35; leaflets obovate to oblanceolate or elliptic (5–25 mm long, 2–15 mm wide), infrequently lanceolate; apex obtuse or acute, sometimes mucronate; base obtuse or acute; margins entire; surfaces villous-tomentose; hairs long, silky, basifixed, without bulbous bases; stipules triangular (to 1.5 cm long), distinct, membranous, silky. INFLORESCENCES: racemes about equaling the height of the subtending leaves, oblong (4–10 cm long, may be longer in fruit), terminal; flowers 4–45; peduncles naked (5–20 cm long), becoming prostrate with maturity. FLOWERS: perfect, papilionaceous; calyx tube cylindric or subcylindric (5–10 mm long), slightly oblique, lobes 5; lobes long-acuminate (2–5 mm long), pubescent; corolla purple to reddish-purple (rarely yellow or white), often drying blue; banner slightly to moderately reflexed (1.7–2.2 cm long); wings 1.5–2 cm long, claws 7–11 mm long; keel 1.5–1.8 cm long, claws 8–11 mm long; stamens 10, diadelphous. FRUITS: legumes ascending or spreading, plump-ellipsoid to oblong-ellipsoid or lanceolate (1.4–2.5 cm long including the beak, 4–9 mm wide), often broadest near the base; usually slightly curved; bilocular for most of the length; surfaces usually glabrous except at the apex; coriaceous to woody; seeds many. SEEDS: reniform to cordiform (2–3 mm long), compressed, smooth to rough, indurate. $2n=22$.

HABITAT: Woolly locoweed is locally common in dry prairies, hillsides, valley floors, and sagebrush flats. It is most abundant in sandy or rocky soils. USES AND VALUES: Generally, it is unpalatable to livestock, but animals will eat it if other forage is not available. Some animals, especially horses, may become habituated to these plants and refuse to eat better feed. The toxic principle is an indolizidine alkaloid (swainsonine). It remains toxic when dry, but pastures can be grazed after the plants have withered away. It does not accumulate selenium.

VARIETIES: This highly variable species has been divided into nearly a dozen varieties. SYNONYMS: *Astragalus mogollonicus* Greene, *A. simulans* Cockerell, *A. thompsoniae* S. Watson, *Phaca villosa* S. Watson. OTHER COMMON NAMES: crazyweed, locoweed, loco, rattlebox weed, stemmed loco, Texas locoweed, Thompson loco, woolly loco, woolly milkvetch.

SIMILAR SPECIES: *Astragalus australis* (L.) Lam., Indian milkvetch, grows in the northwestern part of the region. Its corollas are white or ochroleucous and has notched wings and large, veined, semicoriaceous lower stipules. *Astragalus purshii* Douglas, Pursh milkvetch, also has white to ochroleucous corollas and grows in the northwest portion of the region. It has villous hairs (to 5 mm long) on the fruits. Corollas of *Astragalus shortianus* Nutt., Short milkvetch, are pink to purple, and fruits are strigulose. It is uncommon in the western Great Plains.

Pulse milkvetch
Astragalus multiflorus (Pursh) A. Gray

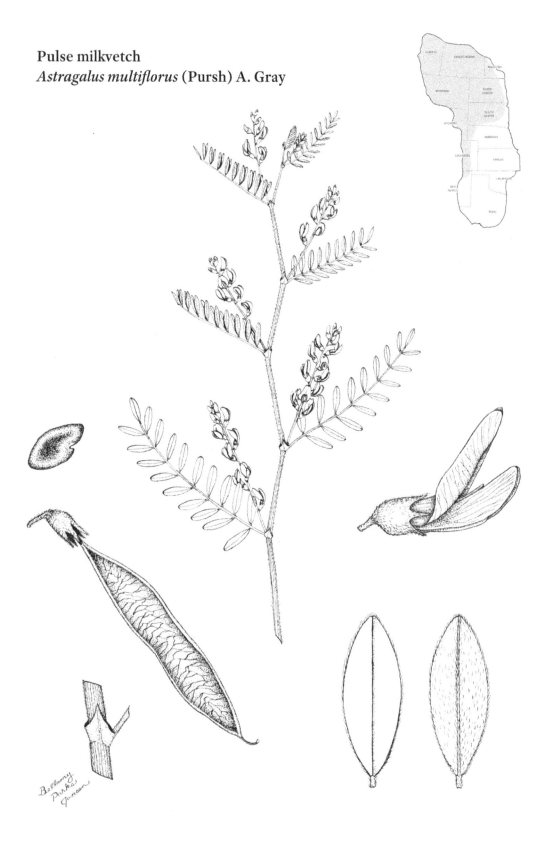

Proceedings of the American Academy of Arts and Sciences 6:226. 1864.
multus (Lat.): many; +*floris* (Lat.): flowers, in reference to the numerous inflorescences.

GROWTH FORM: forb, flowers May to July, reproduces from seeds. LIFE-SPAN: perennial. ORIGIN: native. HEIGHT: 10–50 cm. STEMS: erect to ascending or sprawling, several to many from a superficial or slightly subterranean caudex surmounting a woody taproot, clustered, branched or spurred at most nodes above a leafless base; thinly strigose with basifixed hairs to nearly glabrous. LEAVES: alternate, odd-pinnately compound (3–10 cm long), leaflets 9–27; leaflets elliptic to linear-oblong (7–25 mm long), flat or folded; apex acute to mucronate; base obtuse to acute; margins entire; upper surface glabrate; lower surface thinly strigose, hairs basifixed; upper leaves sessile, lower leaves petiolate; lower stipules connate; upper stipules less connate to distinct, deltoid (1.5–9 mm long), blackening on drying. INFLORESCENCES: racemes 3–6 cm long, numerous, axillary, occasionally 2 per node, lax to remotely flowered; flowers 7–28; flowers first ascending and then declining; peduncles slender or filiform (2–50 mm long). FLOWERS: perfect, papilionaceous; calyx tube campanulate (2–4 mm long), strigulose with white and black hairs or only black hairs, lobes 5; lobes subulate (1–3 mm long), strigulose; corolla ochroleucous to white (rarely pinkish-lavender or pinkish-purple), often with a purple-spotted keel; banner 7–9 mm long; wings 6–9 mm long, claws 2–3 mm long; keel 4–7 mm long, claws 2–3.5 mm long; stamens 10, diadelphous; pedicels 0.5–2.5 mm long in flower. FRUITS: legumes stipitate to exserted-stipitate (stipe to 7 mm long), pendulous, elliptic-oblong to oblanceolate (8–18 mm long, 3–5 mm wide), beak 3–6 mm long; strongly laterally compressed; unilocular; valves papery, green and mottled when immature, black when mature; reticulate, glabrate or strigose, not sulcate, tardily dehiscent; seeds usually 2–5. SEEDS: reniform to cordiform or nearly oval (2–3 mm long), notched, compressed, brown, often purple-spotted, somewhat shiny, indurate. $2n=24, 32$.

HABITAT: Pulse milkvetch grows in prairies, rangelands, sagebrush flats, bluffs, badlands, and roadside cuts, especially in alkaline clay and calcareous soils. USES AND VALUES: Palatability of pulse milkvetch is variable, but it is eaten by cattle, horses, sheep, pronghorn, elk, deer, and bighorn sheep. Small mammals eat the seeds and legumes, and ground-foraging birds eat the seeds. It has not been reported to be poisonous or to accumulate high levels of selenium. It is not planted for erosion control, and its uses in landscaping are limited. OTHER: It was first collected by Meriwether Lewis "on the banks of the Missouri" River in 1804.

SYNONYM: *Astragalus tenellus* Pursh. OTHER COMMON NAMES: loose-flowered milkvetch, looseflower milkvetch, red-stemmed milkvetch.

SIMILAR SPECIES: *Astragalus vexilliflexus* E. Sheld., bentflowered milkvetch, is similar and grows in the northwestern Great Plains. However, its corollas are pinkish-purple (in the northern portion of its range) to white (in the southern portion of it range). Its legumes are sessile, or nearly so, instead of being stipitate.

Nuttall milkvetch
Astragalus nuttallianus DC.

Prodromus Systematis Naturalis Regni Vegetabilis 2:289. 1825.
nuttallianus: named for Thomas Nuttall (1786–1859), English-born naturalist who lived
and collected plants in North America.

GROWTH FORM: forb, flowers April to June, reproduces from seeds. LIFE-SPAN: annual or
winter annual. ORIGIN: native. HEIGHT: 3–40 cm. STEMS: ascending or erect, sometimes
prostrate and forming mats, simple or basally branched from a taproot, glabrate or
silvery-strigose. LEAVES: alternate, odd-pinnately compound (1–7 cm long), leaflets
7–25; leaflets obovate or ovate-cuneate to elliptic or oblong (2–15 mm long, 1–5 mm
wide); apex truncate-emarginate or retuse to acute; base obtuse to acute; margins
entire; surfaces strigose, hairs basifixed; petioled or subsessile above; stipules distinct
(1–6 mm long). INFLORESCENCES: racemes, subcapitate or shortly racemose, axillary,
erect or incurved-ascending; flowers 1–8; peduncles 5–90 mm long. FLOWERS: perfect,
papilionaceous; calyx tube 1.2–3 mm long, thinly or conspicuously black to white
strigulose, hairs 0.5–1 mm long; lobes 5 (1–3 mm long); corolla white to pinkish-purple
or violet to dull blue, often drying yellowish-white; banner 5–9 mm long; wings 5–7
mm long; keel obtuse (4–6 mm long); stamens 10, diadelphous; pedicels commonly
arching (0.5–1 mm long in flower, 0.8–1.6 mm long in fruit). FRUITS: legumes spreading
or declined, oblong to linear-oblanceolate (1–2.5 cm long, 2–3.5 mm wide); initially
subtriquetrous-compressed then subterete, incurved to nearly straight, glabrous to
strigulose, dorsally sulcate, bilocular; valves papery to thinly coriaceous, green when
immature, becoming reticulate and dark brown, splitting from the apex; seeds 10–24.
SEEDS: reniform to cordiform (2–3 mm long), notched, yellow to brown, sometimes
with purple spots, somewhat wrinkled, shiny, compressed, indurate. $2n=22$.

HABITAT: Nuttall milkvetch is scattered to common on sandy and gravelly prairies and
rangelands. USES AND VALUES: It has relatively low palatability and is only occasionally
eaten by cattle, sheep, goats, and deer. Small mammals eat the legumes and seeds.
Quail and other ground-foraging birds eat the seeds. Nuttall milkvetch accumulates
selenium and contains several alkaloids and a small amount of swainsonine. However,
it is rarely toxic to livestock or wildlife.

VARIETIES: Two varieties occur in the Great Plains. Var. *nuttallianus* has glabrous legumes,
and the leaf apices are retuse or truncate-emarginate. It grows on mesic sites from
southern Kansas to southern Texas. Var. *austrinus* (Small) Barneby has strigulose
or glabrous legumes, and the leaf apices are not retuse or truncate-emarginate. It
grows farther west and southwest than var. *nuttallianus*, but the ranges of the two
overlap. SYNONYMS: *Astragalus micranthus* Nutt., *Hamosa nuttalliana* (DC.) Rydb.,
Tragacantha micrantha Kuntze. OTHER COMMON NAMES: annual astragalus, Menzies
loco, smallflowered milkvetch.

SIMILAR SPECIES: *Astragalus americanus* (Hook.) M.E. Jones, American milkvetch, is
caulescent and has white to ochroleucous corollas and distinct stipules. It grows in
the northwestern Great Plains. *Astragalus lindheimeri* Engelm. *ex* A. Gray, Lindheimer
milkvetch, is an annual or winter annual. It has a large purple banner (1.2–1.9 cm long)
and is found in the southern Great Plains.

Narrowleaf milkvetch
Astragalus pectinatus (Hook.) Douglas *ex* G. Don

A General History of the Dichlamydeous Plants 2:257. 1832.
pectinatus (Lat.): comblike, in reference to the leaves.

GROWTH FORM: forb, flowers May to June, reproduces from short rhizomes and seeds.
LIFE-SPAN: perennial. ORIGIN: native. HEIGHT: 10–60 cm. STEMS: ascending to prostrate
with tips ascending, from a short rhizome or a caudex surmounting a thick taproot,
forming mats, diffusely branched at or below the middle, leafless below, strigulose with
basifixed hairs. LEAVES: alternate, odd-pinnately compound (4–11 cm long), leaflets
5–21; leaflets linear to linear-oblanceolate (1.3–7 cm long, 1–2.5 mm wide), somewhat
sickle-shaped, terminal leaflet longest and continuous with the rachis, stiff; apex
acuminate; base rounded to obtuse; margins entire; upper surfaces glabrate; lower
surfaces strigulose, hairs appressed and basifixed; sessile; petioles absent or present
(to 5 mm long); stipules deltate (1–10 mm long), connate and sheathing at lower
nodes, distinct or connate at upper nodes, glabrous. INFLORESCENCES: racemes 3–17 cm
long, axillary, flowers 6–30, becoming loose; peduncles stout (2–12 cm long). FLOW-
ERS: perfect, papilionaceous; calyx tube cylindric (6–10 mm long), strigulose; hairs
black or white or mixed; lobes 5; lobes triangular or short-acuminate (1–3 mm long);
corolla ochroleucous, drying yellow; banner strongly to moderately reflexed (1.7–2.4
cm long), deeply notched; wings 1.7–2 mm long, claws 7–9 mm long; keel 1.5–1.7 cm
long, claws 8–9 mm long; stamens 10, diadelphous; pedicels 1–5 mm long. FRUITS:
legumes sessile, eventually drooping, ovoid to ellipsoid (1–2.5 cm long including the
beak, 6–8 mm wide), straight or slightly curved, slightly compressed dorsiventrally,
unilocular; valves fleshy when immature, tan to brown and woody when mature,
glabrous, rugulose, suture prominent, dehiscent; seeds 20–30. SEEDS: cordiform or
reniform (3–4 mm long), notched, light or pinkish-brown, pitted, indurate. $2n=22$.

HABITAT: Narrowleaf milkvetch is scattered to locally abundant on dry rangeland and
prairie hillsides and hilltops, flats, eroded badlands, and roadsides. It is most common on clay or shale soils, and is an indicator of the presence of selenium in the soil.
USES AND VALUES: Narrowleaf milkvetch is relatively unpalatable, and the foliage is
infrequently eaten by cattle, horses, and sheep. It accumulates selenium, and animals
consuming single massive amounts of the herbage may exhibit lethargy, difficulty in
breathing, increased pulse, blindness, wandering, excitement, and depression. These
symptoms may be followed by coma, respiratory failure, and death. Grazing wildlife
usually ignore narrowleaf milkvetch. However, the legumes and seeds are eaten by
small mammals, and the seeds are eaten by ground-foraging birds and songbirds
without apparent poisoning. It is pollinated by bumblebees and other long-tongued
bees. The flowers are visited by butterflies and skippers. It is not used for erosion
control. The selenium accumulated in the plant causes it to emit a urinelike odor,
making it unacceptable for landscape plantings.

SYNONYMS: *Cnemidophacos pectinatus* (Hook.) Rydb., *Ctenophyllum pectinatum* (Hook.)
Rydb., *Phaca pectinata* Hook., *Tragacantha pectinata* (Hook.) Kuntze. OTHER COM-
MON NAMES: narrowleaved milkvetch, narrowleaf poisonvetch, tineleaf milkvetch.

Platte River milkvetch
Astragalus plattensis Nutt.

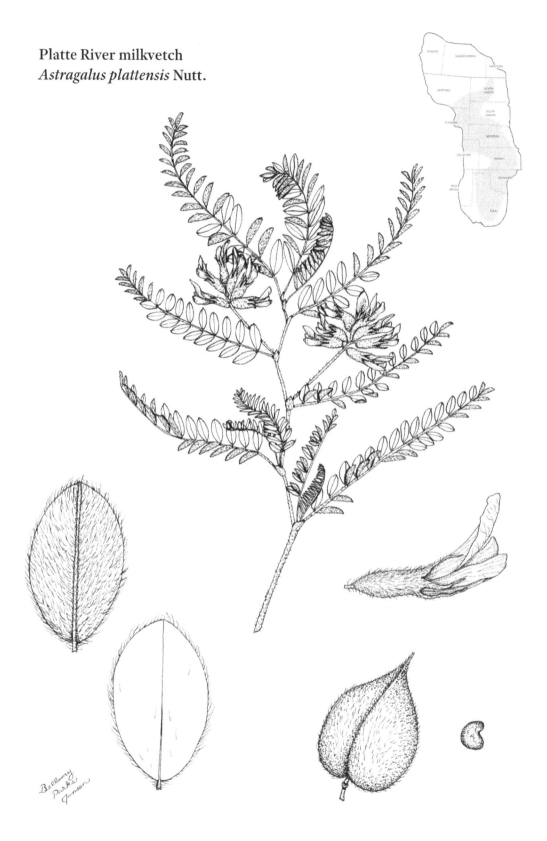

A Flora of North America 1(2):332. 1838.
plattensis: of or from the Platte River region.

GROWTH FORM: forb, flowers March to July, reproduces from rhizomes and seeds. LIFE-SPAN: perennial. ORIGIN: native. LENGTH: 10–40 cm. STEMS: prostrate or decumbent with ascending tips, usually several, sometimes solitary; caulescent from elongated rhizomes and a woody taproot; branched basally, diffuse or spreading, villous with basifixed hairs. LEAVES: alternate, odd-pinnately compound (3–12 cm long), leaflets 9–35; leaflets oblong to elliptic (4–18 mm long, 2.5–8 mm wide); apex acute or obtuse (rarely truncate); base obtuse to acute; margins entire; upper surface glabrous; lower surface villous, hairs basifixed; petiolules short; short-petioled below, subsessile above; stipules lanceolate to ovate-acuminate (1.5–7 mm long), membranous, connate and sheathing on lower nodes, connate or distinct on upper nodes, glabrous or glabrescent. INFLORESCENCES: racemes 4 cm long or more, usually shorter than the subtending leaves, dense, axillary; flowers 3–15, ascending; peduncles stout (4–8 cm long). FLOWERS: perfect, papilionaceous; calyx tube cylindric (6–9 mm long), slightly oblique, pubescent; hairs white or black or mixed; lobes 5, long-acuminate (2.5–6 mm long); corolla pink or pinkish-purple to lavender, drying whitish-blue; banner becoming moderately to strongly reflexed (1.2–2 cm long), deeply notched; wings 1.2–1.8 cm long, claws 6–9 mm long; keel 1.2–1.5 cm long, claw 6–8 mm long; stamens 10, diadelphous; pedicels 2–3 mm long. FRUITS: legumes sessile to shortly stipitate, ascending or often spreading on the soil surface, ovoid or ellipsoid (1–2 cm long excluding the 1–3.5 mm beak, 1–1.3 cm wide), sharply beaked, fleshy, inflated, bilocular; ventral face distinctly sulcate, often mottled with red or purple, villosulous, becoming coriaceous or woody; tardily dehiscent; seeds 25–45. SEEDS: reniform (2.5–3.5 mm long), pinkish-brown to black, smooth, compressed, indurate. $2n=22$.

HABITAT: Platte River milkvetch is infrequent to common on sandy and rocky prairies, rangelands, open woodlands, gullies, and bluffs. USES AND VALUES: It is highly palatable and is eaten by all classes of livestock. Platte River milkvetch decreases with continuous heavy use. Platte River milkvetch is grazed by pronghorn, elk, and deer. Small mammals eat the legumes and seeds, and birds eat the seeds. It is visited by numerous bees and other pollinating insects. It has been used as a specimen plant in rock gardens. ETHNOBOTANY: The legumes were collected, boiled, and eaten by members of some tribes of Plains Indians.

SYNONYMS: *Astragalus pachycarpus* Torr. & A. Gray, *Geoprumnon pachycarpum* (Torr. & A. Gray) Rydb., *G. plattense* (Nutt.) Rydb., *Phaca plattensis* MacMill., *Tragacantha plattensis* (Nutt.) Kuntze. OTHER COMMON NAMES: ground plum, Nebraska milkvetch, Platte milkvetch.

SIMILAR SPECIES: *Astragalus neglectus* (Torr. & A. Gray) E. Sheld., Cooper milkvetch, also, is caulescent, but it has erect stems and white to yellow corollas. It is found in the northeastern Great Plains. *Astragalus praelongus* var. *ellisiae* (Rydb.) Barneby, stinking milkvetch, is caulescent with erect stems and ochroleucous corollas. These plants have a strong, ill-scented, urinelike odor. It grows in the extreme southwestern Great Plains.

Racemed poisonvetch
Astragalus racemosus Pursh

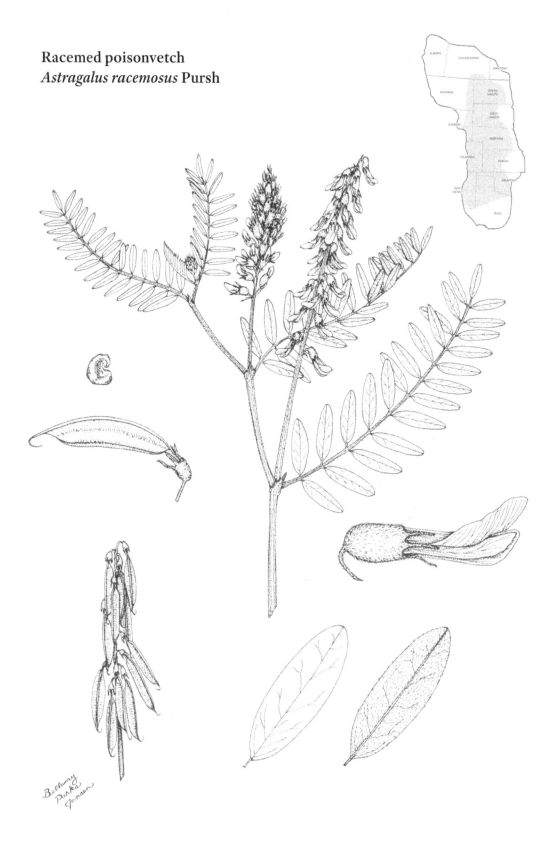

Flora Americae Septentrionalis 2:740. 1814.

racemosus (Lat.): full of clusters, in reference to the inflorescence.

GROWTH FORM: forb, flowers May to July, reproduces from seeds. LIFE-SPAN: perennial. ORIGIN: native. HEIGHT: usually 20–90 cm, rarely to 1 m. STEMS: erect to ascending, few from a short caudex surmounting a taproot, caulescent, usually branched from the base or above, strongly strigulose, hairs basifixed. LEAVES: alternate, odd-pinnately compound (4–15 cm long), leaflets 9–31; leaflets paired or irregularly arranged, linear to narrowly oblong (1–4 cm long, 2–9 mm wide); apex obtuse to acute, sometimes mucronate; base acute; upper surface glabrous; lower surface strigose, hairs basifixed; margins entire, rolled upward when dry; short-petioled below, often sessile above; petiolules short; stipules membranous, those of the lower nodes connate, those of the upper nodes distinct (3–12 mm long), glabrous. INFLORESCENCES: racemes axillary (4–10 cm long) lax; flowers 12–70; peduncles erect (3–11 cm long), stout. FLOWERS: perfect, papilionaceous; calyx tube campanulate to cylindric (5–9 mm long), slightly oblique, sparsely strigose or glabrous; lobes 5, acuminate to subulate (2–6 mm long), green; corolla white or cream, usually tinged with pink or purple; banner moderately to strongly reflexed (1.3–2 cm long); wings 1.2–1.9 cm long, often light purple or tipped with purple; keel 1.1–1.6 cm long, usually purple; stamens 10, diadelphous; pedicels in flower 2–3.5 mm long, in fruit 3–8 mm long. FRUITS: legumes stipitate (stipe 3–7 mm long), pendulous, ellipsoid to linear (1–3 cm long, 3–8 mm wide), triangular in cross section, straight or slightly curved, acute on upper suture, sulcate on lower suture and not grooved ventrally; unilocular; valves papery, glabrous; seeds several to many. SEEDS: reniform to cordiform (2.2–3 mm long), brown, often purple-spotted, somewhat shiny, indurate. $2n=24$.

HABITAT: Racemed poisonvetch grows on dry sandy or rocky prairie uplands, rangelands, hillsides, eroded slopes, and disturbed roadsides. It is an indicator of selenium in the soil. USES AND VALUES: It is rarely eaten by livestock or grazing wildlife. It accumulates selenium, making it a toxic plant. Ingestion of less than 1 kg of herbage by cattle and 0.25 kg by sheep may cause acute selenium poisoning. Chronic poisoning occurs when small amounts are eaten over an extended period. Death may result, and there is no antidote for selenium poisoning. ETHNOBOTANY: Indians in the Great Plains did not use this plant and recognized it as a toxic plant. The Lakota name śuŋkléja hu translates as "horse urine stem" because the plants have a pungent odor resembling urine. It is not used for erosion control or landscaping.

SYNONYMS: *Astragalus galegoides* Nutt., *Tium brevisetum* (M.E. Jones) Rydb., *T. racemosum* (Pursh) Rydb., *Tragacantha racemosa* (Pursh) Kuntze. OTHER COMMON NAMES: alkali milkvetch, cream milkvetch, creamy milkvetch, locoweed, pežuta ska hu (Lakota), racemed milkvetch, racemose milkvetch, śuŋkléja hu (Lakota).

SIMILAR SPECIES: *Astragalus wrightii* A. Gray, Wright milkvetch, has similar flower color, but it is an annual with stiffly erect peduncles, capitate inflorescences, and petals that scarcely exceed the villous calyx. It only grows in Texas at the southern extreme of the Great Plains.

Tufted milkvetch
Astragalus spatulatus E. Sheld.

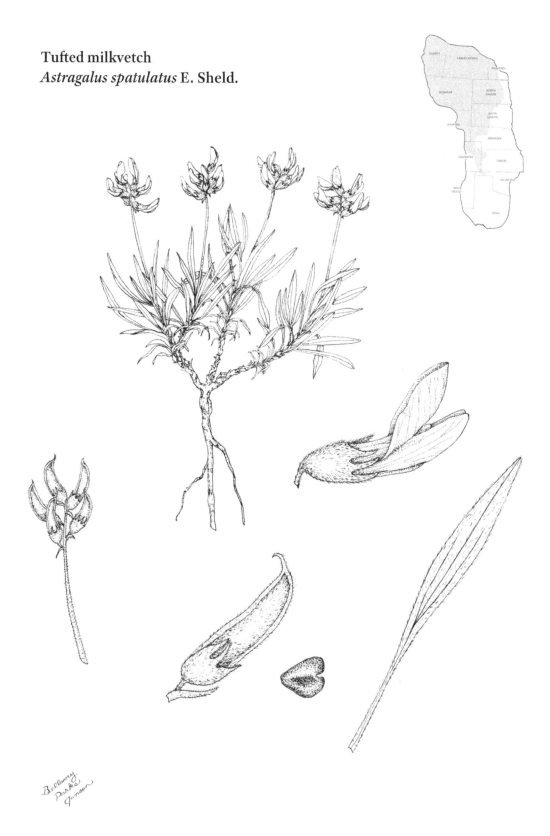

Minnesota Botanical Studies 1(1):22. 1894.

spatula (Lat.): spoon, in reference to the shape of the leaflets.

GROWTH FORM: forb, flowers May to July, reproduces from seeds and a repeatedly forking caudex. LIFE-SPAN: perennial. ORIGIN: native. HEIGHT: 6–10 cm. STEMS: herbaceous, ascending to decumbent; 1 to several from a repeatedly forking caudex surmounting a taproot, subcaulescent, cespitose-tufted, forming mats and domed cushions (up to 30 cm in diameter); surfaces silvery-strigose (hairs up to 1 mm long), some hairs dolabriform; covered with imbricate stipules; most stems reduced to crowns. LEAVES: alternate, clustered, mostly basal, some compound but mostly consisting of simple phyllodes; upper leaves 1.5–6 cm long, leaflets 2–4; leaflets oblong-lanceolate; lower leaflets narrowly spatulate to oblanceolate or linear (5–15 mm long); blades flat to folded or involute; apex subacute to acute, sometimes mucronate; base attenuate; margins entire; surfaces pubescent; petioles well-developed; stipules connate (2–7 mm long), overlapping, membranous, flat or folded to involute, persistent; those of the short outer phyllodes the smallest. INFLORESCENCES: racemes subcapitate (2–10 cm long), terminal; flowers 2–11, sometimes single; peduncles and rachis elongating in fruit (5–70 mm long); bracts ovate to lance-attenuate (0.5–5 mm long), scarious. FLOWERS: perfect, papilionaceous; calyx tube obconic-campanulate (2–3.5 mm long), strigulose with white or black hairs; lobes 5, long-acuminate to subulate (1.5–2.5 mm long); corolla bright pinkish-purple (rarely cream or yellowish) with white-tipped wings; banner slightly to moderately reflexed (6–10 mm long, 4.5–6.5 mm wide), veins suffused with dull lilac, claw cuneate; wings 6–8 mm long, claws 2–3 mm long; keel 4–6 mm long, claws 2–3 mm long; stamens 10, diadelphous; pedicels about 1.5 mm long in flower, about 3 mm long in fruit. FRUITS: legumes subsessile, erect to ascending, sometimes humistrate, narrowly oblong (5–16 mm long, 1–4 mm wide), acuminate, slightly curved or straight, laterally compressed, rounded at the base, unilocular; valves solid green or purplish; strigose, thinly coriaceous, becoming papery; seeds 4–12. SEEDS: cordiform to reniform (1.5–2.5 mm long), slightly compressed, rugulose, purplish-brown to olive, indurate. $2n=24$.

HABITAT: Tufted milkvetch is scattered to locally common in prairies, rangelands, and shallow or eroded soils of hilltops and bluffs. USES AND VALUES: Its palatability is poor for livestock, and it is rarely eaten. It provides poor to fair forage for pronghorn and deer. The legumes are eaten by small mammals, and the seeds are eaten by ground-foraging birds and small mammals. The flowers are visited by various bees and other insects. It protects the soil from erosion, but growth is too slow to warrant planting as a conservation treatment. It is a common and important plant in rock gardens.

SYNONYMS: *Astragalus caespitosus* (Nutt.) A. Gray, *A. simplex* Tidestr., *Homalobus brachycarpus* Nutt., *H. caespitosus* Nutt., *H. canescens* Nutt., *H. uniflorus* Rydb., *Tragacantha caespitosa* (Nutt.) Kuntze. OTHER COMMON NAMES: draba milkvetch, low milkvetch, prairie milkvetch, sessileflower milkvetch, silky milkvetch.

Baptisia Vent.

baptisis (Gk.): a dipping, as in dyeing, referring to some species being used as a source of dye.

Perennial herbs from thick rhizomes; stems herbaceous, ascending to erect, usually much-branched, lateral branches often surpassing the central axis in length; leaves palmately trifoliate (rarely simple); petiolules often short making the leaves appear as pinnately trifoliate; glabrous, glaucous, or pubescent (hairs basifixed); margins entire; generally turning black on drying; inflorescences long or short racemes terminating the central axis and often also the lateral branches, flowers conspicuous; calyx bilabiate, the upper lip entire, notched or lobes 2; lower lip lobes 3, deeply lobed; corolla white, yellow, or blue to purple; banner reniform to nearly circular, not longer than the wings, its sides usually reflexed; wings and keel nearly equal, straight, oblong; keel petals nearly separate; stamens 10, distinct; ovary stipitate; fruits legumes, globose to cylindric or thick-lenticular, inflated, papery to woody, terminated with a long or short curved beak, stalked within the persistent calyx; seeds many. $x=9$.

About 30 species are recognized in North America. Most grow in the eastern and southern parts of the United States. Five species are found in the Great Plains, and 3 are common.

1. Leaves, ovaries, and legumes glabrous; corollas white, blue, or tinged with purple
    2. Stipules slender, shorter than the petioles, caducous; corollas white or tinged with purple; stipe 2–3 times the length of the calyx. .... *Baptisia alba*
    2. Stipules broad, longer than the petioles, persistent; corollas blue; stipe 1–2 times the length of the calyx. .... *Baptisia australis*
1. Leaves, ovaries, and legumes pubescent (at least when young); corollas ochroleucous to cream or pale yellow. .... *Baptisia bracteata*

White wildindigo
Baptisia alba (L.) R. Br.

Hortus Kewensis 3:6. 1811.
alba (Lat.): white, in reference to the color of the corolla.

GROWTH FORM: forb, flowers May to July, reproduces from short rhizomes and seeds. LIFE-SPAN: perennial. ORIGIN: native. HEIGHT: usually 0.5–1.5 m, rarely to 2 m. STEMS: erect to ascending, often solitary, from a caudex and taproot or from short rhizomes, branching widely, herbaceous but appearing shrublike, glabrous to glaucous. LEAVES: alternate, palmately trifoliate; leaflets obovate to oblanceolate (2.5–8 cm long, 7–25 mm wide); apex obtuse to acute; base acute to acuminate; margins entire; surfaces glabrous; petiole 6–25 mm long; petiolule 0.5–1 mm long; stipules lanceolate to ovate (5–24 mm long), slender, small, mostly caducous (some persisting until anthesis). INFLORESCENCES: racemes 20–60 cm long, 1 to few, erect, terminal, showy; flowers numerous; peduncles 5–20 cm long. FLOWERS: perfect, papilionaceous; calyx tube 7–9 mm long, densely pubescent within, persistent, lobes 4; upper lip entire or notched; lobes of lower lip deltoid (2–4 mm long); corolla white to creamy-white with a tinge of purple on the banner; banner 1–1.7 cm long; wings and keel 2–2.5 cm long; pedicels 3–15 mm long; ovaries glabrous. FRUITS: legumes stipitate (stipe 2–3 times the length of the calyx), ellipsoid to oblong (2.5–4 cm long, 8–12 mm wide), glabrous, glaucous, inflated, woody, abruptly narrowed to a short beak; green and then turning brown to black with maturity; seeds many. SEEDS: reniform (3.5–5 mm long, about 3 mm wide), notched, compressed, olive to black, glossy, covered with small resinous dots; indurate. $2n=18$.

HABITAT: White wildindigo is infrequent to common in prairies, open woodlands, ravines, alluvial soils along rivers, and wet meadows and valleys. USES AND VALUES: Animals seldom graze white wildindigo. It contains alkaloids that are emetic and cathartic. Young animals, particularly horses, are most susceptible to the toxins. Small mammals eat the legumes and seeds, and birds eat the seeds without poisoning. It attracts bumblebees, butterflies, and other pollinating insects. Caterpillars of butterflies, skippers, and moths eat the leaves. White wildindigo is difficult to grow from seed, and it is slow to establish. However, it is used in landscaping and for prairie restorations. Insect damage to wild harvested seed is usually severe. Stems and legumes are used in dried floral arrangements. ETHNOBOTANY: A poultice of white wildindigo was used by members of several tribes of Plains Indians for swellings, rheumatism, sores, wounds, and hemorrhoids. OTHER: When plants are mowed, regrowth has more persistent stipules and larger leaves. This foliage can be confused with that of *Baptisia australis*. In autumn, the stems break near the soil surface, and the wind rolls the plants like tumbleweeds scattering the seed.

VARIETIES: Var. *macrophylla* (Larisey) Isely grows in the Great Plains. Var. *alba* has larger flowers, and the legumes have thin walls and are brittle. It grows in the southeastern United States. SYNONYMS: *Baptisia lactea* (Raf.) Thiret, *B. leucantha* Torr. & A. Gray. OTHER COMMON NAMES: Atlantic wildindigo, large whiteindigo, largeleaf wildindigo, prairie indigo, white falseindigo, wild indigo.

Blue wildindigo
Baptisia australis (L.) R. Br.

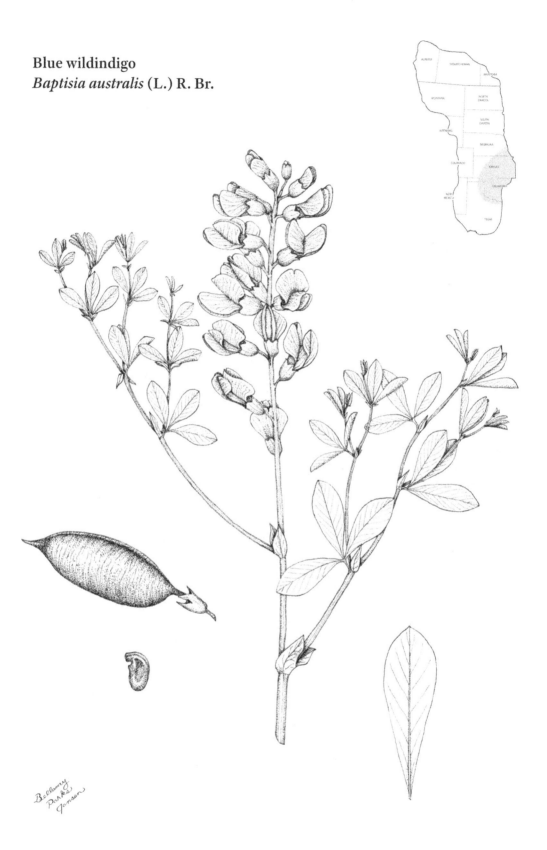

Hortus Kewensis 3:6. 1811.

australis (Lat.): southern, in reference to its primary distribution.

GROWTH FORM: forb, flowers April to June, reproduces from rhizomes and seeds. LIFE-SPAN: perennial. ORIGIN: native. HEIGHT: 0.4–1.5 m. STEMS: single primary stem erect, branches ascending to spreading, glabrous to somewhat glaucous. LEAVES: alternate, palmately trifoliate; leaflets obovate to oblanceolate or elliptic (1.5–5 cm long, 4–20 mm wide); apex rounded or acute; base cuneate; margins entire; surfaces glabrous to somewhat glaucous; petioles 5–18 mm long or subsessile; petiolules 0.5–3 mm long; stipules lanceolate to ovate-lanceolate (6–20 mm long), broad, persistent (at least on some parts of the plant). INFLORESCENCES: racemes 10–50 cm long, terminal, erect; bracts lanceolate (6–15 mm long), caducous. FLOWERS: perfect, papilionaceous; calyx 5–12 mm long, lobes 4; lobes of upper lip slightly notched or entire, lobes of lower lip ovate to triangular (2–5 mm long, usually shorter than the tube); corolla blue to purple or rarely white (2.5–3.5 cm long), the only blue-flowered *Baptisia*; banner 2–2.8 cm long, claw 2–5 mm long; wings 2.5–3 cm long; keel 2.5–3 cm long; pedicels 5–30 mm long; ovaries glabrous. FRUITS: legumes stipitate (stipe 1–2 times the length of the calyx), ellipsoid to globose (2–6 cm long, 1.5–25 cm in diameter), inflated, asymmetrical, beaked, glabrous, glaucous, purplish-black, ascending at maturity, seeds many. SEEDS: reniform to cordiform (3.5–5 mm long), compressed, notched, brown with resinous dots, indurate. $2n=18$.

HABITAT: Blue wildindigo is infrequent to common in sandy or rocky prairies, rocky open woodlands, limestone glades and stream valleys. USES AND VALUES: It contains several quinolizidine alkaloids giving the herbage an unpleasant taste to livestock and may cause anorexia and diarrhea. Birds and small mammals eat the seeds. Blue wildindigo is frequently grown as an ornamental. The legumes are used in dried flower arrangements. ETHNOBOTANY: Members of the Osage tribe made an eyewash from the foliage. Some Plains Indian children used the mature legumes as rattles. Some Cherokees made a cold tea from blue wildindigo to prevent vomiting and a hot tea as a purgative. A pulverized root was held over a tooth to relieve toothache. Also, they used it as a source of indigo dye for clothing. The true indigo plant (*Indigofera tinctoria* L.) is native to the Indian subcontinent.

VARIETIES: Var. *minor* (Lehm.) Fernald is common in the Great Plains, and var. *australis* grows in the eastern and southeastern portions of the United States. Var. *minor* has shorter stems, more spreading branches, and smaller leaflets. Also, the stipe of the fruit is usually twice the length of the calyx in var. *minor* versus scarcely longer, or not at all longer, than the calyx in var. *australis*. Var. *aberrans* (Larisey) M.G. Mendenh. plants are petite and grow in alkaline glades and barrens in the southeastern United States. SYNONYMS: *Baptisia caerulea* Eaton & Wright, *B. confusa* Sweet *ex* G. Don, *B. exaltata* Sweet, *B. versicolor* Lodd., *Podalyria australis* (L.) Willd., *P. coerulea* Michx., *Ripasia caerulea* (Trew) Raf., *Sophora australis* L., *S. caerulea* Trew. OTHER COMMON NAMES: blue fasleindigo, blue indigo, blue rattlebush, false indigo, spiked indigo, wild indigo.

Plains wildindigo
Baptisia bracteata Muhl. *ex* Elliott

A Sketch of the Botany of South-Carolina and Georgia 1(5):469. 1817.
bracta (Lat.): referring to the thin, scalelike leaves subtending the flowers.

GROWTH FORM: forb, flowers from April to June, reproduces from seeds. LIFE-SPAN: perennial. ORIGIN: native. HEIGHT: 20–80 cm. STEMS: ascending, branching from the base, branches divergent, villous-pilosulous. LEAVES: alternate, palmately trifoliate; leaflets oblanceolate to elliptic (3–8 cm long, 8–15 mm wide); apex acute to obtuse; base cuneate; margins entire; surfaces pubescent; petioles 2–5 mm long or subsessile; lower stipules lanceolate to ovate (2–5 cm long, giving the appearance of 5-foliate leaves), persistent, foliaceous; upper stipules reduced. INFLORESCENCES: racemes 10–20 cm long, usually solitary, terminal, sometimes declined and secund, showy; bracts lanceolate to oblong (1–3 cm long), veins reticulate; pedicels 1–4 cm long. FLOWERS: perfect, papilionaceous; calyx 8–11 mm long (tube 4–6 mm long), pubescent, lobes 5; upper 2 lobes nearly completely fused, notched apically; 3 lower lobes deltoid (3–4 mm long); corollas ochroleucous to cream or pale yellow (2–3 cm long); banner 1.5–2 cm long; wings and keel 2–2.8 cm long; pistil densely hairy; ovary pubescent. FRUITS: legumes stipitate (stipe equaling the calyx in length), ovoid to ellipsoid (2–5 cm long, 1.5–2.5 cm in diameter), pubescent, woody, black when mature, veins reticulate; tapering to a long beak; seeds many. SEEDS: reniform to cordiform (4–5 mm long), verrucose, olive to brown, indurate. $2n=18$.

HABITAT: Plains wildindigo is common on prairies and in dry, open woodlands. It is most abundant on sandy and gravelly soils. USES AND VALUES: It contains several quinolizidine alkaloids giving the herbage an unpleasant taste. It may be toxic to cattle and horses if a large quantity is consumed. The legumes and seeds are eaten by small mammals and birds without poisoning. It is used occasionally in landscaping but not for erosion control. ETHNOBOTANY: Some Plains Indians made and consumed an infusion from the roots as a treatment for scarlet fever and typhoid. They also consumed a decoction from leaves as a stimulant and applied it to cuts and wounds. Omaha-Ponca boys used the legumes as rattles. The Pawnees pulverized the seeds, mixed the powder with bison fat, and rubbed the ointment on the abdomen to treat colic.

VARIETIES: Var. *glabrescens* (Larisey) Isely and var. *leucophaea* (Nutt.) Kartesz & Gandhi grow in the Great Plains. Var. *laevicaulis* (A. Gray *ex* Canby) Isely is a glabrous plant with rhombic-cuneate leaflets that grows outside the southeast edge of the Great Plains in southeast Texas and western Louisiana. SYNONYMS: *Baptisia saligna* Greene, *Podalyria bracteata* Muhl. OTHER COMMON NAMES: black rattlepod, bracted falseindigo, cream wildindigo, gasatho (Omaha-Ponca), largebract wildindigo, longbract wildindigo, pira-kari (Pawnee), tdika shande nuga (Omaha-Ponca).

SIMILAR SPECIES: *Baptisia nuttalliana* Small, Nuttall wildindigo, grows in Texas in extreme southeastern Great Plains. It has caducous bracts and pale yellow to yellow corollas. *Baptisia sphaerocarpa* Nutt., yellow wildindigo, grows in eastern Oklahoma and Texas. It is woody and has caducous bracts and smaller bright yellow corollas.

Caragana Fabr.

caragan (Lat. from Turkic): Mongolian name of a tree.

Deciduous shrubs or small trees; primary leaves alternate on new terminal growth; secondary leaves clustered on spurs from the previous year's growth; primary leaves even-pinnately compound, leaflets 2–18; leaflets small, margins entire; rachis often persistent and spiny; stipules small, deciduous or persistent and spiny; inflorescences intercalary from spurs; flowers solitary or 2–5 in fascicles; calyx campanulate or cylindric, lobes 5; lobes nearly equal, upper 2 sometimes smaller; flowers papilionaceous, corolla yellow (rarely white or pink); banner recurved, claws long; wings with long claws; keel straight, obtuse; stamens 10, diadelphous; ovary sessile; rarely stipitate; fruits legumes, linear to oblong, terete or inflated, usually pointed, dehiscent; seeds several. $x=8$.

A genus of about 50 species from southern Russia to China, mostly in central Asia. A single species has been introduced into the Great Plains.

Siberian peashrub
Caragana arborescens Lam.

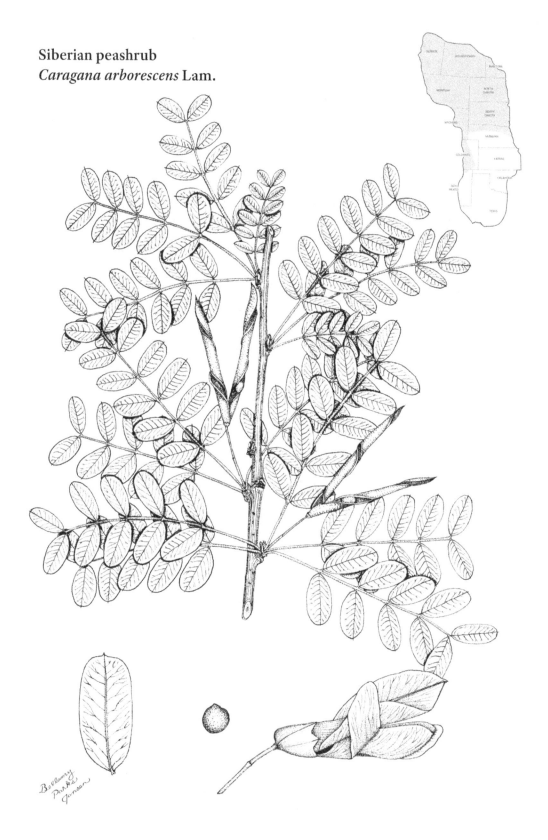

Encyclopédie Méthodique, Botanique 1(2):615. 1785.
arboris (Lat.): a tree, referring to its growth form.

GROWTH FORM: shrub or small tree, flowers May to June, reproduces from seeds. LIFE-SPAN: perennial. ORIGIN: introduced (from Eurasia). HEIGHT: to 6 m. TWIGS: branches several, ascending, pubescent when young, generally unarmed or weakly prickly, smooth, shiny, yellowish-brown to olive-green, remain green for several years. TRUNKS: older bark dark gray. LEAVES: alternate or whorled at the tips of new growth of branchlets, even-pinnately compound (5–10 cm long), leaflets 6–14; leaflets oblong-elliptic to ovate (1–3 cm long); apex obtuse to truncate, mucronate; base obtuse; margins entire; surfaces appressed villosulous when young, becoming glabrescent with maturity; upper surface bright to dark green; lower surface lighter in color; rachis sometimes extending as a short spine; petiolate; stipules acicular to linear (5–9 mm long), inconspicuous, sometimes persisting and forming spines. INFLORESCENCES: flowers borne alone or in fascicles of 2–5, in axils of leaves on numerous lateral branches; flowers ascending or pendant. FLOWERS: perfect, papilionaceous, fragrant; calyx tube broadly campanulate (5–7 mm long), silky pubescent, lobes 5; lobes triangular (about 1 mm long), pubescent; corolla yellow to bright yellow; banner (1.8–2.1 cm long, 1.7–1.9 mm wide) notched apically, narrowed abruptly to a claw (4–6 mm long); wings 1.8–2.1 cm long; keel 1.5–1.8 cm long; stamens 10, diadelphous; pedicels 1–4.5 cm long. FRUITS: legumes linear to lanceolate (3–5 cm long, 4–5 mm wide), straight, compressed, thinly coriaceous, yellow to brown, twisting at maturity; valves spiral after dehiscing, persistent; beak narrow or slender; seeds usually 3–6. SEEDS: nearly globose (4–5 mm long, 2.5–3 mm wide), dark reddish-brown, smooth, lustrous, indurate. $2n=16$.

HABITAT: Siberian peashrub has escaped from shelterbelts to pastures, waste places, abandoned fields, roadsides, and woodlands. It is difficult to control and is recognized as one of the most significant threats to maintenance of biodiversity in natural areas within its range. It grows best in sandy soils. It is tolerant of alkaline soils, saline environments, infertile soils, drought, and high winds. USES AND VALUES: Palatability to livestock is relatively low. Siberian peashrub provides woody cover for wildlife. Its legumes and seeds are eaten by small mammals, and its seeds are eaten by birds. It is tolerant of extremely cold temperatures and is most commonly planted as a windbreak border. It is also planted in buffer strips and is occasionally used as an ornamental. Several cultivars of Siberian peashrub are available from plant nurseries. ETHNOBOTANY: Siberian peashrub contains flavonoids and lectins. In Asia extracts have been used to treat headaches, cough, nosebleeds, strain-induced fatigue. It is thought to have the potential to treat many additional maladies.

SYNONYMS: *Aspalathus arborescens* Amman, *Caragana sibirica* Medik., *Robinia caragana* L. OTHER COMMON NAMES: black karagana, caragana, common caragana, pea tree, Siberian peatree, yellow acacia.

Crotalaria L.

krotalon (Gk.): a rattle, in reference to the inflated legumes with loose seeds after ripening.

Annual (in the Great Plains), or perennial herbs, slightly suffrutescent, shrubs in the tropics; leaves simple or rarely palmately trifoliate; leaflets broadly ovate to linear; stipules obsolescent or conspicuous and decurrent on the stem; inflorescences racemes, terminal or axillary, flowers many; calyx tube campanulate, lobes obscurely bilabiate, upper lip less deeply cleft; flowers papilionaceous; corolla yellow, commonly dark-striate, sometimes suffused with reddish-orange, not exceeding the calyx lobes; banner short-clawed, wings not articulate at the base, keel longer than the other petals, connivent on both margins; stamens 10, alternating lengths, monadelphous below the middle; ovary typically sessile; fruits legumes, subglobose to cylindric or ellipsoid, subcoriaceous or papery, greatly inflated, dehiscent; seeds several to many. x=7, 8, 16.

A genus of about 600 species. Most are tropical and occur in both hemispheres. About 12 grow in the United States, and a single species is common in the Great Plains. It may have been brought to the region in hay from the southeastern United States. A second species is becoming more common as its use as a cover and green manure crop increases in the region.

Rattlebox
Crotalaria sagittalis L.

Species Plantarum 2:714. 1753.

sagitta (Lat.): an arrow, in reference to the shape of the stipules of the upper leaves.

GROWTH FORM: forb, flowers May to September, reproduces from seeds. LIFE-SPAN: annual in the Great Plains but sometimes a short-lived perennial in southern states. ORIGIN: native to the southeastern United States but not native to the Great Plains. HEIGHT: 10–40 cm. STEMS: erect to ascending, simple or branched above, strigose or pilose. LEAVES: alternate, simple, upper blades broadly elliptic to lanceolate or linear (1.5–8 cm long, 6–15 mm wide); apex rounded to acute; base rounded to truncate; margins entire; surfaces strigose or pilose; basal leaves elliptic to oval, smaller; petiole 0–3 mm long; stipules inversely sagittate (2–3 mm wide), decurrent on the stem for half or more of the internode, persistent at least on the upper leaves. INFLORESCENCES: racemes, terminating stems and branches, flowers 2–4; peduncle 1–5 cm long. FLOWERS: perfect, papilionaceous; calyx campanulate (7–20 mm long), villous, lobes 5; lobes linear to lanceolate, unequal, upper lobe largest (8–10 mm long); corolla yellow, petals usually shorter than the calyx; banner about 8 mm long; wings oblong; keel oblong and curved upward; stamens 10, 5 shorter than the others; ovary glabrous; pedicels 1–10 mm long. FRUITS: legumes sessile in the calyx, oblong (2–4 cm long, about 1 cm wide), inflated, dull brown becoming black with maturity, glabrous, seeds 10–20. SEEDS: obliquely reniform or cordiform (2–3 mm wide), compressed, notched, greenish- to grayish-brown, glossy to lustrous, indurate. $2n=32$.

HABITAT: Rattlebox may be locally common on dry, sandy or gravelly soils on disturbed sites, waste areas, prairie glades, and open woodlands. USES AND VALUES: Livestock generally avoid rattlebox, but some develop a taste for it. It has been classified as a poisonous plant for more than a century. The poisoning was called "Missouri bottom disease" because horses grazing the Missouri River bottoms were affected while horses grazing the adjacent uplands were not poisoned. The toxic substance is likely one or more pyrrolizidine alkaloids. Seeds are the most poisonous plant part, and the toxic substances are present in fresh and dried herbage. Horses are more readily poisoned than cattle. Symptoms of poisoning include slow emaciation, weakness, and stupor. It may take a few weeks or even a few months before the animal dies from degeneration of the liver and spleen. Birds, especially quail, consume the seeds with no obvious ill effect. Rattlebox has little value for erosion control or landscaping.

SYNONYMS: *Anonymos sagittalis* Walter, *Crotalaria belizensis* Lundell, *C. fruticosa* Mill., *C. lunulata* Raf., *C. matthewsana* Benth., *C. parviflora* Roth, *C. pilosa* Raf., *C. platycarpa* Link, *C. pringlei* A. Gray, *C. tuerckheimii* H. Senn. OTHER COMMON NAMES: arrow crotalaria, arrow rattlebox, arrowleaf crotalaria, Virginia rattlebroom, rattleweed, wild pea.

SIMILAR SPECIES: *Crotalaria juncea* L., Sunn hemp, is a tropical species originating in India. It is a shrubby annual with yellow corollas and may reach a height of 3 m. Its use as a cover crop and a green manure crop is increasing in the southern and eastern Great Plains.

Dalea L.

Dalea: named after Samuel Dale (1659–1739), English botanist.

Annual or perennial herbs or shrubs, stems leafy; leaves alternate, small, odd-pinnately compound or trifoliate; glandular-punctate; leaflet margins entire; stipules subglandular or herbaceous, usually inconspicuous; inflorescences of terminal spikes; flowers papilionaceous or not conventionally papilionaceous; flowering from the bottom upward; flowers small, subtended by a deciduous or persistent bract; bract usually glandular-punctate; calyx tube campanulate or cylindric, ribs 10, glandular at intervals, lobes 5; lobes deltate to aristate, about equal; corollas of various colors; stamens 5–10, monadelphous; fruits legumes, indehiscent, included in a persistent calyx or partially exserted, thin-walled, usually somewhat compressed; seeds 1. x=7, 14.

About 160 species of *Dalea* have been described. They are most abundant in Mexico and the southern and southwestern United States. Eighteen occur in the Great Plains, and 10 may be considered common or abundant in the region.

1. Plants shrubs. .... *Dalea formosa*
1. Plants annuals or herbaceous perennials
  2. Plants annual. .... *Dalea leporina*
  2. Plants perennial
    3. Stems prostrate. .... *Dalea lanata*
    3. Stems erect or ascending to decumbent
      4. Stems and leaves mainly glabrous
        5. Spikes sparingly flowered (flowers 5–25). .... *Dalea enneandra*
        5. Spikes densely flowered
          6. Corollas purple to rose. .... *Dalea purpurea*
          6. Corollas white or yellowish-white
            7. Spikes subglobose, 1.5 cm long or less. .... *Dalea multiflora*
            7. Spikes cylindric, more than 1.5 cm long
              8. Spikes mostly less than 6 cm long; calyx tube glabrous or with hairs less than 0.5 mm long. .... *Dalea candida*
              8. Spikes mostly more than 6 cm long; calyx tube densely pilose with hairs more than 1.5 mm long. .... *Dalea cylindriceps*
      4. Stems and leaves softly villous-tomentose, silky-pilose, or densely villous
        9. Calyx lobes greater than 3.5 mm long. .... *Dalea aurea*
        9. Calyx lobes less than 1.5 mm long. .... *Dalea villosa*

Silktop dalea
Dalea aurea Nutt. *ex* Pursh

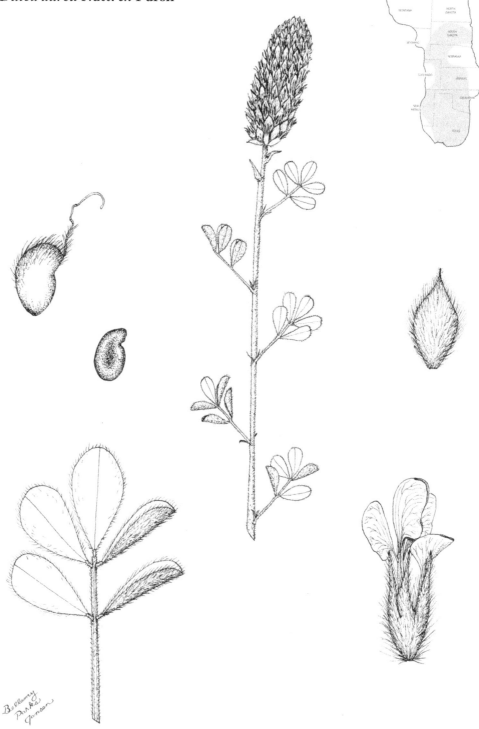

Flora Americae Septentrionalis 2:740–41. 1814.
aureus (Lat.): golden, in reference to the color of the corollas.

GROWTH FORM: forb, flowers June to September, reproduces from seeds. LIFE-SPAN: perennial. ORIGIN: native. HEIGHT: 20–80 cm. STEMS: herbaceous, erect or ascending from a short caudex surmounting a deep, woody taproot, clustered, stout, simple or branched above, pubescence silky-pilose, hairs appressed (hairs to 1.5 mm long), inconspicuously glandular. LEAVES: alternate, odd-pinnately compound (1.5–4 cm long), leaflets 3–9 (commonly 5); leaflets elliptic to oblanceolate or ovate (5–20 mm long, 2–8 mm wide), reduced in size above; apex obtuse or rarely acute, occasionally mucronulate; base acute; margins entire; upper surface deep green, often glabrescent or glabrous; lower surface villous-tomentose; foliage is somewhat sparse, leaves usually deciduous as fruit ripens; petioles glandular-punctate (6–25 mm long); stipules subulate to narrowly lanceolate (0.8–3 mm long), thinly pilosulous or glabrate, drying early, fragile, stramineous. INFLORESCENCES: Spikes cylindric (1–6 cm long, 1.2–2 cm wide), terminal, densely flowered; axis pilosulous, not visible in flower; bracts ovate to lanceolate (2.5–5.5 mm long), peduncled or sessile, pilose, persistent. FLOWERS: perfect, not conventionally papilionaceous; calyx tube turbinate (2.2–3 mm long), silky externally (hairs to 2.5 mm long), ribs prominent, glands 3–4 between the ribs; lobes 5, lanceolate-aristate (3.5–5 mm long), subequal; corollas yellow, pale, clear, remaining yellow on drying; banner sagittate (6–9 mm long), claws 3.5–5 mm long; wings oblong (5–6 mm long), claws about 1 mm long; keel oblong (5–9 mm long), united valvately; stamens 10 (1–1.2 cm long), occasionally 9, longer filaments free for 2.5–3 mm. FRUITS: legumes 3–4 mm long; upper portion silky-villous, may be glabrous (or nearly so) below; seeds 1. SEEDS: broadly ellipsoid to reniform (2–2.5 mm long), yellow to dark brown, smooth, indurate. $2n=14$.

HABITAT: Silktop dalea is infrequent to common on dry prairies, open woodlands, brushy hillsides, and ravines. It grows in all types of soil but is most common in gravelly and calcareous soils. USES AND VALUES: Its forage is fair for cattle and good for deer. It is readily eaten by livestock and decreases with continuous grazing. Deer and pronghorn eat the leaves and inflorescences. Its seeds are eaten by ground-foraging birds and small mammals. Silktop dalea attracts bees, butterflies, and small birds. It is not used in erosion control and is used sparingly in landscaping. It is used often in prairie restorations. ETHNOBOTANY: Members of some Plains Indian tribes consumed an infusion of the leaves for colic and diarrhea.

SYNONYMS: *Cylipogon capitatum* Raf., *Parosela aurea* (Nutt. *ex* Pursh) Britton, *Petalostemon capitatus* (Raf.) DC., *Psoralea aurea* (Nutt. *ex* Pursh) Poir. OTHER COMMON NAMES: golden dalea, golden parosela, golden prairieclover, pezhuta pa (Dakota).

SIMILAR SPECIES: *Dalea nana* Torr. *ex* A. Gray, dwarf dalea, is similar in appearance, except that dwarf dalea is seldom taller than 35 cm. Its stems are diffusely branched, and its yellow corollas are small and fade to pink or brown on drying. It grows in the southwestern Great Plains and is uncommon.

White prairieclover
Dalea candida Willd.

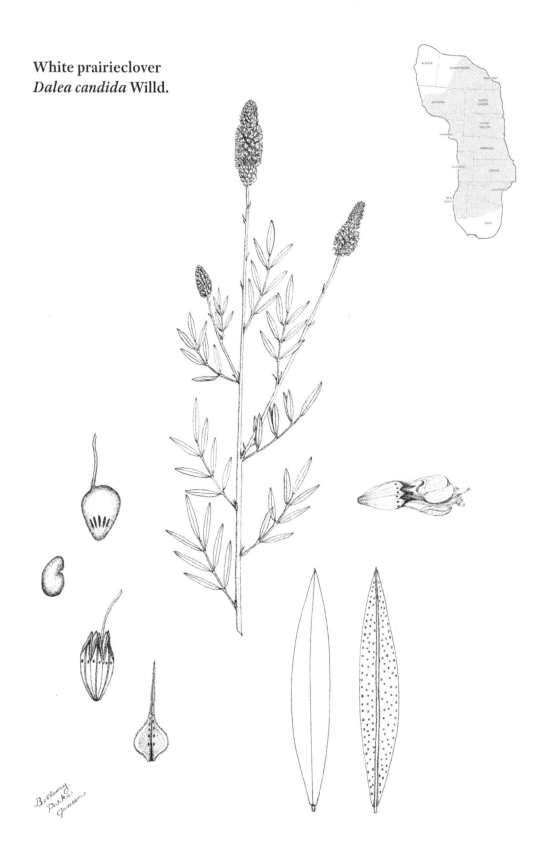

Species Plantarum. Editio quarta 3(2):1337. 1802.

candidus (Lat.): dazzling white, in reference to the color of the corollas.

GROWTH FORM: forb, flowers May to August, reproduces from seeds. LIFE-SPAN: perennial. ORIGIN: native. HEIGHT: 0.3–1 m. STEMS: herbaceous, erect to ascending, simple or branched above, branches stiffly spreading, 1 to several stems from a woody caudex surmounting a taproot, glabrous. LEAVES: alternate, odd-pinnately compound (1–6 cm long), leaflets 5–9, commonly 7; leaflets narrowly oblanceolate to narrowly elliptic or linear (1–3 cm long, 2–6 mm wide); apex acute to obtuse, often mucronate; base acute to obtuse; margins entire; upper surface glabrous; lower surface glandular-punctate; often folded along the midrib; stipules subulate, glabrous. INFLORESCENCES: spikes cylindric to ovoid (1.5–5.5 cm long, 6–10 mm wide), 1 to few, terminal, densely flowered, axis usually not visible; bracts oblanceolate, tapering to acuminate, glabrous, glandular-punctate, caducous, ciliate, slender tip surpassing the calyx lobes. FLOWERS: perfect, not conventionally papilionaceous; calyx tube 2–3 mm long, ribs 10, ring of glands near the top, glabrous to finely ciliate; lobes 5, lanceolate to triangular (0.5–1.8 mm long), slightly unequal, lowest one the longest, glabrous or with fine, short ciliate hairs (hairs under 0.5 mm long); corolla white; banner 4–6 mm long; wings 3–5.5 mm long; lower petals that would normally form the keel not united (3–5.5 mm long); stamens 5, monadelphous. FRUITS: legumes ovate (2.5–4.5 mm long), usually exserted, walls thin, glandular, upper portion glabrate or pilose; seeds 1. SEEDS: asymmetrically reniform (1.5–2 mm long), brown, smooth, indurate. $2n=14$.

HABITAT: White prairieclover is common in dry prairies, rangelands, meadows, and rocky upland woodlands. USES AND VALUES: It is palatable to all classes of livestock and adds quality to prairie hay. It decreases with continuous heavy grazing. Deer, elk, bighorn sheep, pronghorn, and wild turkeys eat the foliage. Small mammals and birds eat the seeds. It has the potential to stabilize erosive soils and is an attractive landscape plant. ETHNOBOTANY: Some Lakotas chewed the roots for their pleasant taste and made tea from the leaves. Other Plains Indians bruised the leaves and steeped them in water for application to fresh wounds.

VARIETIES: The axis of spikes is concealed in var. *candida*, external surface of the calyx is glabrous, and largest leaflets are 1.3–4 cm long. It grows in the moist eastern Great Plains. The spike axis of var. *oligophylla* (Torr.) Shinners is visible, the calyx tube is generally pubescent, and largest leaflets are 5–25 mm long. It grows in the drier western part of the region. SYNONYMS: *Kuhnistera candida* (Willd.) Kuntze, *K. occidentalis* A. Heller *ex* Britton & Kearney, *Petalostemon candidus* (Willd.) Michx., *P. gracilis* Nutt., *P. occidentalis* (A. Heller *ex* Britton & Kearney) Fernald, *P. virgatus* Nees & Schwein., *Psoralea candida* (Willd.) Poir. OTHER COMMON NAMES: bloka (Lakota), pussyfoot, slenderwhite prairieclover, western prairieclover.

SIMILAR SPECIES: *Dalea phleoides* (Torr. & A. Gray) Shinners, small-leaf dalea, also has white corollas. However, it has up to 40 leaflets per leaf, and the calyx is oblique with the lobes placed on the lower side. It grows in the southeastern Great Plains.

Largespike prairieclover
Dalea cylindriceps Barneby

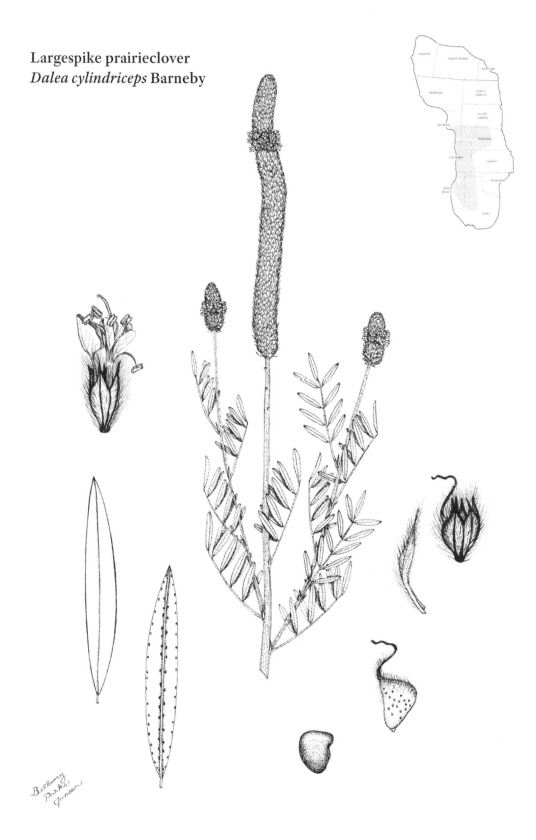

Memoirs of the New York Botanical Garden 27:227–28. 1977.
kylindros (Gk.): a cylinder, in reference to the shape of the inflorescence.

GROWTH FORM: forb, flowers May to September, reproduces from seeds. LIFE-SPAN: perennial (short-lived). ORIGIN: native. HEIGHT: 20–80 cm. STEMS: herbaceous, erect to ascending, usually simple or few-branched, 1 to several from a woody caudex surmounting a taproot, glabrous to the base of the inflorescence, glandular-punctate. LEAVES: alternate, odd-pinnately compound (3–8 cm long), leaflets 5–9; leaflets oblong-elliptic to oblanceolate (1–2.5 cm long, 1–5 mm wide); apex acute to acuminate; base cuneate to attenuate; margins entire; upper surface glabrous; lower surface sparsely glandular-punctate; stipules lanceolate to subulate. INFLORESCENCES: spikes cylindric (6–24 cm long, 8–12 mm wide), terminal, frequently recurved near the apex, dense, axis not visible; long-peduncled; bracts long-attenuate, longer than the calyx, usually pilose with silky-villous hairs, apex dark-colored, caducous but mostly held in place by the surrounding flowers. FLOWERS: perfect, not conventionally papilionaceous; calyx tube campanulate (1.9–2.3 mm long), densely pilose externally (hairs over 1.5 mm long), ribs 10, row of small glands between the ribs; lobes 5, deltoid-acuminate to ovate (1.5–2.5 mm long); corollas white to yellowish-white (rarely light pink), fading to light yellow; banner 4.5–6.3 mm long, claw 3–4 mm long; wings 2–5 mm long; lower petals that would normally form the keel not united (2–5 mm long); stamens 5, monadelphous. FRUITS: legumes ovate (2.5–3 mm long), contained within the calyx, pilose, glandular-punctate; seeds 1. SEEDS: cordiform to reniform (1.5–2.5 mm long), compressed, smooth, indurate. $2n=14$.

HABITAT: Largespike prairieclover is uncommon to scattered in well-drained sandy and gravelly soils of sandsage prairies and rangelands, woodlands, and stream valleys. Its long, thick spikes set it apart from the other *Daleas*, but it is uncommon and seldom collected. USES AND VALUES: It is grazed by cattle, horses, and sheep but is not sufficiently abundant to be an important forage species. It decreases with continuous heavy grazing. Largespike prairieclover is grazed by deer, elk, and pronghorn. Wild turkeys eat the leaves. Small mammals and birds eat the seeds. It attracts bees, butterflies, skippers, and moths. Largespike prairieclover serves as a host of the southern dogface caterpillar (*Zerene cesonia*). Greenish-white eggs are laid on the underside of the host plant leaves by the butterfly. The larvae are green with a white stripe running down each side of their bodies. The green chrysalis hangs upright surrounded with a silken girdle. Largespike prairieclover is not used for erosion control. It is an attractive plant and has the potential to be used in landscaping. It is often included in prairie restorations. However, high-quality seed is not readily available. ETHNOBOTANY: Some Lakotas steeped the dry leaves in water to make tea.

SYNONYMS: *Petalostemon compactus* sensu auct., *P. macrostachyus* Torr. OTHER COMMON NAMES: Andean prairieclover, compact prairieclover, cylindrical prairieclover, denseflowered prairieclover, massivespike prairieclover, sandsage prairieclover.

Nineanther dalea
Dalea enneandra Nutt.

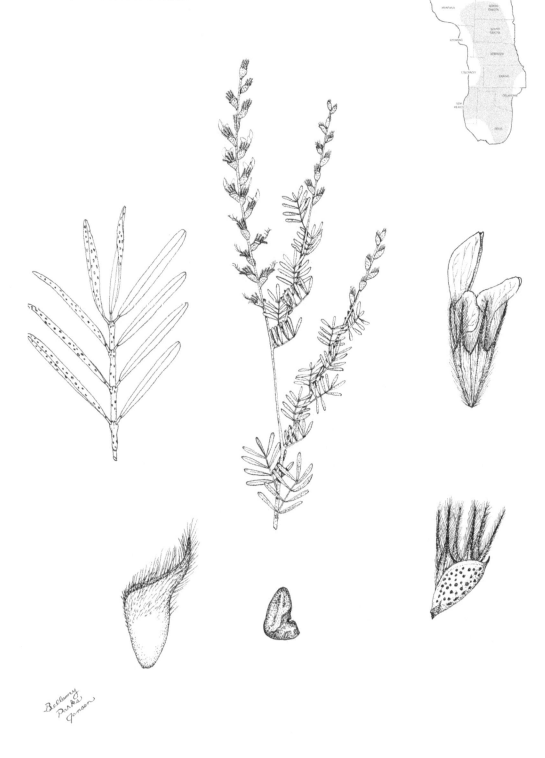

Catalogue of New and Interesting Plants Collected in Upper Louisiana No. 30. 1813. *ennea* (Gk.): nine; + *andos* (Gk.): man, in reference to the 9 stamens.

GROWTH FORM: forb, flowers June to September, reproduces from seeds. LIFE-SPAN: perennial. ORIGIN: native. HEIGHT: 0.4–1.5 m. STEMS: herbaceous, erect to ascending, 1–3 from a woody caudex surmounting an orangish-yellow taproot, unbranched below, many spreading branches above, glabrous to glabrate to the inflorescence, glandular-punctate. LEAVES: alternate, odd-pinnately compound (1–2.5 cm long), subsessile; primary stem leaves usually absent by flowering; leaflets 5–13, commonly 9; leaflets linear to narrowly oblong (3–12 mm long, 0.5–1.5 mm wide), sometimes involute; apex obtuse to rounded; base obtuse to cuneate; margins entire; upper surface glabrous; lower surface glandular-punctate. INFLORESCENCES: spikes (3–10 cm long), terminal, flexuous, loose, open; sparingly flowered (flowers 5–25), appearing 2-ranked; bracts broadly ovate to obovate or truncate (3–4 mm long), enfolding, cuspidate, persistent, glandular-punctate; margins prominent, membranous, pale. FLOWERS: perfect, not conventionally papilionaceous; calyx tube campanulate (3–4 mm long), silky pilosulous; lobes 5, setaceous to filiform above, triangular below (3.5–4.5 mm long), longer than the tube, conspicuously white-pilose, persistent; corollas primarily white; banner white (6–8 mm long, claw 2.5–3.5 mm long), slightly reflexed; wings white (3–5 mm long); keel pale yellow (9–12 mm long); stamens 9, monadelphous. FRUITS: legumes about equal to the calyx tube (3–4 mm long), upper portion pubescent; lower portion glabrous, thin, transparent, yellow; seeds 1. SEEDS: narrowly ovoid to reniform (2.5 mm long), beaked, yellow to brown, smooth, indurate. $2n=14$.

HABITAT: Nineanther dalea is infrequent to locally abundant in dry prairies, stream valleys, and roadsides. It is most common in calcareous, rocky, or sandy soils. USES AND VALUES: It is palatable to cattle, horses, and sheep but does not grow in sufficient densities to furnish much forage to livestock. It is grazed by deer, elk, and pronghorn, and small mammals and birds eat the seeds. High-quality seed is rarely available, and it has only limited potential for soil stabilization and landscaping. ETHNOBOTANY: The Dakotas said that the roots contained a narcotic or poisonous substance, but the presence of this substance has not been substantiated. Kiowas made small arrow shafts from the stems, tipped each with a cactus thorn, and used them to hunt birds, rabbits, and small mammals.

VARIETIES: Var. *enneandra* is most common in the Great Plains. Var. *pumila* (Shinners) B.L. Turner is smaller (20–30 cm tall), with a shorter and more dense inflorescence. It has been collected in Texas. SYNONYMS: *Cylipogon virgatum* Raf., *Dalea laxiflora* Pursh, *Parosela enneandra* (Nutt.) Britton, *Petalostemon laxiflorus* Steud., *Psoralea laxiflora* (Pursh) Poir. OTHER COMMON NAMES: bigtop dalea, nineanther prairieclover, plume dalea, slender dalea.

SIMILAR SPECIES: *Dalea pogonanthera* A. Gray, bearded prairieclover, grows in the extreme southern Great Plains. Its corollas are pink to violet or purple rather than white. Its spikes are compact rather than loose. The well-defined keel is often deciduous, and the stamen column becomes exserted.

Featherplume
Dalea formosa Torr.

Annals of the Lyceum of Natural History of New York 2:177–78. 1827.
formosus (Lat.): beautiful, in reference to the flowers.

GROWTH FORM: shrub, flowers April to September, reproduces from seeds. LIFE-SPAN: perennial. ORIGIN: native. HEIGHT: 0.2–1 m. STEMS: woody, divaricately branched; branches zigzag from node to node, glabrous; young twigs glandular-punctate; bark light gray to brown. LEAVES: alternate, odd-pinnately compound (3–15 mm long), aromatic, slightly resinous, glabrous, thick, deciduous and soon inconspicuous or absent; leaflets 5–15, oblanceolate to narrowly obovate or obovate-cuneate (1–7 mm long, 1–2 mm wide), usually folded or rolled with the upper surface on the inside; apex emarginate to obcordate; base obtuse; margins entire; upper surface glabrous, dark green to grayish-green; lower surface glandular-punctate. INFLORESCENCES: spikes subcapitate (1–1.5 cm in diameter), terminal on branches and spurs, bracts caducous, flowers 2–10; may flower twice in a single year if sufficient soil moisture is available. FLOWERS: perfect, papilionaceous, fragrant; calyx tube campanulate (3–5 mm long), longitudinally ribbed, abundantly pilose; hairs 1–3 mm long, glandular-punctate; lobes 5, filiform to aristate (4–10 mm long), longer than the tube, plumose; corollas bicolored; banner yellow (7–9 mm long), becoming reddish to purple with age, claw 4–5 mm long; wings purple to rose-purple (8–10 mm long); keel rose-purple (8–12 mm long); stamens 10, monadelphous. FRUITS: legumes obovate (3–3.5 mm long), compressed, contained within the persistent calyx; seeds 1. SEEDS: yellow to brown, smooth, indurate. $2n=14, 28, 42$.

HABITAT: Featherplume is infrequent to common on prairie hills, rangelands, open areas, shrubby barren areas, cedar breaks, canyon margins, and rocky hillsides. Featherplume can grow in poor, dry limestone-based soils, caliche, and sandy to clayey soils. USES AND VALUES: It provides valuable browse for cattle, horses, and goats. Pronghorn and deer browse the leaves and young twigs. Rabbits browse the foliage and twigs. Small mammals and ground-foraging birds eat the seeds. It is pollinated by various bees and visited by many other nectar-seeking insects. This attractive shrub is used in landscape plantings. The fine leaves give it a delicate texture, and the yellow and reddish-purple flowers can be spectacular. It is most showy if it is cut back in the spring. ETHNOBOTANY: Some southwestern Plains Indians took an infusion of the leaves as an emetic, cathartic, and strengthener before a long run or other strenuous activity. Some dried the flowering branches to make a sweet tea to relieve aches and pains. Featherplume was used to treat influenza and viral infections such as the common cold.

SYNONYM: *Parosela formosa* (Torr.) Vail. OTHER COMMON NAMES: feather dalea, indigo-bush, limoncillo (Spanish), peabush, yerba de Alonso Garcea (Spanish).

SIMILAR SPECIES: *Dalea frutescens* A. Gray, black dalea, is another shrub and is often thicket-forming. Its calyx lobes are much shorter than the tube, banners are white while the wings and keel are purple. The leaves are 1–2 cm long. The ranges of *Dalea frutescens* and *Dalea formosa* overlap in the southern and southwestern Great Plains.

Woolly prairieclover
Dalea lanata Spreng.

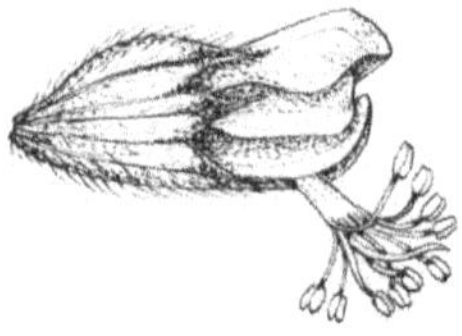

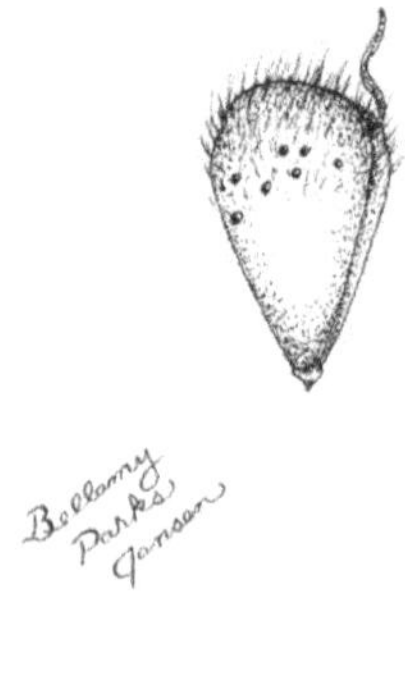

Bellamy
Parks
Jansen

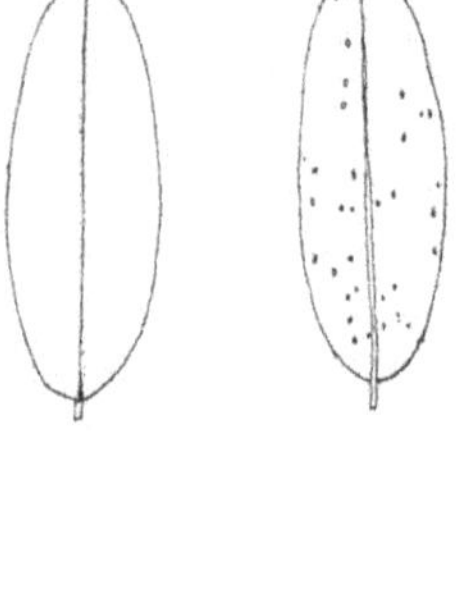

Systema Vegetabilium, editio decima sexta 3:327. 1826.

lanatus (Lat.): woolly, describing the pubescence on the leaves and fruit.

GROWTH FORM: forb, flowers July to September, reproduces from seeds. LIFE-SPAN: perennial. ORIGIN: native. LENGTH: 0.2–1 m. STEMS: herbaceous, prostrate, trailing or ascending from a caudex surmounting a taproot, 1 to several, divaricately branching from the base, arching, occasionally forming mats, densely lanate or short-villous-tomentose, sparsely glandular-punctate or eglandular. LEAVES: alternate, odd-pinnately compound (1.5–3 cm long); leaflets 5–21, obovate to oblanceolate (3–10 mm long, 1.5–6 mm wide); apex obtuse; base obtuse; margins entire; upper surface short-villous; lower surface densely short-villous to glabrate, glandular-punctate; petioled or subsessile; subtended by a gland on the rachis; stipules bristlelike to subulate to acicular (1–2.5 mm long), inconspicuous to caducous to sometimes absent. INFLORESCENCES: spikes cylindric (2–9 cm long, 6–9 mm wide), terminal, loosely flowered, axis visible; bracts subtending flowers obovate, glandular-punctate, persistent; peduncles 4–40 mm long. FLOWERS: perfect, not conventionally papilionaceous; calyx tube campanulate (2–2.5 mm long), villous externally, glandular-punctate or with a glandular spot; lobes 5, triangular to lanceolate (1.5–2.5 mm long), usually acuminate; corollas reddish-violet to purple or magenta; petals with a few scattered glandular dots; banner cordiform (3–4 mm long); wings 2–4 mm long, claws short; keel petals 3.5 mm long, not forming an obvious keel; stamens 8–10, monadelphous. FRUITS: legumes oblong or elliptical (2.5–3 mm long), pilose, glandular-punctate, contained within the persistent calyx; seeds 1. SEEDS: reniform to cordiform (1.8–2.5 mm long), olive or brown to black, smooth, indurate. $2n=14$.

HABITAT: Woolly prairieclover is infrequent to common in dry sandy soils of flood plains, prairies, rangelands, and roadsides. It may be found growing in a broad range of soils, but it grows best in sandy soils. USES AND VALUES: It is grazed by cattle, horses, sheep, and goats. Deer and pronghorn graze the foliage. The legumes and seeds are eaten by small mammals and birds. ETHNOBOTANY: Some Great Plains Indians made the foliage into a paste and applied it to insect bites and centipede bites to reduce swelling and itching. The roots were used as a sweetener and eaten raw.

VARIETIES: Var. *lanata* grows in the Great Plains. Its calyx tubes are villosulous, and the apices of the lobes are usually acuminate. Var. *terminalis* (M.E. Jones) Barneby has glabrous calyx tubes, and the lobes are deltate or shortly lanceolate. It grows in the southwest portion of the region and into Mexico. Ranges of the two infrequently overlap. SYNONYMS: *Dalea glaberrima* S. Watson, *D. lanuginosa* Nutt. *ex* Torr. & A. Gray, *D. terminalis* M.E. Jones, *Parosela lanata* (Spreng.) Britton. OTHER COMMON NAMES: woolly parosela, woolly dalea.

SIMILAR SPECIES: *Dalea jamesii* (Torr.) Torr. & A. Gray, James dalea, stems are erect (1–1.2 m tall), leaves are mainly trifoliate, and the corollas are yellow. Its range overlaps on the Great Plains with the range of *Dalea lanata* in western Texas and Oklahoma and eastern Colorado and New Mexico.

Foxtail dalea
Dalea leporina (Aiton) Bullock

Bulletin of Miscellaneous Information, Royal Gardens, Kew 1939(4):196. 1939.
leporinus (Lat.): of hares, in reference to the developing silky inflorescence, which resembles a rabbit's foot.

GROWTH FORM: forb, flowers July to September, reproduces from seeds. LIFE-SPAN: annual. ORIGIN: native. HEIGHT: 0.2–1.5 m. STEMS: herbaceous, erect, diminutive to robust, multiple from the base, simple to much-branched from midstem or near the base, branches ascending, glandular-punctate, glabrous below the inflorescence. LEAVES: alternate, odd-pinnately compound (3–10 cm long); leaflets 13–49 (fewest on upper leaves), elliptic to lanceolate or obovate (3–12 mm long, 1–3 mm wide), subtended by a gland; apex acute to rounded; base acute; margins entire, sometimes reddish; upper surface darker green, glabrous; lower surface lighter in color, glabrous, glandular-punctate; glands often showing through to the upper surface; petioles 0–8 mm long; stipules lanceolate to subulate (1–3 mm long). INFLORESCENCES: spikes obovoid, becoming cylindric (1.5–9 cm long, 8–15 mm wide), terminal, erect, dense, axis not visible; bracts obovate, long-acuminate, margins thin and membranous, caducous; peduncles 3–9 cm long. FLOWERS: perfect, not conventionally papilionaceous; calyx tube 1.7–2.9 mm long, slit ventrally on the upper side, villous, glandular-punctate, ribs 10, hyaline intervals between the ribs with 1 or 2 rows of small glands; lobes 5, lanceolate to acuminate (1–2 mm long), shorter than the tube; corolla white, purple, or blue, sometimes pinkish; banner 3.5–6 mm long; wings 2–3 mm long; petals 1.5–3 mm long, not forming an obvious keel; stamens 9 or 10, monadelphous; anthers shortly exserted. FRUITS: legumes obovoid (1.5–3 mm long), papery, pubescent above, glandular-punctate, style beak positioned to one side, indehiscent; seeds 1. SEEDS: reniform to cordiform (1.6–2.5 mm long), gray to brown, shiny, smooth, indurate. $2n=14$.

HABITAT: Foxtail dalea is scattered to common in moist, alluvial, sandy soils of open areas, disturbed sites, roadsides, prairies, rangelands, edges of fields, wooded areas, and streambanks. USES AND VALUES: Foxtail dalea is grazed by cattle, sheep, goats, and deer, but it has limited forage value. The legumes and seeds are eaten by small mammals and ground-foraging birds. It has been seeded as a cover crop and plowed under as a green manure crop. Since it is an annual, it has little value for erosion control. Foxtail dalea has limited landscaping potential. ETHNOBOTANY: In Mexico it was reported to act against tumors. However, the plant parts used and treatment methods are unknown.

SYNONYMS: *Dalea alba* Michx. *ex* Roem., *D. alopecuroides* Willd., *D. bigelovii* (Rydb.) B.L. Turner, *D. exserta* (Rydb.) Gentry, *D. lagopus* (Cav.) Willd., *D. leporina* (Aiton) Kearney & Peebles, *D. linnaei* Michx., *D. oreophila* (Cory) Cory, *Parosela bigelovii* Rydb., *P. costaricana* Rydb., *P. lagopus* Cav., *P. leporina* (Aiton) Rydb., *Petalostemon alopecuroides* (Willd.) Pers., *P. oreophilum* Cory, *Psoralea alopecuroides* (Willd.) Poir., *P. lagopus* Cav., *P. leporina* Aiton. OTHER COMMON NAMES: cola de zorra (Spanish), foxtail prairieclover, harefoot dalea, parosela, pink parosela.

Roundhead prairieclover
Dalea multiflora (Nutt.) Shinners

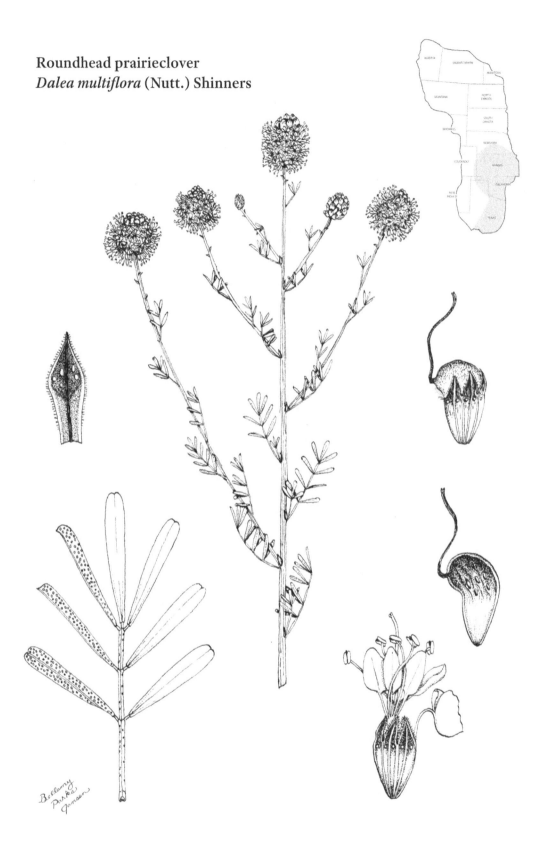

Field & Laboratory 17(3):82. 1949.
multus (Lat.): many; + *florus* (Lat.): flower, in reference to the numerous spikes of flowers.

GROWTH FORM: forb, flowers May to August, reproduces from seeds. LIFE-SPAN: perennial. ORIGIN: native. HEIGHT: 20–80 cm. STEMS: herbaceous, erect to ascending, several from a caudex and taproot, much-branched from the base; branches ascending, glandular-punctate, glabrous, sometimes with reddish stripes. LEAVES: alternate, odd-pinnately compound (1.5–4 cm long), somewhat fragrant when crushed; leaflets 3–13 (commonly 7 or 9), elliptic-oblanceolate to narrowly oblong or linear (5–15 mm long, 1–2 mm wide), often folded at the midrib, involute when dry; apex acute to usually obtuse, sometimes mucronate; base cuneate; margins entire; upper surface glabrous; lower surface glandular-punctate; stipules subulate to lanceolate (1–2 mm long), reddish-purple with a green base, thickened at the base, sometimes fringed below the midpoint. INFLORESCENCES: spikes subglobose to short-ovoid (9–15 mm long, 7–14 mm wide), numerous, terminal, axis not visible; bracts tapering to a short-acuminate apex, shorter than the calyx, caducous, but the internal ones held in place by the crowded flowers; flowers 15–40, dense and overlapping; peduncle slender (1–2.5 cm long), erect, glandular-punctate. FLOWERS: perfect, not conventionally papilionaceous; calyx tube 1.5–2.5 mm long; glabrous to finely pubescent, ribs 10; hyaline intervals between the ribs with 1–3 glands; glands aging reddish; lobes 5, triangular to lanceolate (1–1.5 mm long), unequal, lowest one longest, dark reddish-purple, ciliate; corolla white; banner 4–6 mm long, claw 2–3 mm long; wings 3.5–5 mm long; lowest petals not united to form a keel (3.5–5 mm long); stamens 5, monadelphous; external filaments alternating with 4 exserted petals (2 wing and 2 keel petals). FRUITS: legumes obliquely obovoid (2.5–5 cm long), partially exserted from the calyx, nearly glabrous, elongated glands on the sides, indehiscent; seeds 1. SEEDS: asymmetrically obovoid to reniform (1.5–2.3 mm long), brown or tan, smooth. $2n=14$.

HABITAT: Roundhead prairieclover is infrequent to common on dry hills, prairies, range-lands, brushy hillsides, and roadsides. It is most abundant on rocky, limestone soils of undisturbed prairies. USES AND VALUES: It is grazed by cattle, horses, sheep and goats. It is seldom abundant and is not an important forage for livestock. It is a more important forage resource for deer. Roundhead prairieclover decreases with continuous heavy grazing. The seeds and legumes are eaten by small mammals and ground-foraging birds such as quail. Roundhead prairieclover is pollinated by various bees and visited by many nectar-seeking insects. It has limited potential as a soil stabilizer, but its numerous spikes make it an attractive plant for landscaping. ETHNOBOTANY: Members of some Plains Indian tribes made tea from the leaves to be used as a preventative medicine for an unknown malady. They chewed the roots for their pleasant taste. The stems were tied together for brooms.

SYNONYMS: *Kuhnistera candida* var. *multiflora* (Nutt.) Rydb., *K. multiflora* (Nutt.) A. Heller, *Petalostemon multiflorus* Nutt. OTHER COMMON NAMES: prairie clover, round-headed prairieclover.

Purple prairieclover
Dalea purpurea Vent.

Description des Plantes Nouvelles . . . Jardin de J.M. Cels pl. 40. 1801.
purpureus (Lat.): purple of various shades, in reference to the color of the corollas.

GROWTH FORM: forb, flowers June to August, reproduces from seeds. LIFE-SPAN: perennial. ORIGIN: native. HEIGHT: 20–90 cm. STEMS: herbaceous, erect to ascending or spreading, few to many from a woody caudex surmounting a taproot, glabrous to thinly pilosulous, striate-ribbed, with scattered brownish-black glands. LEAVES: alternate, odd-pinnately compound (1–4 cm long), somewhat fragrant when crushed; leaflets 3–7 (often 5), linear (5–28 mm long, 0.5–1.5 mm wide); apex obtuse; base obtuse to cuneate; margins entire, involute; upper surface usually glabrous; lower surface glabrate to occasionally pubescent, glandular-punctate; petiole similar to the leaflets; stipules absent or small, inconspicuous, bristlelike. INFLORESCENCES: spikes cylindric (1–7 cm long, 7–14 mm wide), dense, axis not visible; bracts (2–8 mm long, 1–2 mm wide) abruptly contracted into a recurved or erect tail, persistent until fruit fall. FLOWERS: perfect, not conventionally papilionaceous, calyx tube (1.7–3 mm long) densely villous externally, areas between ribs glandless; lobes 5, lanceolate to ovate (1–3 mm long), upper pair broadest; corolla purple to rose (may be reddish-purple, magenta, or lilac); banner 4.5–7 mm long, separate; others attached at the separation of the filaments; stamens 5, monadelphous. FRUITS: legumes obliquely ovate (2–2.5 mm long), enclosed by the persistent calyx, pilosulous and glandular-punctate; seeds 1. SEEDS: reniform to cordiform (1.5–2 mm long), indurate. $2n=14$.

HABITAT: Purple prairieclover is common on dry prairies throughout the region. USES AND VALUES: It is highly palatable and produces excellent forage for all classes of livestock and wildlife. It decreases with continued heavy grazing. Its use in landscaping is increasing. ETHNOBOTANY: Plains Indians ate fresh and boiled leaves. Bruised leaves were steeped in water and applied to fresh, open wounds. Some Comanches and Poncas chewed the roots for their pleasant flavor and made tea from the leaves.

VARIETIES: Two varieties grow in the Great Plains. Var. *purpurea* is most common and has spikes 9–14 mm in diameter and some pubescence. Var. *arenicola* (Wemple) Barneby spikes are 7–10 mm in diameter and the plants are primarily glabrous. It grows in the western and southwestern Great Plains. SYNONYMS: *Dalea violacea* (Michx.) Willd., *Kuhnistera violacea* (Michx.) Aiton *ex* Steud., *Petalostemon pubescens* A. Nelson, *P. purpureus* (Vent.) Rydb., *P. standleyanus* Rydb., *P. violaceus* Michx., *Psoralea purpurea* (Vent.) MacMill. OTHER COMMON NAMES: ba'sibûgûk (Chippewa), khats-pidpatski (Pawnee), kiha piliwus hawastat (Pawnee), makan skithe (Omaha-Ponca), prairie clover, purple clover, red tasselflower, thimbleweed, tokala tapežuta hu wiŋjela (Lakota), violet clover, violet prairieclover, wanah'cha (Dakota).

SIMILAR SPECIES: *Dalea compacta* Spreng., compact prairieclover, grows in the southeastern Great Plains. Its spikes are 4–7 cm long and 1–1.5 cm in diameter. Its corollas are reddish-purple. *Dalea tenuis* (J.M. Coult.) Shinners, pinkglobe prairieclover, grows in the southern part of the region. It has pinkish or pinkish-purple corollas, wiry stems, and short compact spikes on long peduncles.

Silky prairieclover
Dalea villosa (Nutt.) Spreng.

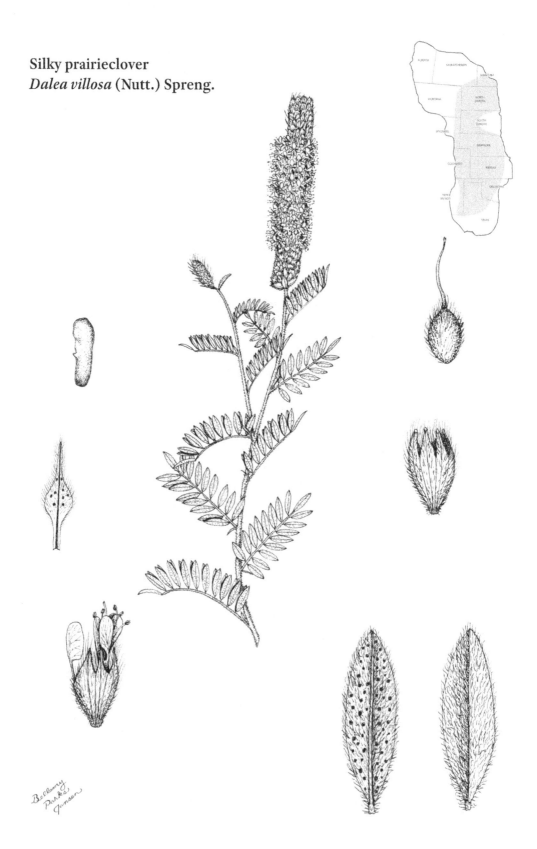

Systema Vegetabilium, editio decima sexta 3:326. 1826.

villous (Lat.): hairy or shaggy, referring to the densely villous covering on the plant.

GROWTH FORM: forb, flowers July to August, reproduces from seeds. LIFE-SPAN: perennial. ORIGIN: native. HEIGHT: 15–90 cm. STEMS: herbaceous, erect or ascending to decumbent, arising from a woody caudex surmounting a reddish-orange taproot, densely villous, grayish-green, simple below, usually branching above; branches ascending or lax. LEAVES: alternate, odd-pinnately compound (1.5–4 cm long); leaflets 9–23, elliptic to elliptic-lanceolate (5–14 mm long, 1–4 mm wide); apex acute; base obtuse to cuneate; margins entire; both surfaces densely villous, grayish-green; lower surface glandular-punctate; usually flat, even when dry; stipules bristlelike, subulate or acicular, inconspicuous, caducous or absent. INFLORESCENCES: spikes cylindric (2–14 cm long, 7–9 mm wide), terminal, erect, sometimes bent or curved, moderately dense, axis eventually visible; bracts narrowly lanceolate (1.5–5 mm long), surpassing the calyx, villous, caducous; peduncles up to 2.5 cm long. FLOWERS: perfect, not conventionally papilionaceous; calyx tube 2–3 mm long, ribs 10, obscured by pubescence; lobes 5, lanceolate to subulate (1–1.5 mm long), upper pair ovate to triangular; corolla pale rose to rose-purple, rarely white; banner separate (4.5–5.5 mm long, claw 2–3 mm long); other petals (2.5–4.5 mm long) attached at the separation of the filaments; stamens 5, monadelphous; anthers orange, conspicuous. FRUITS: legumes obliquely ovate (2.5–3 mm long), densely villous; seeds 1. SEEDS: narrowly ovoid (2–2.5 mm long), enclosed in the persistent calyx; olive, brown, or black, smooth, indurate. 2*n*=14.

HABITAT: Silky prairieclover is most common on sandy prairies, floodplains, and open woodlands. It can be common on the edges of sandy blowouts. USES AND VALUES: It has only fair to moderate palatability, but it is grazed by cattle, horses, pronghorn, elk, bighorn sheep, and deer. Ground-foraging birds eat the seeds, and small mammals eat the seeds and legumes. Silky prairieclover is pollinated by honeybees, bumblebees and wasps. It is promising for erosion control. The attractive flowers and grayish-green foliage make it valuable for landscaping. ETHNOBOTANY: Some Lakotas ate the leaves and flowers to reduce swelling inside the throat. They consumed a decoction made from the roots as a laxative.

VARIETIES: Var. *villosa* is the most common in the region. Var. *grisea* (Torr. & A. Gray) Barneby is 50–80 cm tall, erect, and thinly pubescent. It grows near the southeastern edge of the Great Plains. SYNONYMS: *Kuhnistera villosa* (Nutt.) Kuntze, *Petalostemon villosus* Nutt. OTHER COMMON NAMES: bláye zitká tačaη hu stóla (Lakota), camsú huȟolȟóta (Lakota), downy prairieclover, hairy prairieclover, waptáya huȟolȟóta (Lakota).

SIMILAR SPECIES: *Dalea tenuifolia* (A. Gray) Shinners, slimflower prairieclover, has fewer leaflets per leaf (3–11) than silky prairieclover. Its stems and foliage are rarely villous. It is not found in loose sand. It grows in the southwestern portion of the Great Plains, and its range overlaps with the range of silky prairieclover.

Desmodium Desv.

desmos (Gk.): a bond or connection, referring to the segmented fruit.

Perennial herbs, erect to prostrate or trailing, usually with uncinate hairs; leaves usually alternate, pinnately trifoliate; margins entire; terminal leaflet on a petiolule; stipules ovate to subulate, nerves striate; axils pubescent, some with uncinate hairs; inflorescences racemes or panicles of racemes, terminal; calyx tube slightly oblique; lobes 5, more or less bilabiate, upper 2 lobes connate for all or most of their length, lower 3 separate, middle lobe often longest; flowers perfect, papilionaceous; corolla pink or white to purple or violet, sometimes marked with yellow, small; banner oblong to nearly orbicular, narrowed at the base; wings oblong; keel nearly straight; stamens 10, diadelphous or occasionally monadelphous, persistent; median upper filament free; ovary short; fruits loments, indehiscent, segments 2–7; elevated on a stipe above the persistent calyx, eventually separating into 1-seeded segments, covered with uncinate hairs, sutures between segments usually more shallow on the upper margin than on the lower margin. x=11.

Nearly 300 species grow in North America, South America, Africa, Asia, and Australia. About 45 species grow in North America. Fifteen grow in the Great Plains, but only 7 are common.

1. Calyx lobes less than half as long as the tube; stipe of loment at least 3 times as long as the calyx tube, longer than the persistent remains of stamens; stamens monadelphous; upper margin of loment glabrous
    2. Flowering stems simple, usually leafless. .... *Desmodium nudiflorum*
    2. Flowering stems branched, leafy. .... *Desmodium glutinosum*
1. Calyx lobes more than half as long as the tube; stipe of loment less than 2 times as long as the calyx, shorter than the persistent remains of the stamens; stamens diadelphous; upper margin of loment pubescent
    3. Petioles up to 3 mm long. .... *Desmodium sessilifolium*
    3. Petioles 3 mm long or more
        4. Leaflets with uncinate hairs on lower surface
            5. Stem usually unbranched with 1 inflorescence; lower leaf surfaces reticulate-veined. .... *Desmodium illinoense*
            5. Stem usually much-branched with several inflorescences; lower leaf surfaces not reticulate-veined. .... *Desmodium canescens*
        4. Leaflets without uncinate hairs on the lower surface
            6. Leaflets about 3 times as long as wide; lower margin of loment segments obtuse or rounded, segments more or less semicircular. ..... *Desmodium canadense*
            6. Leaflets up to 5 times as long as wide; lower margin of loment segments abruptly curved, segments more or less triangular. .... *Desmodium paniculatum*

Canada tickclover
Desmodium canadense (L.) DC.

Prodromus Systematis Naturalis Regni Vegetabilis 2:328. 1825.
canadense: of or from Canada.

GROWTH FORM: forb, flowers July to September, reproduces from seeds. LIFE-SPAN: perennial. ORIGIN: native. HEIGHT: 0.4–2 m. STEMS: herbaceous, erect, usually simple, from a caudex surmounting a taproot, branched above, uncinulate-puberulent to glabrate. LEAVES: alternate, pinnately trifoliate (6–15 cm long, including the petiole); leaflets lanceolate to lance-oblong or oblong (terminal leaflet 4–10 cm long, 1–3 cm wide); apex acute to attenuate; base obtuse to broadly acute; margins entire, ciliate; upper surface appressed-pilose to glabrate; lower surface appressed-pilose; petiole 5–25 mm long, petiole and leaf rachis combined are one-sixth to one-half as long as the terminal leaflet; stipules linear to subulate (4–10 mm long, 1–1.5 mm wide at the base), caducous; stipels lanceolate (2–4 mm long), persistent. INFLORESCENCES: panicle of racemes, densely flowered, terminal and axillary; bracts conspicuous, ovate-lanceolate (4–10 mm long). FLOWERS: perfect, papilionaceous; calyx tube campanulate (1.5–2 mm long); lobes 5, bilabiate, upper lobe 4.5–5 mm long, middle tooth of lower lobe 5–7 mm long; corolla 1–1.5 cm long, pink or rose-purple to reddish-violet (rarely white), fading to purple or blue; banner usually with 2 yellow spots near the base; stamens diadelphous. FRUITS: loments straight to slightly convex on the upper margin, obtuse or rounded on the lower margin, breaking apart, segments 2–5; segments more or less semicircular (5–7 mm long, 4–5 mm wide), uncinate-pubescent; stipe 2–3 mm long; seeds one per segment. SEEDS: reniform or ovoid to broadly ellipsoid (3–4 mm long), compressed, smooth, indurate. $2n=22$.

HABITAT: Canada tickclover is infrequent to common in prairies, rangelands, open thickets, open woodlands, old fields, wet meadows, roadsides, and stream banks. USES AND VALUES: Forage quality of immature plants is good, and they are grazed by cattle, horses, sheep, and goats. The plants decrease with continued heavy grazing. The foliage is eaten by deer, elk, rabbits, and small mammals. Many kinds of birds and small mammals eat the seeds. Upland birds, such as quail and pheasants, use stands of Canada tickclover for escape cover. Its flowers are visited by bumblebees, honeybees, and other pollinating insects. It has little value for erosion control or landscaping. ETHNOBOTANY: The Pawnees and Cherokees recognized its importance for wildlife. East of the Great Plains, the Iroquois made a decoction of the roots for a gastrointestinal aid.

SYNONYM: *Meibomia canadensis* (L.) Kuntze. OTHER COMMON NAMES: beggarlice, beggarticks, bush trefoil, Canadian tickclover, Canadian ticktrefoil, hoary tickclover, hoary ticktrefoil, sanfoil, showy ticktrefoil, sticktight, ticktrefoil, wókaȟtaŋ blaskáska (Lakota).

SIMILAR SPECIES: Two other species in the region have fruits similar to those of Canada tickclover. *Desmodium marilandicum* (L.) DC., Maryland tickclover, grows on the eastern edge of the Great Plains and has small (4–7 mm long) reddish flowers. It is similar to *Desmodium ciliare* (Muhl. *ex* Willd.) DC., slender tickclover, which grows in the southeastern Great Plains. It is difficult to separate the two, though slender tickclover has pink or white flowers.

Hoary tickclover
Desmodium canescens (L.) Poir.

Dictionnaire des Sciences Naturelles, 2nd ed., 13:110. 1819.

canescens (Lat.): becoming gray, referring to the color created by the pubescence.

GROWTH FORM: forb, flowers July to September, reproduces from seeds. LIFE-SPAN: perennial. ORIGIN: native. HEIGHT: 0.5–2 m. STEMS: herbaceous, robust, erect to ascending, usually much-branched, 1–10 from a caudex surmounting a taproot, more or less pubescent. LEAVES: alternate, pinnately trifoliate (6–16 cm long, including the petiole), scattered; leaflets ovate to ovate-lanceolate, reticulate, moderately coriaceous, terminal leaflet largest (3–12 cm long, 1.5–6 cm wide); apex acute to acuminate, sometimes mucronate; base rounded to cuneate; margins entire, ciliate; upper surface uncinate-pubescent to glabrous, uncinate hairs on veins; lower surface uncinate-puberulent with a few villous hairs, not reticulate-veined; lateral leaflets smaller, inconspicuously asymmetrical; petioles of principal leaves 3–12 cm long, nearly as long as the lateral leaflets; stipules ovate to lanceolate or deltate (5–12 mm long, 3–6 mm wide), apex acuminate, margins ciliate, surfaces glabrous, base partially clasping, commonly reflexed, persistent; stipels linear (2–7 mm long), persistent. INFLORESCENCES: panicles of a few to several racemes; raceme axis villous or hirsute; bracts ovate (3–6 mm long). FLOWERS: perfect, papilionaceous; calyx tube campanulate (1.5–2.5 mm long), sparsely to densely villous; lobes 5, bilabiate (3–5.5 mm long), longer than the tube; corolla pink to pinkish-white (8–13 mm long), drying to purple; stamens 10, diadelphous; pedicels slender (5–15 mm long), pubescent, subtended by bracts; bracts ovate-lanceolate (3–7 mm long), caducous. FRUITS: loments straight or slightly curved, uncinate-pubescent, segments 2–6; segments triangular to semirhomboidal (7–13 mm long, 4–5 mm wide), margins pubescent; upper margin convex to straight, more deeply indented and obtuse on the lower side; stipe 1–2 mm long; seeds 1 per segment. SEEDS: reniform (3.5–4.5 mm long, about 2 mm wide), compressed, brown, smooth, indurate. $2n=22$.

HABITAT: Hoary tickclover is widely scattered in moist or dry soils of open woodlands, prairies, rangelands, roadsides, stream banks, and railroad rights-of-way. It is most common on sandy soils. USES AND VALUES: Forage value is fair to good. Cattle, horses, sheep, and goats graze the immature plants. Deer eat the foliage, and wild turkeys eat the flowers. Birds and small mammals eat the loments and seeds. Hoary tickclover has little value for erosion control or landscaping.

VARIETIES: This highly variable species has been divided into vars. *canescens, hirsutum* (Hook.) B.L. Rob., *paleaceum* (Poir.) DC., and *villosissimum* Torr. & A. Gray. SYNONYM: *Meibomia canescens* (L.) Kuntze. OTHER COMMON NAMES: beggar's lice, hoary ticktrefoil, tick trefoil, ukwalága (Lakota), ûnistilûistĭ (Lakota).

SIMILAR SPECIES: *Desmodium viridiflorum* (L.) DC., velvetleaf ticktrefoil, grows most commonly in cutover woodlands near the extreme eastern edge of the Great Plains. Its corollas are purple or pink to lavender, and it is usually less than 1 m tall. The reticulum of the leaves is largely obscured by the pubescence, and the stems are villous with uncinate hairs.

Largeflower tickclover
Desmodium glutinosum (Muhl. *ex* Willd.) Alph. Wood

A Class-book of Botany 2:120. 1845.

glutinous (Lat.): sticky, referring to the uncinate hairs covering the loments.

GROWTH FORM: forb, flowers July to September, reproduces from seeds. LIFE-SPAN: perennial. ORIGIN: native. HEIGHT: 0.4–1 m. STEMS: herbaceous, erect, usually simple from a caudex and taproot, branched, leafy, minutely uncinate-puberulent to sparsely pilose or glabrous; flowering stems branched, leafy. LEAVES: alternate, pinnately trifoliate (15–30 cm long, including the petiole), crowded into a single pseudo-whorl near the top of the stem; terminal leaflet ovate to broadly ovate (7–15 cm long, 5–12 cm wide), larger than the lateral leaflets, apex abruptly acuminate, base rounded; lateral leaflets slightly asymmetrical (3–11 cm long, 2–8 cm wide), apex acuminate to abruptly acuminate, base rounded, margins entire, sometimes slightly ciliate; upper surface glabrous; lower surface sparsely pubescent, especially along the veins; petioles 6–14 cm long; stipules linear 8–13 mm long, semipersistent. INFLORESCENCES: racemes or panicles of racemes (30–80 cm long) terminal, not densely flowered; erect, ascending, or inclined to one side; elevated above the leaves on a naked peduncle; bracts 5–10 mm long, caducous. FLOWERS: perfect, papilionaceous; calyx tube 1.5–2.5 mm long, slightly irregular; lobes 5, less than half the length of the tube; corolla pink to purple, rarely white; banner rounded with a notch in the middle at the fold; wings and keel slightly longer or equaling the banner; stamens 10, monadelphous, apices yellow. FRUITS: loments uncinate-puberulent, seldom with more than 3 segments; segments semiobovate (7–9 mm long, 4–6 mm wide); upper margin straight or concave, glabrous; lower margin convex or angular-convex, initially green, turning brown with maturity; stipe 5–10 mm long. SEEDS: irregular (6–7 mm long, 4–4.5 mm wide), narrow at one end, compressed, yellow to brown. 2*n*=22.

HABITAT: Largeflower tickclover is common in rich woodlands, woodland borders, and wooded valleys. USES AND VALUES: Immature plants are grazed by cattle, horses, sheep, goats, and deer. Birds, small mammals, and rabbits eat the loments and seeds. It is an especially important food for quail and wild turkeys. Loments cling to animals enhancing seed dispersal. It is pollinated by bumblebees and several other species of long-tongued bees and visited by many kinds of insects. Caterpillars of several butterfly and skipper species feed on the foliage. It is not used for erosion control. Largeflower tickclover has little value for landscaping because the plants are rank, the flowers are not showy, and the uncinate foliage clings to clothing. ETHNOBOTANY: East of the Great Plains, Iroquois made a cold infusion of the roots for use as "basket medicine." The specific meaning of "basket medicine" is unknown. Records of use by Plains Indians were not found.

SYNONYMS: *Desmodium acuminatum* (Michx.) DC., *D. glutinosum* (Muhl. *ex* Willd.) Schindl., *Hedysarum acuminatum* Michx., *H. glutinosum* Muhl. *ex* Willd., *Meibomia acuminata* (Michx.) S.F. Blake. OTHER COMMON NAMES: clusteredleaf ticktrefoil, pointed ticktrefoil, pointedleaf ticktrefoil, tickseed, ticktrefoil.

Illinois tickclover
Desmodium illinoense A. Gray

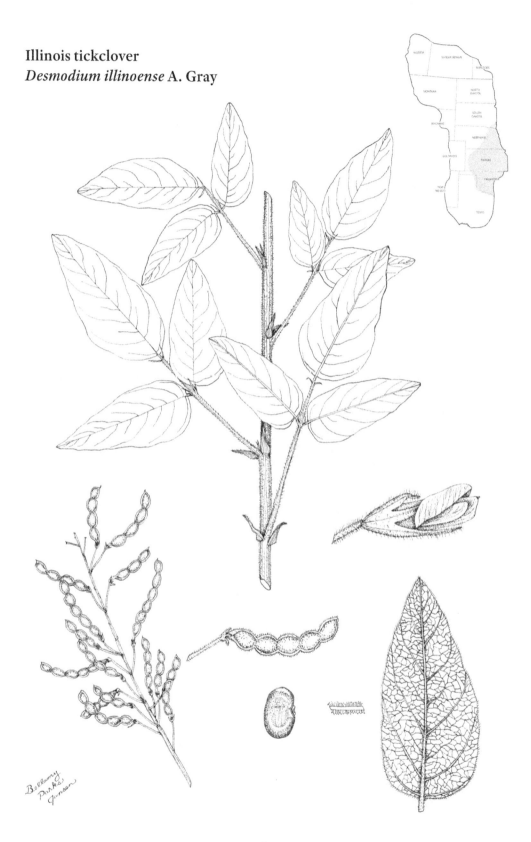

Bellamy
Parks
Jansen

Proceedings of the American Academy of Arts and Sciences 8:289. 1870.
illinoense: of or from Illinois.

GROWTH FORM: forb, flowers June to September, reproduces from seeds. LIFE-SPAN: perennial. ORIGIN: native. HEIGHT: 1–2 m. STEMS: Herbaceous, erect, stout, unbranched or rarely sparingly branched, from a caudex surmounting a taproot, thinly to densely uncinate-puberulent. LEAVES: alternate, pinnately trifoliate (7–15 cm long, including petiole), scattered; leaflets lanceolate-ovate or lanceolate-oblong, terminal leaflet largest (3–10 cm long, 1.5–6 cm wide), lateral leaflets 2–6 cm long and 1–4 cm wide; apex acute to obtuse, often mucronate; base rounded to truncate; margins entire, sometimes slightly sinuate, ciliate; upper surface uncinate-puberulent; lower surface whitish-green, uncinate-puberulent, also with glandular hairs, strongly reticulate-veined; petioles 3–14 cm long, nearly as long to longer than the terminal leaflet, puberulent-uncinate; petiolules of lateral leaflets very short; stipules ovate-acuminate (9–20 mm long, 2–6 mm wide), margins ciliate, upper surfaces pubescent, lower surface glabrous, semipersistent. INFLORESCENCES: usually a single raceme, terminal, few flowers in bloom at the same time, occasionally a simply branched panicle of a few racemes; axis villous to hirsute; bracts lanceolate (about 1 cm long), caducous. FLOWERS: perfect, papilionaceous; calyx tube campanulate (1–2 mm long), bilabiate; upper lobe 2–4.5 mm long, lower lobe 3–6 mm long; corolla pink to purple to white (8–10 mm long), fading to purple; 2 green or yellow spots near the base of the banner; stamens 10, diadelphous. FRUITS: loments, straight to slightly curved, segments 3–7; segments 4–6 mm long, rounded on both margins; margins pubescent with uncinate hairs; stipe less than 1 mm long; seeds 1 per segment. SEEDS: ovoid to reniform (3–3.5 mm long, about 2 mm wide), slightly compressed, smooth, golden-brown, indurate. $2n=22$.

HABITAT: Illinois tickclover is scattered to locally common in rich soils of ravines and hillsides of prairies, glades, and rangelands. This drought-tolerant species also grows in waste areas, roadsides, and woodlands. USES AND VALUES: Cattle, horses, and sheep graze the immature plants, and goats graze them at any time of the year. It decreases with continuous heavy grazing. Deer, elk, and rabbits eat the foliage. Ring-necked pheasants, quail, and wild turkeys eat the seeds and use Illinois tickclover for cover. Small mammals eat the loments and seeds. Its flowers are visited by bumblebees, honeybees, leaf-cutting bees, and many other pollinating insects. It is a good honey plant. It serves as a host to butterfly, skipper, and moth larvae. ETHNOBOTANY: Immediately east of the Great Plains, the Meskwaki (Fox) combined it with other plants to make a "powerful medicine" that was used as an adjuvant.

SYNONYM: *Meibomia illinoense* (A. Gray) Kuntze. OTHER COMMON NAMES: Illinois tick-trefoil, tick trefoil.

SIMILAR SPECIES: *Desmodium tweedyi* Britton, Tweedy ticktrefoil, has uncinate pubescence on its stems and both leaf surfaces. However, its leaflets generally are white-blotched along the midrib. Its loment joints are convex above and angled or rounded below. It is scattered in the southern Great Plains.

Barestem tickclover
Desmodium nudiflorum (L.) DC.

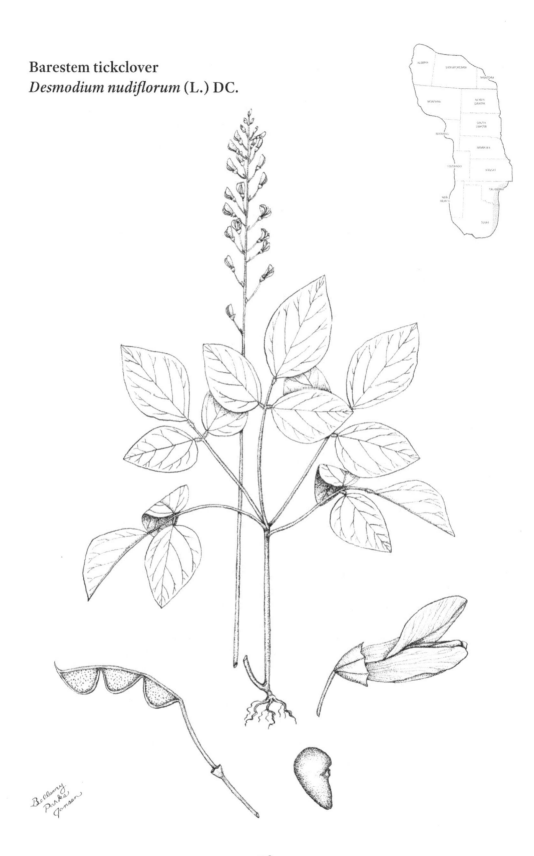

Prodromus Systematis Naturalis Regni Vegetabilis 2:330. 1825.
nudus (Lat.): naked; +*florus* (Lat.): flower, in reference to the sterile stem.

GROWTH FORM: forb, flowers July to September, reproduces from seeds. LIFE-SPAN: perennial. ORIGIN: native. HEIGHT: 0.4–1 m. STEMS: herbaceous, erect to ascending, ultimately divergent or spreading, forked from the base; stems of 2 types, sterile and fertile; sterile stem (10–30 cm tall) naked below, glabrate or sparsely pilose, light green to reddish-green, bearing a crowded cluster of 5–6 tightly spaced leaves in a pseudo-whorl at the summit; fertile branches 1–4, arising from slightly below ground level, scapose, ascending, about 3 times the height of the sterile stem, usually without leaves. LEAVES: alternate, pinnately trifoliate; terminal leaflet elliptic to ovate (4–10 cm long), apex acute to short-acuminate, base rounded to cuneate, margins entire; upper surface medium green, short-pubescent to glabrous; lower surface pale-green, glabrous; lateral leaflets ovate-oblong to ovate, smaller; petioles 10–15 cm long; stipules linear-subulate, caducous. INFLORESCENCES: usually a raceme, occasionally a panicle of few racemes, terminal, leafless; flowers widely spaced; bracts linear-lanceolate (4–6 mm long), caducous. FLOWERS: perfect, papilionaceous; calyx tube campanulate (1.5–2.5 mm long); lobes 5, short, hardly evident, blunt, lower tooth larger than the others, sparsely to minutely hairy; corolla pink to pale lavender and rarely white (6–9 mm long), petals nearly equal in length; banner erect; wings spreading; keel horizontal; stamens 10, monadelphous, filaments white; pedicels capillary (1.2–2.5 cm long). FRUITS: loments with 2–4 segments; segments semi-obovate (8–11 mm long, 4–5 mm wide), margins glabrous; upper margin concave to nearly straight, glabrous; asymmetrical and convex or rhombic below; surfaces uncinate-puberulent; stipe glabrous (8–12 mm long); seeds 1 per segment. SEEDS: variable (5–6 mm long), compressed, yellow to brown, smooth, indurate. $2n=22$.

HABITAT: Barestem tickclover grows in rich moist soils of ravines, woodlands, and wooded slopes. It is less abundant in dry woodlands. It is not abundant in the Great Plains. USES AND VALUES: It is not an important forage species because it usually does not grow where cattle and horses graze. Deer and rabbits eat the foliage, and numerous birds, including wild turkeys and quail, and small mammals eat the loments and seeds. The loment segments stick to hair, and it is likely that animals such as deer are the primary seed dispersers. It has little value for erosion control and landscaping. It is pollinated by bumblebees and several other species of long-tongued bees. Barestem tickclover serves as host to caterpillars of several species of butterflies, skippers, and moths. It produces insect-repellent substances. ETHNOBOTANY: Some Cherokees chewed the root to treat mouth inflammation and bleeding gums. They also used an infusion from the roots as a wash for cramps.

SYNONYMS: *Hedysarum nudiflorum* L., *Meibomia nudiflora* (L.) Kuntze, *Pleurolobus nudiflorus* (L.) MacMill. OTHER COMMON NAMES: nakedflower ticktrefoil, scapose tickclover, tickseed, únistilû′istĭ-yû′ (Lakota).

Panicled tickclover
Desmodium paniculatum (L.) DC.

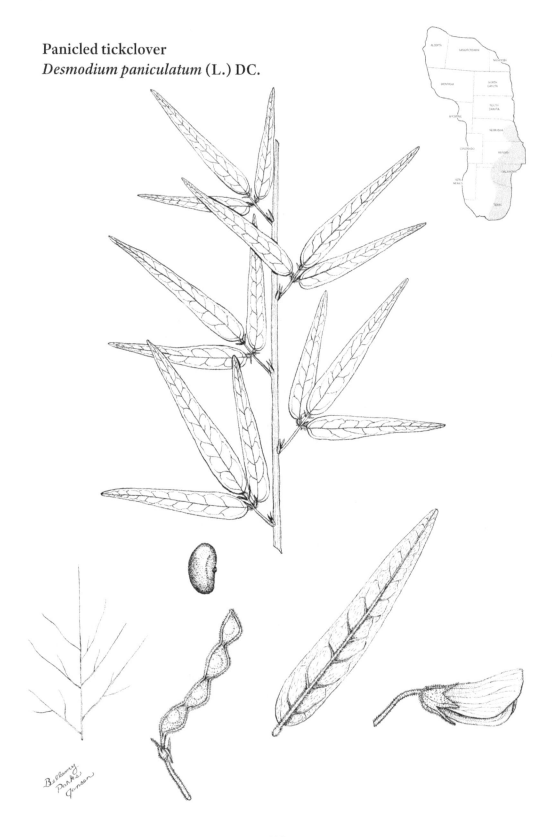

Prodromus Systematis Naturalis Regni Vegetabilis 2:329. 1825.
panicula (Lat.): panicle, referring to the inflorescence.

GROWTH FORM: forb, flowers July to September, reproduces from seeds. LIFE-SPAN: perennial. ORIGIN: native. HEIGHT: 0.5–1.4 m. STEMS: herbaceous, erect to ascending, slender, ridged, 1 to several from a caudex and taproot, usually branching above, uncinate-puberulent to appressed-pubescent or pilose. LEAVES: alternate, pinnately trifoliate (2.5–15 cm long), scattered; leaflets variable, narrowly lanceolate to narrowly oblong to linear; terminal leaflet 2–10 cm long, 1–3 cm wide; lateral leaflets 2–8 cm long, 5–20 mm wide; apex obtuse to nearly acute; base rounded to truncate; margins entire; upper surface appressed-pubescent; lower surface more densely hairy or tomentose, not uncinate-puberulent; lateral veins arcuate-ascending; petioles of principal leaves 3–5 cm long; stipules lanceolate-subulate (2–6 mm long), caducous to semipersistent. INFLORESCENCES: panicles of racemes (10–40 cm long), much-branched, terminal and axillary. FLOWERS: perfect, papilionaceous; calyx tube 1.5–2 mm long, bilabiate; upper lobe 1.5–3.5 mm long, lower lobe 2.5–5.5 mm long; corolla lavender to reddish-purple, rarely white (6–8 mm long), fading to blue; stamens 10, diadelphous; pedicels 4–11 mm long. FRUITS: loments usually straight, segments 3–6, each more or less triangular to rhombic (5–10 mm long, 3.5–4.5 mm wide), convex above, abruptly curved below; margins and surfaces uncinate-puberulent; stipe exceeds the length of the persistent calyx; seeds 1 per segment. SEEDS: reniform to elliptic (3.5–4.5 mm long), reddish-brown or tan, smooth, indurate. $2n=22$.

HABITAT: Panicled tickclover is infrequent to locally common on roadsides, prairies, rangelands, and dry woodlands, especially if the soil is sandy or rocky. USES AND VALUES: Its forage quality is only fair, but it is eaten while immature by cattle, horses, sheep, and deer. Birds and small mammals eat the seeds. It has little value for landscaping or erosion control. ETHNOBOTANY: Outside of the region, members of the Houma nation in Louisiana consumed a root infusion in whiskey to treat weakness and cramps.

VARIETIES: Plants of this species are highly variable, especially in pubescence and leaflet shape. It has been divided into seven varieties. Ronald L. McGregor, in the *Flora of the Great Plains*, said that the following two may be recognized in the region: *dillenii* (Darl.) Isely with terminal leaflets generally 1.5–3 times longer than wide and var. *paniculatum* with terminal leaflets 3–8 times longer than wide. SYNONYMS: *Desmodium dichromum* Shinners, *Hedysarum paniculatum* Michx., *Meibomia chapmanii* (Britton) Small, *M. paniculata* (L.) Kuntze, *M. pubens* (Torr. & A. Gray) Rydb. OTHER COMMON NAMES: panicleleaf tickclover, panicleleaf ticktrefoil.

SIMILAR SPECIES: Three additional species grow in the eastern and southeastern Great Plains. Only careful observations of stipules, stem and leaf pubescence, and measurements of plant parts will separate *Desmodium cuspidatum* (Muhl. *ex* Willd.) DC. *ex* G. Don, longleaf tickclover; *D. glabellum* (Michx.) DC., Dillenuis ticktrefoil; and *D. perplexum* B.G. Schub., perplexed ticktrefoil.

Sessileleaf tickclover
Desmodium sessilifolium Torr. & A. Gray

A Flora of North America 1(3):363. 1840.

sessilis (Lat.): pertaining to sitting; + *folium*: a leaf, referring to the close attachment of the leaf blade and stem.

GROWTH FORM: forb, flowers July to September, reproduces from seeds. LIFE-SPAN: perennial. ORIGIN: native. HEIGHT: 0.8–1.5 m. STEMS: herbaceous, erect to ascending, unbranched or sparingly so, usually simple to the inflorescence, 1 to several from a caudex surmounting a taproot, densely uncinate-puberulent, striped green and tan. LEAVES: alternate, pinnately trifoliate; leaflets narrowly oblong to oblong-lanceolate; terminal leaflet 5–9 cm long and 5–18 mm wide; lateral leaflets 3–9 cm long; apex acute to obtuse, usually mucronate; base rounded to obtuse; upper surface dull green, sparsely pubescent or pilose to occasionally glabrous; lower surface pale green, sparsely uncinate-puberulent along the veins, reticulate-veined; petioles 1–3 mm long; stipules subulate to lanceolate (3–5 mm long), semipersistent to caducous. INFLORESCENCES: sparingly branched panicle of racemes or a single raceme (10–30 cm long), terminal; branches ascending; axis uncinate-puberulent; bracts ovate (2.5–3.6 mm long). FLOWERS: perfect, papilionaceous; calyx tube 1–1.5 mm long, light green to occasionally purplish, bilabiate, lobes 5, often longer than the tube, upper lobe entire or notched; corolla pink, lavender, yellowish-green, or white (4–6 mm long); banner with a patch of yellow at the base; stamens 10, diadelphous. FRUITS: loments; segments 1–4, each 4–6.5 mm long and 3–4.5 mm wide, compressed, straight to convex above, rounded below; margins and surfaces uncinulate-pubescent; stipe 1–3 mm long; pedicels 0–4 mm long; seeds 1 per segment. SEEDS: narrowly ovoid (2.5–3.5 mm long), slightly compressed, light-tan to olive or brown, smooth, indurate. $2n=22$.

HABITAT: Sessileleaf tickclover is scattered to locally common in dry or sterile soils of open woodlands, prairies, meadows, rangelands, limestone glades, savannas, and roadsides. USES AND VALUES: The quality of the forage is fair to good, and it is grazed by cattle, horses, sheep, goats, rabbits, and deer when it is immature. Forage quality and palatability decrease with maturity resulting in reductions in animal use of the forage. Small mammals and birds, such as quail and wild turkeys, eat the loments and seeds. The uncinate hairs cause the loments and loment segments to stick to animal hair, facilitating seed dispersal. Sessileleaf tickclover flowers are pollinated by bumblebees, long-horned bees, alkali bees, and leaf-cutting bees. It is visited by many other insects. It is a good honey plant. It serves as a host for larvae of several butterflies, skippers, and moths. It has little value for erosion control or landscaping.

SYNONYM: *Meibomia sessilifolia* (Torr. & A. Gray) Kuntze. OTHER COMMON NAMES: sessile tickclover, sessile ticktrefoil, sessileleaf ticktrefoil.

SIMILAR SPECIES: *Desmodium rotundifolium* DC., prostrate tickclover, has prostrate stems and forms mats (1.5–2.5 m in diameter). Pubescence on the leaves and stems is appressed to spreading. Its corollas are usually purple. Loment segments are 3–6. It grows in the extreme eastern Great Plains.

Glycyrrhiza L.

glukus (Gk.): sweet; + *rhiza* (Gk.): root, referring to the taste of the root.

Perennial herbs from rhizomes; leaves odd-pinnately compound, leaflets few to several; leaflets glandular-punctate; stipules small, caducous; inflorescences dense axillary racemes or spikes, bracts caducous; calyx tube campanulate to shortly cylindric; lobes 5, filiform to lanceolate, bilabiate, upper 2 lobes fused part of their length; corolla ochroleucous to bluish, petals usually acute; banner oblong-obovate, tapering to the base, slightly recurved; wings and keel shorter, oblong, clawed at the base; stamens diadelphous for half their length, often irregular in length; anthers alternately large and small; ovary cylindric, glandular or glabrous; fruits legumes, ovate to oblong, tardily dehiscent to indehiscent, somewhat inflated, slightly compressed, glandular, enveloped with hooked prickles; seeds few. $x=8$.

About 20 species grow in Eurasia, Africa, and the New World. Two grow in the United States, and only 1 grows in the Great Plains. The licorice of commerce is *Glycryrrhiza glabra* L. It is grown commercially in several areas of the world, including the western United States.

Wild licorice
Glycyrrhiza lepidota Pursh

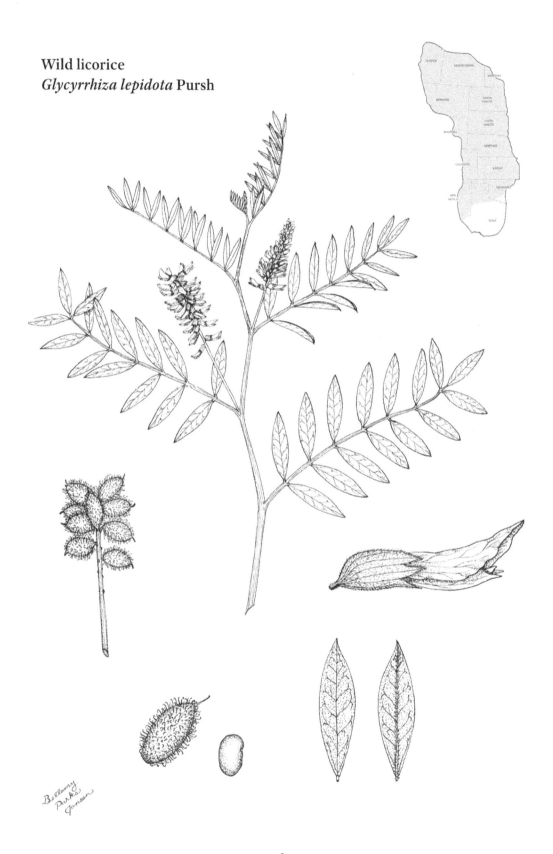

Flora Americae Septentrionalis 2:480. 1814.

lepidotos (Gk.): scaly, referring to the leaves being minutely scaly when young.

GROWTH FORM: forb, flowers June to September, reproduces from rhizomes and seeds. LIFE-SPAN: perennial. ORIGIN: native. HEIGHT: 0.3–1.2 m. STEMS: herbaceous, erect to ascending, 1 to several from long rhizomes or a caudex surmounting an aromatic taproot; simple or little-branched, glabrous or glandular-pubescent. LEAVES: alternate, odd-pinnately compound, leaflets 7–21; leaflets ovate to lanceolate to elliptic (1–5 cm long, 4–18 mm wide); apex acute, mucronate; base acute to cuneate; margins entire; surfaces glandular-punctate; midrib of lower surface often puberulent; rachis not extended into a tendril; petioles 5–50 mm long; stipules lance-acuminate (3–10 mm long), caducous. INFLORESCENCES: racemes axillary (2.5–5 cm long), spikelike, dense, erect, shorter than the subtending leaves; peduncle 1–7 cm long. FLOWERS: perfect, papilionaceous; calyx tube campanulate (5–7 mm long), glandular-punctate, pubescent; lobes 5, bilabiate, unequal; corolla white to yellowish-white or ochroleucous; banner oblong to obovate (9–15 mm long), tapering to the base; wings narrowly oblong; keel narrowly oblanceolate (8–12 mm long), apex acute, clawed at the base; stamens 10, diadelphous; anthers alternately large and small; pedicels 0.5–1 mm long. FRUITS: legumes oblong to elliptic (1–2 cm long, 5–8 mm wide), slightly compressed, covered with hooked prickles, burrlike, reddish-brown at maturity, indehiscent to tardily dehiscent; stipe absent; seeds usually 3–5. SEEDS: short-reniform (2.5–4 mm long), plump but slightly compressed, olive to brown, smooth, indurate. $2n=16$.

HABITAT: Wild licorice is common on moist prairies, pastures, rangelands, rich shores, meadows, ravines, railroad rights-of-way, and waste places. USES AND VALUES: Palatability of the herbage to livestock is low, but it is readily eaten in dry hay. Deer, pronghorn, elk, and bighorn sheep consume the foliage. Birds eat the seeds, small mammals eat the legumes and seeds, and plains pocket gophers eat the roots. It has limited application in erosion control and landscaping. The Eurasian *Glycyrrhiza glabra* L. is the source of licorice flavoring. ETHNOBOTANY: Wild licorice was widely used as medicine by Great Plains Indians. A poultice was applied to open wounds to stop bleeding. The same poultice was applied to horses' backs to relieve soreness. Some Lakotas used a decoction of the roots as a fever remedy. Steeped leaves were places in ears to relieve earache. Roots were chewed for their pleasant flavor and held in the mouth to relieve toothache and sore throats. The roots and leaves were used to treat upset stomach, coughs, and chest pain. Roots were eaten either raw or baked. The root was also used to protect pregnant women from spiritual harm.

VARIETIES: Var. *lepidota* is common to the Great Plains. Var. *glutinosa* (Nutt.) S. Watson is generally found west of the region and has stalked glands on the petioles, upper stems, and calyces. SYNONYM: *Glycyrrhiza glutinosa* Nutt. OTHER COMMON NAMES: American licorice, buffalobur, deseret weed, licorice, pithahatusakitstsuhast (Pawnee), sweetroot, sweetwood, wi-nawizi (Dakota), winawizi čik'ala (Lakota).

Kummerowia Schindl.

kummerowia: named for Jochen Kummerow (1927–2004), German-born botanist who worked in Chile and at San Diego State University.

Annual, herbs, low or prostrate plants; stems much-branched; leaves alternate, palmately trifoliate (appearing to be pinnately compound), leaflet margins entire to slightly serrulate, terminal leaflet with a petiolule, rachis not extended into a tendril, petioled or subsessile; leaflets transverse-striate with parallel and unbranched secondary veins; stipules conspicuous, chartaceous, striate, persistent; inflorescences short terminal racemes or solitary flowers in axils along the stem; flowers papilionaceous; chasmogamous and cleistogamous intermixed or in separate inflorescences; calyx tube campanulate; lobes broad, upper pair partly or nearly fused; corolla of chasmogamous flowers bicolored, pinkish-purple and white; fruits legumes, elliptic, reticulate-veined, papery, sessile, subtended by the persistent calyx, indehiscent; seeds 1. x=10, 11.

Genus of two species. Both were introduced into the Great Plains and are native to Asia. They are cultivated in both hemispheres.

1. Petioles of the main leaves 4–10 mm long; leaflets emarginate and conspicuously ciliate; stems antrorsely appressed-pubescent; calyx covering one-third to one-half of the fruit. .... *Kummerowia stipulacea*
1. Petioles of the main leaves 1–4 mm long, appearing subsessile; leaflets not emarginate or conspicuously ciliate; stems retrorse-strigose; calyx covering one-half to four-fifths of the fruit. .... *Kummerowia striata*

Korean clover
Kummerowia stipulacea (Maxim.) Makino

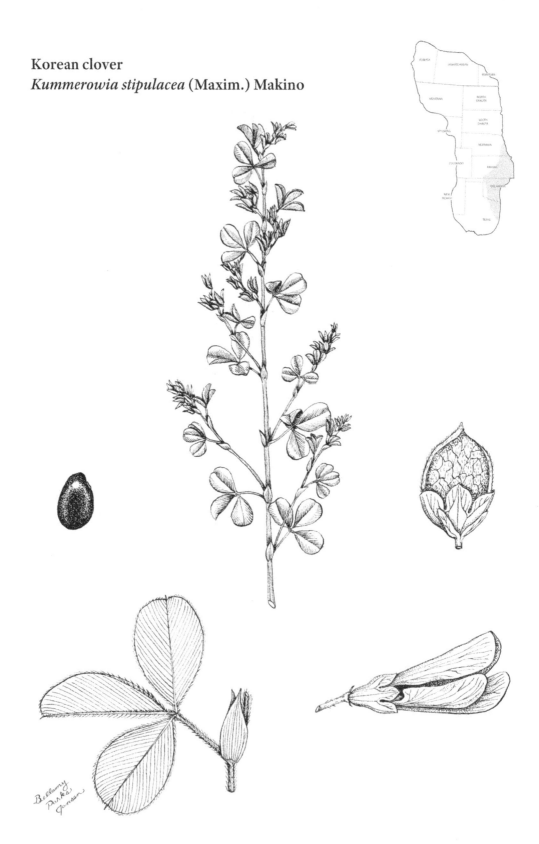

Botanical Magazine, Tokyo 28(328):107. 1914.

stipula (Lat.): a small stalk, referring to the stipules at the base of the petioles.

GROWTH FORM: forb, flowers July to September, reproduces from seeds. LIFE-SPAN: annual. ORIGIN: introduced (from Asia). HEIGHT: 10–60 cm. STEMS: herbaceous, erect or ascending from a shallow taproot, diffusely branched from the base, sparingly pubescent with antrorsely appressed hairs. LEAVES: alternate, palmately trifoliate (appearing pinnately trifoliate); terminal leaflet with a petiolule; rachis not extending into a tendril; leaflets broadly obovate (6–20 mm long, 4–10 mm wide); apex emarginate, mucronate; base acute to obtuse; margins entire, ciliate; surfaces glabrous or glabrate; lower surface sparsely pubescent on the midrib, especially on younger leaves; petioles 4–10 mm long; petiolules 0.5–1 mm long; stipules ovate-lanceolate (3–6 mm long), brown, scarious, many-veined, persistent. INFLORESCENCES: short racemes, axillary and near the branch tips, dense, chasmogamous and cleistogamous flowers intermixed; flowers 1–3. FLOWERS: chasmogamous flowers, perfect, papilionaceous (7–8 mm long); calyx tube campanulate (about 1 mm long); lobes 5, ovate, shorter or about equaling the tube; upper lobe nearly completely connate; corolla bicolored, pinkish-purple (6–8 mm long), usually with a dark purple or red spot at the base of the banner and tip of the keel; banner broadly ovate; wings oblong, white, shorter than the keel; keel petals oblanceolate, white, often with dark red tips; stamens 10, diadelphous; cleistogamous flowers without petals, inconspicuous. FRUITS: legumes elliptic to oval or obovate (2.5–3 mm long, 1.5–2.5 mm wide), one-third to one-half enclosed in the persistent calyx, strongly reticulate, sparsely appressed-pubescent; seeds 1. SEEDS: ellipsoid to orbiculate (1.5–2 mm long), slightly compressed, smooth, brown. $2n=22$.

HABITAT: Korean clover has been planted in pastures and has escaped to disturbed prairies, heavily grazed rangelands, open woodlands, roadsides, and waste areas. It is most common in sandy soils. It can spread aggressively and is often classified as a weed. USES AND VALUES: Korean clover was introduced to North America from Korea in 1919 for pasture and soil improvement. Korean clover is commonly planted with cool-season grasses to provide much-needed summer grazing when higher temperatures reduce the rate of growth of cool-season grasses. It is occasionally cut for hay, and is palatable to all classes of livestock and is readily eaten. It does not cause bloat, but instances of hemorrhagic syndrome have been reported. The condition is similar to sweetclover poisoning and occurs after animal eats moldy Korean clover hay. The foliage is eaten by deer and rabbits. Seeds are eaten by songbirds, gamebirds, and small mammals. It is pollinated by bees and visited by many kinds of insects. Korean clover is used for soil conservation but has virtually no value for landscaping.

SYNONYMS: *Lespedeza stipulacea* Maxim., *L. striata* var. *stipulacea* (Maxim.) Debeaux, *Microlespedeza stipulacea* (Maxim.) Makino. OTHER COMMON NAMES: Korean bush-clover, Korean lespedeza.

Japanese clover
Kummerowia striata (Thunb.) Schindl.

Repertorium Specierum Novarum Regni Vegetabilis 10(257/259):403. 1912.
stria (Lat.): furrowed or marked with fine longitudinal lines, in reference to the surface of the stipules.

GROWTH FORM: forb, flowers July to September, reproduces from seeds. LIFE-SPAN: annual. ORIGIN: introduced (from eastern Asia). HEIGHT: 10–50 cm. STEMS: herbaceous, erect to prostrate and sprawling, from a shallow taproot, sparingly branched to much-branched, sparsely pubescent, hairs retrorse-strigose. LEAVES: alternate, palmately trifoliate (appearing pinnately trifoliate); terminal leaflet with a petiolule; leaflets oblong-obovate to oblong-elliptic (6–18 mm long, 3–9 mm wide); apex retuse or rounded, not emarginate, mucronate; base acute; margins entire to slightly serrated, slightly appressed-ciliate; upper surface glabrous, medium to dark green; lower surface sparsely pubescent on the midrib, otherwise glabrous, pale green; petioles 1–4 mm long; petiolules 0.5–1 mm long; stipules ovate-lanceolate (4–6 mm long), striate, longitudinal veins many, scarious, becoming brown, persistent. INFLORESCENCES: short racemes in upper axils, chasmogamous and cleistogamous flowers intermixed; flowers sessile or on pedicels to 2 mm long; flowers 1–5. FLOWERS: chasmogamous flowers perfect, papilionaceous; calyx tube campanulate (1–2 mm long), glabrous to sparsely pubescent; lobes 5, oblong, about as long as the tube, nearly equal, reticulate-veined, glabrous to sparsely pubescent; corolla bicolored, purple to pink or white (5–7 mm long); banner predominantly pink or purple with several dark purple veins near the base; wings white; keel white except for a purplish-black outer edge. FRUITS: legumes obovate (3–4 mm long), beak acute, inconspicuously reticulate, densely appressed puberulent, one-half to four-fifths covered with the persistent calyx; seeds 1. SEEDS: obovate to reniform (2.5–3 mm long), compressed, black, often mottled, lightly pitted, indurate. $2n=22$.

HABITAT: Japanese clover has been planted for forage and escaped from cultivation to roadsides, waste places, fields, and upland woodlands. It is most abundant in well-drained fertile soils. It can be a problematic weed in turf. USES AND VALUES: It produces high-quality forage for all classes of livestock. It tolerates close grazing, and is sometimes cut for hay. Cattle eating moldy Japanese clover hay may suffer from internal hemorrhaging. It can be important in the diets of white-tailed deer. Wild turkeys eat the foliage, while songbirds, gamebirds, and small mammals eat the seeds. Seeds are especially important in the diets of bobwhite quail. Japanese clover has been grown in North America since 1846, and several cultivars have been developed. It has limited applications for soil stabilization and is not used in landscaping. ETHNOBOTANY: In Asia, a decoction of the whole plant, sometimes mixed with other plants, has been used as a diuretic, to reduce fever, and to treat diarrhea, headache, and vertigo.

SYNONYMS: *Desmodium striatum* (Thunb.) DC., *Hedysarum striatum* Thunb., *Lespedeza striata* (Thunb.) Hook. & Arn., *Meibomia striata* (Thunb.) Kuntze, *Microlespedeza striata* (Thunb.) Makino. OTHER COMMON NAMES: common lespedeza, Japanese bushclover, Japanese lespedeza, wild clover.

Lathyrus L.

lathyros (Gk.): ancient name of some leguminous plant.

Annual or perennial herbs; generally rhizomatous; stems decumbent, erect, or ascending, often twining, sometimes winged; leaves alternate, even-pinnately compound, usually terminated by a tendril or a bristle; stipules smaller than the leaflets, hastate to obliquely or semisagittately lobed, persistent; inflorescence an axillary raceme of few to many flowers; calyx regular or irregular, tube campanulate to somewhat turbinate; lobes 5, equal or upper 2 shorter; corolla papilionaceous, purple to reddish-purple or white to yellow; banner broadly obovate to rotund; claw nearly as wide as the blade; wings free or nearly so, narrowly to broadly obovate; keel upwardly curved; stamens 10, diadelphous; ovary sessile or stipitate; style twisted or straight, compressed, bearded down the inner face; fruits legumes, sessile or substipitate, oblong to oblong-lanceolate, terete to compressed, papery or coriaceous, thin-walled, dehiscent; seeds 2 to several. $x=5, 6, 7$.

A genus of about 150 species found primarily in the northern temperate zone and in South America. Of the 7 species growing in the Great Plains, 3 are common.

1. Leaflets on mature leaves only 2; stem winged. .... *Lathyrus latifolius*
1. Leaflets on mature leaves 4 or more; stems angled, not winged
    2. Rachis extended into a tendril; tendril sparingly branched. .... *Lathyrus venosus*
    2. Rachis extended into a bristle; bristle simple. .... *Lathyrus decaphyllus*

Prairie vetchling
Lathyrus decaphyllus Pursh

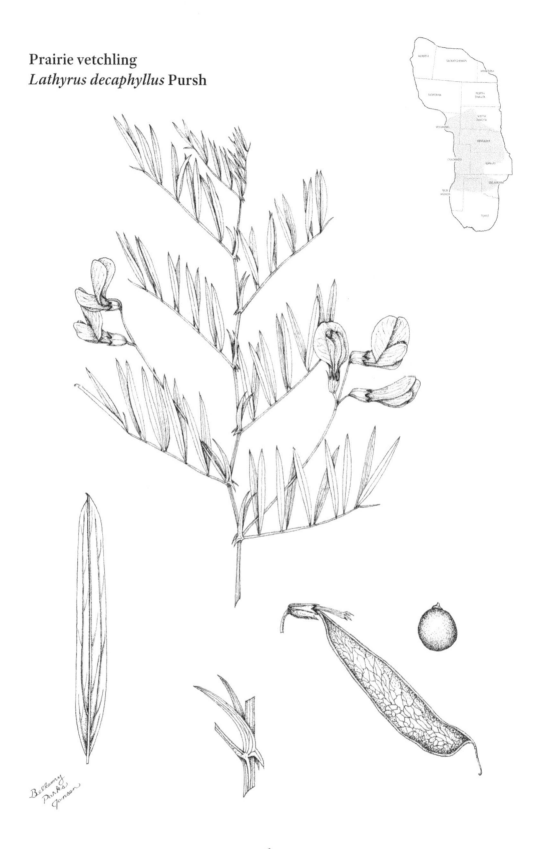

Flora Americae Septentrionalis 2:471. 1814.

deca (Lat.): ten; + *phyllus* (Lat.): leaves or foliage, in reference to the number of leaflets.

GROWTH FORM: forb, flowers May to June, reproduces from rhizomes and seeds. LIFE-SPAN: perennial. ORIGIN: native. HEIGHT: 10–50 cm. STEMS: herbaceous, erect to ascending from rhizomes or a branching caudex, glabrous to pilose or villous, angled, not winged. LEAVES: alternate, evenly pinnately compound, glabrous or rarely pubescent, rachis extended to a simple curved or straight bristle, lacking tendrils; leaflets 4–12 (commonly 10), linear-lanceolate to linear-elliptic (1.5–5 cm long, 1–6 mm wide), scattered or paired; apex acuminate to attenuate, asymmetric; base cuneate to attenuate; margins entire; surfaces usually glabrous but may be pilose to villous; petioles wingless (2–16 mm long); stipules half-sagittate (7–21 mm long, 1–4 mm wide), lower lobe lanceolate to triangular, upper lobe lanceolate. INFLORESCENCES: racemes axillary; flowers 2–8, showy; peduncles 6–8 cm long. FLOWERS: perfect, papilionaceous, fragrant; calyx campanulate (6–12 mm long), tube 4–6 mm long; lobes 5, upper lobes 2–3 mm long, lower lobe 4–6 mm long, lateral lobes 3–4 mm long, persistent; corolla bicolored, violet or purple to reddish-pink; banner purple (2–3 cm long), clawless; wings purple or blue to white (1.5–2.5 cm long); keel 1.5–2.5 cm long; pedicels 3–9 mm long. FRUITS: legumes narrowly oblong (4–6 cm long, 5–10 mm wide), coriaceous; seeds 3–8. SEEDS: globose (3–6 mm long), brown, smooth. $2n=14$.

HABITAT: Prairie vetchling is locally common on dry, sandy prairies and rocky, open woodlands. It is also found in stream valleys and on sand dunes. USES AND VALUES: It has good forage quality and is eaten by all classes of livestock. Forage quality in dried hay remains good. It decreases with continued heavy grazing. The herbage is eaten by deer, pronghorn, elk, bighorn sheep, and rabbits. Ground-foraging birds eat the seeds and small mammals eat the legumes and seeds. Ingestion of prairie vetchling by horses has been reported to cause lameness. The toxic substance has not been identified. Pollination is by various kinds of bees. The flowers are visited by butterflies, skippers, and many other kinds of insects. It has the potential to prevent erosion of sandy soils. Interest in using prairie vetchling in landscaping is increasing. ETHNOBOTANY: Members of some Great Plains Indian tribes ate the legumes raw or after boiling.

VARIETIES: The three recognized varieties are *decaphyllus*, *incanus* (J.G. Sm. & Rydb.) Broich, and *minor* Hook. & Arn. Most of the plants in the Great Plains are glabrous or nearly so and belong to var. *decaphyllus*. Var. *incanus* has pilose herbage and grows most commonly in Colorado and Wyoming on the western edge of the Great Plains. Var. *minor* grows in California. SYNONYMS: *Lathyrus polymorphus* Nutt., *L. stipulaceus* (Pursh) Butters & H. St. John. OTHER COMMON NAMES: hoary peavine, hoary vetchling, manystem pea, sweet pea, wild pea.

SIMILAR SPECIES: *Lathyrus ochroleucus* Hook., yellow vetchling, differs from *Lathyrus decaphyllus* by having branched tendrils and white or ochroleucous flowers. It grows in the northern Great Plains.

Everlasting pea
Lathyrus latifolius L.

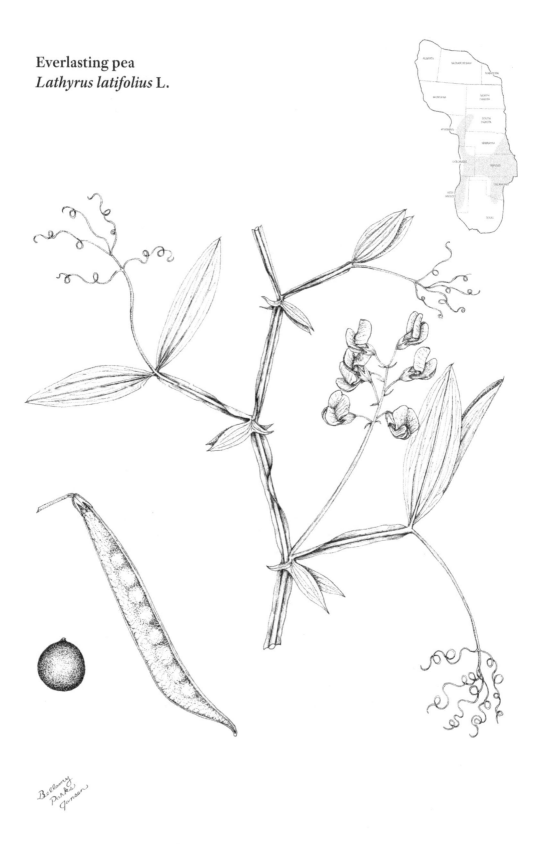

Species Plantarum 2:733. 1753.
latus (Lat.): broad; + *folium* (Lat.): leaf, in reference to the broad leaves.

GROWTH FORM: forb, flowers May to September, reproduces from rhizomes and seeds.
LIFE-SPAN: perennial. ORIGIN: introduced (from southern Europe). LENGTH: 0.5–2 m.
STEMS: herbaceous, climbing or trailing, glabrous, slightly glaucous, broadly winged;
wings 5–10 mm wide. LEAVES: alternate, pinnately compound; leaflets 2, lanceolate to
elliptic (4–15 cm long, 1–5 cm wide); rachis extended into a much-branched tendril;
veins parallel; apex attenuate, mucronate; base attenuate; margins entire; surfaces
glabrous; petiole broadly winged (3–9 cm long, about as wide as the stem); stipules
lanceolate with a basal lobe (1.5–8 cm long, usually wider than the stem), semisagittate, foliaceous. INFLORESCENCES: racemes axillary; flowers 4–12; peduncle 10–20 cm
long; bracts setaceous to linear-subulate (2–6 mm long). FLOWERS: perfect, papilionaceous, not fragrant; calyx tube campanulate (4–6 mm long), persistent; lobes 5,
unequal, upper lobes 2–4 mm long, lateral and lower lobes 4–10 mm long; corollas
purple (less commonly red, white, or pink); banner strongly reflexed (1.5–2.5 cm
long), clawed; wings and keel clawed; stamens 10, diadelphous; pedicels 8–25 mm
long. FRUITS: legumes oblong (6–10 cm long, 7–10 mm wide), compressed, reticulate; seeds usually 6–25. SEEDS: globose or oblong-reniform (3–5 mm long), rugose,
dark brown. $2n=14$.

HABITAT: Everlasting pea has been cultivated in the United States since the early 1700s.
It has escaped from cultivation and is infrequent in fence rows and roadsides. It
is found in vacant lots and waste places near homes and abandoned farmsteads.
USES AND VALUES: This garden plant is not commonly eaten by livestock or wildlife
because of where it grows. The legumes and seeds have been reported to be highly
toxic. They contain a dangerous amino acid. Ingestion of the seeds may result in
neurolathyrism, which is an imbalance in the nervous system causing trouble with
breathing and walking, slow pulse, convulsions, and paralysis followed by death. It
can cause osteolathyrism, resulting in skeletal deformities and aortic rupture. Horses
are especially at risk, but it has poisoned humans, all classes of livestock, poultry,
and dogs. Birds and small mammals eat the seeds seemingly with no ill affect. The
flowers are pollinated by various bees and visited by many kinds of insects. Butterflies
drink the nectar, but are not effective at pollination. It has been planted for erosion
control. Several cultivars are available in garden stores.

SYNONYM: Lathyrus laevigatus Arechav. OTHER COMMON NAMES: everlasting peavine,
perennial pea, perennial sweetpea, sweet pea, wild sweetpea.

SIMILAR SPECIES: Lathyrus pusillus Elliott, low vetchling, also has 2 leaflets per leaf. It is
an annual with wingless petioles and small flowers (less than 1.5 cm long). It grows in
the southeastern part of the Great Plains. *Lathyrus tuberosus* L., tuberous sweetpea,
has 2 leaflets per leaf. The stems climb or sprawl. They are angled but not winged.
The flowers are small (up to 1.6 cm long). It is most common in the southeastern
Great Plains.

Bushy vetchling
Lathyrus venosus Muhl. *ex* Willd.

Species Plantarum. Editio Quatra 3(2):1092. 1802.
venosus (Lat.): conspicuously veined, in reference to the petals.

GROWTH FORM: forb, flowers May to July, reproduces from rhizomes and seeds. LIFE-SPAN: perennial. ORIGIN: native. LENGTH: 0.5–2 m. STEMS: herbaceous, sprawling (occasionally ascending), 4-angled, not winged, glabrous to finely pubescent, coarse, arising from rhizomes, branched. LEAVES: alternate, even-pinnately compound (7–15 cm long); leaflets 8–14, scattered or in pairs, ovate or narrowly to broadly elliptic (3–6 cm long, 1–3 cm wide); apex rounded to obtuse, usually mucronate; base rounded to obtuse; margins entire; surfaces glabrous to finely pubescent; upper surface medium green; lower surface lighter in color; rachis extended into a sparingly branched tendril; petioles 0.5–3 mm long; stipules somewhat sagittate, narrowly lanceolate to ovate-lanceolate (5–35 mm long), pointed at both ends, upper portion longer than the lower portion. INFLORESCENCES: racemes axillary; flowers 5–20, dense or loose; bracts minute, caducous; peduncles stout, often arcuate, usually finely pubescent, shorter or equaling the subtending leaf. FLOWERS: perfect, papilionaceous; calyx tube campanulate (3.5–4.5 mm long), glabrous to shortly pubescent; lobes 5, unequal, lower lobes longer than the upper but shorter than the tube; corollas bicolored; lavender, purple, blue, or reddish-purple; venation is distinctly darker; banner obcordate (1.5–2 cm long), toothed at the top center, flaring out at the sides, claw nearly equaling the blade; wings shorter than the banner, pale pink to nearly white; keel shorter than the wings, slightly darker pink than the wings; pedicels 3–7 mm long. FRUITS: legumes linear-oblong (4–6 cm long, 5–8 m wide), compressed, glabrous to short-pubescent; seeds usually 3–9. SEEDS: nearly oval (4–5 mm long, 2–3 mm wide), dark brown, smooth. $2n=14$.

HABITAT: Bushy vetchling is infrequent to locally common in prairie and rangeland ravines, stream valleys, open woodlands, roadsides, and lake shores. USES AND VALUES: It is grazed by all classes of livestock. Deer eat the foliage, and pocket gophers eat the roots. It is important nesting cover for upland birds. Ground-foraging birds eat the seeds, and small mammals eat the legumes and seeds. It is pollinated by bumblebees and other long-tongued bees. It has few applications for erosion control and landscaping. It aggressively spreads from garden plantings. ETHNOBOTANY: Chippewas (Ojibwas) used a decoction of the foliage as an emetic, stimulant, tonic, and anticonvulsant. Legumes and seeds were eaten raw or cooked.

VARIETIES: At least seven varieties have been described. Var. *intonsus* Butters & H. St. John is finely pubescent and the one most common in the Great Plains. SYNONYMS: *Lathyrus multiflorus* Nutt. *ex* Torr. & A. Gray, *L. oreophilus* Wooton & Standl., *L. rollandii* Vict. & J. Rousseau, *Orobus venosus* P.J. Braun. OTHER COMMON NAMES: bushy vetch, mĭs'nĭsĭno'wûck (Chippewa), veiny pea, veiny peavine, wild pea, wild peavine.

SIMILAR SPECIES: *Lathyrus palustris* L., marsh vetchling, is similar, but it usually has 8 or fewer leaflets, and each raceme has 9 or fewer flowers. Marsh vetchling grows in the northeastern Great Plains. The ranges of bushy vetchling and marsh vetchling overlap.

Lespedeza Michx.

Lespedeza: derived from a misspelling of Céspedes after V. M. de Céspedes (about 1721–94), Spanish governor of Florida, who aided French botanist André Michaux's botanical explorations of North America.

Annual or perennial herbs, some introduced shrubs; often from a woody caudex, erect or ascending, often branched, pubescent or glabrous, not glandular-punctate; leaves alternate, numerous, pinnately trifoliate; leaflets often mucronate, margins entire; stipulate but without stipels; inflorescences spicate or capitate racemes, clusters, or solitary flowers; calyx tube campanulate, short; lobes 5, linear-subulate; corollas purple to yellowish-white; banner suborbicular to oblong-obovate, spreading to erect, sometimes with a purple spot, claws short; wings oblong, straight, clawed, connivent with the keel; keel obliquely obovate; stamens 10, diadelphous; ovary short; fruits legumes, oval to elliptic, indehiscent, persistent; seeds 1; flowers of perennial species of two types, chasmogamous and cleistogamous; in white-flowered species chasmogamous flowers are petalous (papilionaceous), fertile, and most abundant; in purple-flowered species cleistogamous flowers are apetalous, most abundant, and produce more mature fruits. x=9, 10, 11.

About 40 species are native to eastern Asia and North America. Most in North America grow in the eastern part of the continent. Nine species grow in the Great Plains, and 4 are common.

1. Corollas cream to white (may be marked with purple or pink); calyx sometimes equaling or exceeding the mature legume
    2. Flowers borne singly or in axillary racemes of 2–4; wings and keel equal; leaflets 1–2.5 cm long. .... *Lespedeza cuneata*
    2. Flowers numerous in subglobose to short-ovoid racemes; wings longer than the keel; leaflets 2–4.5 cm long. .... *Lespedeza capitata*
1. Corollas purple; calyx shorter than the mature legume
    3. Leaflets 3 or more times as long as wide; peduncles usually shorter than the subtending leaves. .... *Lespedeza virginica*
    3. Leaflets less than 3 times as long as wide; peduncles usually longer than the subtending leaves. .... *Lespedeza violacea*

Roundhead lespedeza
Lespedeza capitata Michx.

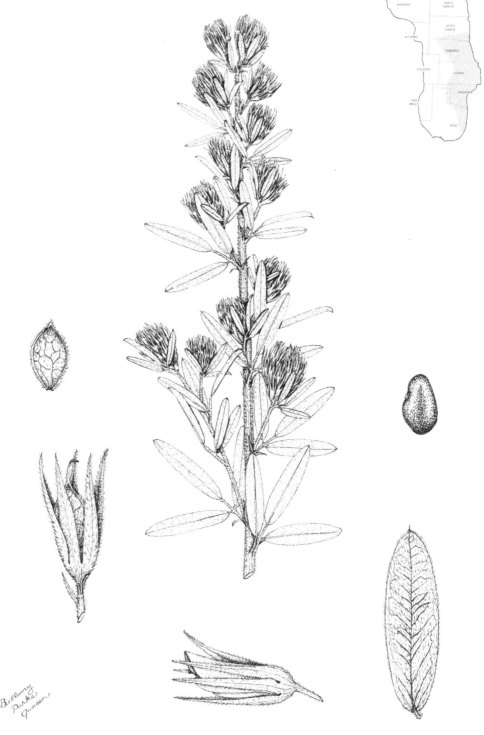

164

Flora Boreali-Americana (Michaux) 2:71. 1803.
capitatus (Lat.): headlike or globose, referring to the flower clusters.

GROWTH FORM: forb, flowers June to September, reproduces from seeds. LIFE-SPAN: perennial. ORIGIN: native. HEIGHT: 0.6–2 m. STEMS: herbaceous, usually erect from a caudex and taproot, rigid, simple or branched above, appressed-pubescent to villous-pubescent, densely leafy. LEAVES: alternate, pinnately trifoliate, subsessile; leaflets elliptic to oblong-lanceolate (2–4.5 cm long, 5–15 mm wide); apex obtuse to acute, mucronate; base rounded to cuneate; margins entire, ciliate; upper surface glabrate; lower surface densely silvery-pubescent; petioles 2–5 mm long, shorter than the stalk of the terminal leaflet; stipules filiform to subulate (3–8 mm long), persistent. INFLORESCENCES: racemes subglobose to short-ovoid (1.2–2.5 cm long), axillary and on short branches at the stem tips, numerous, crowded; flowers 8–40, cleistogamous flowers scattered among the chasmogamous flowers; peduncles rarely longer than the racemes, usually shorter than the subtending leaves. FLOWERS: chasmogamous flowers perfect, papilionaceous; calyx tube 0.5–1 mm long, villous, reddish-brown, persistent; lobes 5 (6–14 mm long), villous; corolla cream to white (8–12 mm long); banner about as long as the calyx, longer than the keel, sometimes reddish-purple at the base; wings longer than the keel; stamens 10, diadelphous; cleistogamous flowers 4–5 mm long; pedicels 1–3 mm long. FRUITS: legumes elliptic to oblong (4–7 mm long), shorter than the calyx, puberulent; seeds 1. SEEDS: ellipsoid, slightly compressed, tan to green or brown to black, smooth, shiny, indurate. $2n=20$.

HABITAT: Roundhead lespedeza is common on prairies, rangelands, dry fields, sandy woodlands, sand dunes, and roadsides. USES AND VALUES: It provides excellent forage for all classes of livestock during the growing season. It can be an important component of prairie hay. It decreases with continued heavy grazing. The foliage is eaten by deer and wild turkeys. The seeds are eaten by upland birds, songbirds, and small mammals. It is important for soil stabilization. Roundhead lespedeza is seldom used for landscaping, but it is included in dry floral arrangements because of its attractive, persistent, reddish-brown calyces. ETHNOBOTANY: Some Comanches made a beverage from the leaves, and some Omaha-Poncas burned pieces of the stem into their flesh as counterirritants to treat rheumatism and neuralgia.

VARIETIES: This variable species has been divided into about ten varieties largely based on pubescence, leaflet proportions, and the degree to which the flowers are masked by the subtending leaves. SYNONYMS: *Hedysarum conglomeratum* Poir., *H. frutescens* Willd., *H. umbellatum* Walter, *Lespedeza bicknellii* House, *L. frutescens* (L.) Elliott, *L. velutina* E.P. Bicknell. OTHER COMMON NAMES: bushclover, dusty clover, parus-as (Pawnee), rabbitsfoot, roundhead bushclover, tall bushclover, te-hunton-hi nuga (Omaha-Ponca).

SIMILAR SPECIES: *Lespedeza hirta* (L.) Hornem., hairy lespedeza, has terminal leaflets that are more than half as wide as long. It is found occasionally in the eastern Great Plains.

Sericea lespedeza
Lespedeza cuneata (Dum. Cours.) G. Don

A General History of the Dichlamydeous Plants 2:307. 1832.
cuneata (Lat.): wedge-shaped, in reference to the shape of the leaflets.

GROWTH FORM: forb, flowers July to October, reproduces from seeds. LIFE-SPAN: perennial. ORIGIN: introduced (from eastern Asia). HEIGHT: 0.5–2 m. STEMS: herbaceous to somewhat woody at the base, erect from a knobby and woody caudex, branched above; branches numerous, virgate, and elongate, sulcate, appressed-pubescent in lines on the ridges of the stem. LEAVES: alternate, pinnately trifoliate; leaflets linear-cuneate (1–2.5 cm long, 2–6 mm wide); apex truncate, mucronate; base cuneate to obtuse; margins entire; upper surface glabrous; lower surface sericeous; petioles 1–5 mm long; stipules setaceous (3–12 mm long). INFLORESCENCES: solitary flowers or short-cylindric racemes with 2–4 flowers (appearing as clusters), axillary; chasmogamous and cleistogamous flowers mingled. FLOWERS: two types; chasmogamous flowers perfect, papilionaceous; calyx tube 0.5–1 mm long, sericeous; lobes 5, lance-subulate (3–5 mm long), sericeous; corolla white, yellowish-white, or cream; banner 6–9 mm long, marked with purple or pink along the veins; wings and keel equal, shorter than the banner; stamens 10, diadelphous; cleistogamous flowers common, scattered among the chasmogamous flowers. FRUITS: legumes oval (2.5–3.5 mm long), shorter than the calyx, glabrate to appressed-pubescent; seeds 1. SEEDS: ellipsoid to ovoid (1.5–2.5 mm long), slightly compressed, brown to olive, often mottled with brown, indurate. $2n=18, 20$.

HABITAT: Sericea lespedeza has escaped from plantings to roadsides, prairies, pastures, stream valleys, open woodlands, thickets, and waste places. It grows most frequently on well-drained clay loam and silt loam soils. It can grow on infertile soils with a low pH. USES AND VALUES: Initially, sericea lespedeza was planted for grazing and control of soil erosion. It has aggressively spread into native prairies causing serious degradation of the natural vegetation. It has become a pernicious weed that is extremely difficult to control. Sericea lespedeza is classified a noxious weed in some states. It produces low-quality forage because it contains condensed tannins that reduce digestibility by inhibiting cellulolytic enzymes. It decreases the value of hay. Low-tannin cultivars are available, but planting this aggressive, invasive species is not recommended in the Great Plains. The foliage is lightly grazed by deer, rabbits and wild turkeys. It is pollinated by various long-tongued bees and is a good honey plant. Butterflies and skippers obtain nectar from the flowers. The seeds are eaten by ground-foraging birds and small mammals. It is still used for erosion control in some southeastern states. It is not used in landscaping.

SYNONYMS: *Anthyllis cuneata* Dum. Cours., *Aspalathus cuneata* (Dum. Cours.) D. Don, *Hedysarum sericeum* Thunb., *Lespedeza argyraea* Siebold & Zucc., *L. juncea* var. *sericea* F.B. Forbes & Hemsl., *L. sericea* Benth., *L. sericea* Miq. OTHER COMMON NAMES: Chinese bushclover, Chinese lespedeza.

SIMILAR SPECIES: *Lespedeza texana* Britton *ex* Small, Texas bushclover, is a perennial native species. Its leaflets are grayish beneath. The racemes contain 8–10 purple flowers, and its stipules are up to 8 mm in length. It grows in central Texas.

Violet lespedeza
Lespedeza violacea (L.) Pers.

Synopsis Plantarum 2(2):318. 1807.
violaceus (Lat.): violet, in reference to the color of the corollas.

GROWTH FORM: forb, flowers July to September, reproduces from seeds. LIFE-SPAN: perennial. ORIGIN: native. HEIGHT: 20–80 cm. STEMS: herbaceous, weakly ascending or trailing to occasionally erect, several from a short-branching caudex surmounting a taproot, much-branched throughout, light green to brown, glabrous to sparsely pubescent with appressed to ascending hairs. LEAVES: alternate, pinnately trifoliate; leaflets elliptic to broadly oblong (1–5 cm long, 5–25 mm wide), less than 3 times as long as wide, terminal leaflet larger than the lateral leaflets, leaflets reduced in size above; apex rounded, mucronate; base rounded to obtuse; margins entire, ciliate; upper surface medium green, glabrous to sparsely pubescent; lower surface light green, appressed-pubescent; petioles slender, nearly as long as the leaflets; stipules filiform (2–6 mm long), persistent. INFLORESCENCES: racemes of chasmogamous flowers, terminal, loose, much exceeding the subtending leaves; flowers 3–12; peduncles longer than the subtending leaves; cleistogamous flowers in short axillary clusters. FLOWERS: chasmogamous flowers perfect, papilionaceous; calyx 3–6.5 mm long, sparsely pubescent; tube 1.5–3 mm long; lobes 5 (2–3.5 mm long, longer than the tube), 2 uppermost lobes connate beyond the middle; corolla violet to purple or rose-purple (7–9 mm long); banner dark purple at the base; wings slightly spreading; banner and wings shorter than the keel; stamens 10, diadelphous; pedicels 2–6 mm long. FRUITS: legumes elliptic to oblong (4–7 mm long), longer than the calyx, compressed, tapering to a short beak, strigose; cleistogamous legumes (3–6 mm long) in clusters, glabrate to sparsely pubescent; seeds 1. SEEDS: ellipsoid (2.5–3 mm long), slightly compressed, olive to purple or brown, smooth, indurate. 2*n*=20.

HABITAT: Violet lespedeza is scattered to locally common in dry or rocky soils of upland woodlands, prairies, rangelands, waste grounds, and roadsides. It can grow in relatively sterile soils. USES AND VALUES: It produces good quality forage for livestock and deer. It decreases with continued heavy grazing. Songbirds, ground-foraging birds, and small mammals eat the seeds. It is pollinated by various bees including leaf-cutting bees, digger bees, and halictid bees. It is visited by many types of insects including butterflies and skippers. It has the potential to control erosion, but its landscaping applications are limited.

VARIETIES: This species has been divided into several varieties. SYNONYMS: *Hedysarum divergens* Muhl. *ex* Willd., *H. frutescens* L., *H. violaceum* L., *Lespedeza divergens* Pursh, *L. prairea* (Mack. & Bush) Britton. OTHER COMMON NAMES: diverging bushclover, prairie lespedeza, slender lespedeza, wand lespedeza.

SIMILAR SPECIES: *Lespedeza procumbens* Michx., trailing bushclover, is another perennial native species. It is mat-forming and spreads to 1 m. It has 6–10 pink to purplish-pink flowers in each raceme. It grows in the southeastern Great Plains. *Lespedeza repens* (L.) W.P.C. Barton, creeping lespedeza, is another trailing species and has 4–6 pink to pinkish-purple flowers in each raceme. These plants have stipules mostly less than 4 mm, and it also grows in the southeastern Great Plains.

Slender lespedeza
Lespedeza virginica (L.) Britton

Transactions of the New York Academy of Sciences 12(2):64. 1893.
virginica: of or from Virginia.

GROWTH FORM: forb, flowers June to October, reproduces from seeds. LIFE-SPAN: perennial. ORIGIN: native. HEIGHT: 0.3–1 m. STEMS: herbaceous, erect or nearly so, several in a cluster from a stout caudex surmounting a taproot, simple or sparingly branched above, appressed-pubescent, usually densely leafy. LEAVES: alternate, pinnately trifoliate, crowded, erect to ascending; leaflets linear to narrowly oblong (6–40 mm long, 2–7 mm wide), about 3–6 times as long as wide; apex rounded to obtuse, mucronate; base rounded to obtuse; margins entire; upper surface short-strigose to glabrous; lower surface silky-hairy, hairs appressed; petioles 3–25 mm long; stipules filiform (3–6 mm long), persistent. INFLORESCENCES: chasmogamous flowers in congested racemes in the upper axils; racemes cylindric to ovoid, flowers 4–14; cleistogamous flowers inconspicuous, usually in small axillary clusters; peduncles usually shorter than the subtending leaves. FLOWERS: chasmogamous flowers perfect, papilionaceous; calyx campanulate (3–6 mm long), shorter than the fruit; lobes 5 (1.7–3 mm long), nearly equal, uppermost connate for one-half to two-thirds of their length; calyces fall with maturity exposing the fruit; calyx lobes of cleistogamous flowers shorter; chasmogamous corolla purple to pink to rarely whitish (6–8 mm long); banner darker purple at the base; wings shorter than the keel; keel occasionally longer than the banner and wings; stamens 10, diadelphous. FRUITS: chasmogamous legumes elliptic to oblong (4–7 mm long), longer than the calyx; cleistogamous legumes smaller; compressed; seeds 1. SEEDS: ellipsoid (2.5–3 mm long), slightly compressed, shiny, olive to tan, indurate. $2n=20$.

HABITAT: Slender lespedeza is scattered to common on prairies, rangelands, upland woodlands, limestone glades, savannas, riverbanks, and roadsides. It is adapted to a broad range of soils including rocky and sandy soils. USES AND VALUES: It has good forage quality and is grazed by all classes of livestock. It decreases with continued heavy grazing. The foliage is eaten by deer, rabbits, and wild turkeys, and the seeds are eaten by upland birds and small mammals. Seeds are especially valuable to bobwhite quail. Slender lespedeza is pollinated by various long-tongued bees and is a good honey plant. The flowers are visited by butterflies and skippers seeking nectar. Caterpillars of some species of skippers feed on the foliage. It is often a pioneer plant on roadcuts helping to stabilize the soil. It has little potential for landscaping. ETHNOBOTANY: Some Great Plains tribes used it as an antidote for unspecified poisons.

SYNONYMS: *Hedysarum reticulatum* Muhl. *ex* Willd., *H. sessiliflorum* Poir., *Lespedeza angustifolia* Darl., *L. reticulata* (Muhl. *ex* Willd.) Pers., *L. sessiliflora* Michx., *Medicago virginica* L. OTHER COMMON NAMES: bush clover, prairie clover, slender bushclover, slenderbush lespedeza, Virginia bushclover.

SIMILAR SPECIES: *Lespedeza stuevei* Nutt., tallbush lespedeza, is similar in appearance and difficult to separate from slender lespedeza. However, it has purple flowers, and its leaflets are oblong to elliptic or ovate, and they are less than 3 times as long as wide. It grows in the southeastern Great Plains.

Lotus L.

lotos (Gk.): ancient name applied to several kinds of plants.

Annual or perennial herbs or suffrutescent plants; leaves odd-pinnately or subpalmately compound, usually with 3–5 leaflets, margins entire; stipules glandlike or wanting; inflorescences umbels or flowers solitary on leafy-bracteate peduncles from the axils; flowers perfect, papilionaceous, small; corollas yellow to orangish-yellow or reddish (sometimes white); calyx campanulate or short-cylindric; lobes 5, nearly equal or not; petals clawed, free from stamens; banner ovate or obtuse, not articulate; keel usually fused at both margins, incurved, beaked; stamens 10, diadelphous; filaments partly dilated at the apices; ovary sessile or stipitate; fruits legumes, linear or oblong, straight or curved, round in cross section, sessile within the calyx, valves papery or coriaceous, dehiscent; seeds several.

About 90 species have been described, and more than half are native to the Mediterranean region. About 40 species are native to western North America. Their range has been expanded by humans. Two species are common in the Great Plains, and a third is occasionally collected. x=5, 6, 7.

1. Leaflets 3; stipules reduced to glands; flowers usually solitary, occasionally in pairs; annual. . . . . **Lotus purshianus**
1. Leaflets 5; stipules wanting; flowers in umbels of 3 or more; perennial. . . . . **Lotus corniculatus**

Birdsfoot trefoil
Lotus corniculatus L.

Species Plantarum 2:775–76. 1753.
cornu (Lat.): horn, referring to the shape of the legume.

GROWTH FORM: forb, flowers May to September, reproduces from seeds. **LIFE-SPAN:** perennial. **ORIGIN:** introduced (from Europe). **LENGTH:** 5–60 cm. **STEMS:** herbaceous, prostrate to ascending or rarely erect, glabrous to lightly pubescent; several from a stout crown and taproot. **LEAVES:** alternate, odd-pinnately compound; leaflets 5, obovate to broadly lanceolate (5–18 mm long, 3–8 mm wide), sessile or nearly so, lower pair basal on the rachis in the stipular position, upper 3 grouped apically; apex obtuse to acute; base cuneate; margins entire; surfaces glabrous to pubescent, hairs basifixed, not glandular-punctate; stipules wanting. **INFLORESCENCES:** umbels, axillary on peduncles (2–10 cm long) exceeding the subtending leaves, subtended by a trifoliate bract; flowers 3–8. **FLOWERS:** perfect, papilionaceous; calyx tube campanulate (2.7–3.5 mm long); lobes 5, linear to triangular, nearly the length of the tube; corolla bright yellow to orangish-red (1.5 cm long); banner as broad as long (1–1.6 cm long), exceeding the oblong wings (9–14 mm long) and oblanceolate keel (1.2–1.4 cm long); stamens 10, diadelphous; filaments unequal, alternating long and short, the 5 longer ones dilated at the apex; pedicels 1–3 mm long, sometimes obsolete. **FRUITS:** legumes linear to narrowly oblong (1.5–4 cm long, 1.5–2.5 mm wide), round in cross section, straight, brown to black at maturity, valves splitting and twisting when mature; seeds 8–15. **SEEDS:** broadly reniform (1.2–2 mm long), compressed, olive to dark brown, often mottled, indurate. $2n=24$.

HABITAT: Birdsfoot trefoil is planted for hay and pasture. It has also been planted for soil conservation in gullies, roadsides, and dunes. It has escaped to roadsides, waste places, and pastures. **USES AND VALUES:** More than 25 cultivars are commercially available. It has limited use for summer pastures and production of hay. It does not cause bloat in cattle because it contains condensed tannins that precipitate the soluble leaf proteins that cause bloat. Birdsfoot trefoil contains potentially toxic amounts of a cyanogenic glucosides which can be hydrolyzed to produce prussic or hydrocyanic acid. However, there are few reports of livestock losses. Upland birds and small mammals eat the seeds. Birdsfoot trefoil is rich in nectar and pollen. It is a good honey plant and attracts bees and a variety of butterflies, such as the common blue and meadow brown butterflies. The larvae of each feed on the foliage. **ETHNOBOTANY:** Plant extracts have been used as an antispasmodic, sedative, to reduce gas, and reduce fever. An orangish-yellow dye has been extracted from the flowers.

VARIETIES: This variable species is divided into dozens of varieties. **SYNONYMS:** *Lotus caucasicus* Kuprian. *ex* Juz., *L. caucasicus* Kuprian., *L. macbridei* A. Nelson, *L. major* Scop. **OTHER COMMON NAMES:** birdseyes, bloomfell, butterjags, cat clover, crowtoes, ground honeysuckle, lady fingers, lady slippers.

SIMILAR SPECIES: *Lotus tenuis* Waldst. & Kit. *ex* Willd., narrowleaved trefoil, is similar in appearance but has linear leaflets that are less than half as broad as long.

American deervetch
Lotus purshianus Clem. & E.G. Clem.

Rocky Mountain Flowers: An Illustrated Guide for Plant-Lovers and Plant-Users 183. 1914. *purshianus*: named for Frederick Pursh (1774–1820) botanist and author of *Flora Americae Septentrionalis* (1814).

GROWTH FORM: forb, flowers May to October, reproduces from seeds. LIFE-SPAN: annual. ORIGIN: native. HEIGHT: 0.2–1.2 m. STEMS: herbaceous, ascending to erect, simple to much-branched, densely pilose to nearly glabrous with maturity. LEAVES: alternate, pinnately trifoliate, nearly sessile, terminal leaflet on a short petiolule; leaflets ovate to narrowly lanceolate (1–2.5 cm long, 2–9 mm wide); apex acute, often mucronate; base rounded to acute; margins entire; surfaces pilose; petiole 0.5–1.5 mm long; stipules glandlike, minute. INFLORESCENCES: flowers solitary or rarely paired, axillary; peduncle (1–2 cm long) about equaling the subtending leaves; usually with a single leaflike bract. FLOWERS: perfect, papilionaceous; calyx tube obconic (6–8 mm long), hirsute; lobes 5, much longer than the tube, equaling or exceeding the corolla; corolla white or ochroleucous with pink veins (rarely yellowish-white), or pink with red veins; banner (5–8 mm long) streaked with red; wings oblong (4–6 mm long); keel crescent-shaped (4–6 mm long); stamens 10, diadelphous, about equal, filaments dilated at the apex. FRUITS: legumes linear (1.5–3.5 cm long, 2–3 mm wide), nearly round in cross section, straight, entirely exserted, spreading or pendant, beak curved (0.8–1.6 mm long), glabrous, dehiscent; seeds usually 3–9. SEEDS: oblong (about 3 mm long, 2 mm wide), plump, slightly compressed, mottled, indurate. $2n=14$.

HABITAT: American deervetch is locally common on moist sandy soils of prairies, rangelands, stream valleys, railroad rights-of-way, abandoned fields, and open woodlands. USES AND VALUES: All classes of livestock eat the foliage, especially in the early growing season. However, forage value has not been determined. It has been cultivated for forage, but the yields were generally unsatisfactory. Deer, pronghorn, and elk eat the foliage. The seeds are eaten by ground-foraging birds and small mammals. It is pollinated by bees and is a good honey plant. It is relatively easy to grow and has been propagated to increase pollinators. ETHNOBOTANY: Lakotas recognized that it provided nutritious feed for their horses.

VARIETIES: It has been divided into four varieties. Var. *helleri* (Britton) Isely is most common in the Great Plains. SYNONYMS: *Acmispon aestivalis* A. Heller, *A. americanus* (Nutt.) Rydb., *A. elatus* (Nutt.) Rydb., *A. floribundus* (Nutt.) A. Heller, *A. glabratus* A. Heller, *A. gracilis* A. Heller, *A. mollis* (Nutt.) A. Heller, *A. pilosus* (Nutt.) A. Heller, *Hosackia americana* (Nutt.) Piper, *H. elata* Nutt., *H. floribunda* Nutt., *H. mollis* Nutt., *H. pilosa* Nutt., *H. purshiana* Benth., *H. sericea* Branner & Colville, *H. unifoliolata* Hook., *Lotus americanus* (Nutt.) Bisch., *L. sericeus* Pursh, *L. unifoliolatus* (Hook.) Benth., *Trigonella americana* Nutt. These plants may best fit in *Acmispon americanus*, and some taxonomists are recognizing *Acmispon* rather than *Lotus* for this species. OTHER COMMON NAMES: Dakota vetch, Nebraska birdsfoot, prairie birdsfoot, prairie trefoil, Spanish clover, wild vetch, ziŋtkála tȟawóte (Lakota).

Lupinus L.

lupus (Lat.): wolf, its use for this genus is uncertain, perhaps from an old belief that these plants destroyed the soil.

Annual or perennial herbs (rarely shrubs), stems erect or ascending, leafy or subscapose, from a taproot or rhizome; leaves alternate, palmately compound; leaflets 5–11, not glandular-punctate, petiolate; stipules distinct or partially adnate to the petiole, caducous or not; inflorescences terminal racemes or spikes; flowers perfect; calyx tube short and asymmetrical, bilabiate, upper lip slightly bulged to saccate or short-spurred at the base; corolla papilionaceous; white, pink, blue, or yellow; sometimes bicolored, banner broad, commonly reflexed, generally glabrous or sometimes strigose on the back; wings commonly connivent by their edges and enclosing the mostly falcate-pointed keel; stamens 10, monadelphous; anthers of 2 kinds, 5 large alternating with 5 smaller; fruits legumes, elliptic to oblong, somewhat flattened and constricted between the seeds, dehiscing along 2 sutures; seeds 2 to many. $x = 6$.

This is primarily a North American genus, but some species are native to all continents except Australia and Antarctica. Nearly 200 species have been described. Plants are highly variable making positive identification difficult. Many specimens are intermediate between species descriptions and hybridization is presumed. Seven species have been recorded in the Great Plains, and several more grow near the western boundary of the region. Five are considered to be common.

1. Plants winter annuals
    2. Legumes usually with 2 seeds; leaflets less than 1 cm wide; lower leaf surface sparsely pilose. .... *Lupinus pusillus*
    2. Legumes usually with 4 or more seeds; leaflets more than 1 cm wide; lower leaf surface densely silky. .... *Lupinus subcarnosus*
1. Plants perennial.
    3. Plants rhizomatous. .... *Lupinus plattensis*
    3. Plants with a short-branching caudex surmounting a taproot
        4. Calyx spurred, upper lip 5 mm long or more, lower lip 6 mm long or more; upper leaf surface densely sericeous. .... *Lupinus caudatus*
        4. Calyx not spurred, upper lip less than 5 mm long, lower lip 6 mm long or less; upper leaf surface glabrous to strigulose. .... *Lupinus argenteus*

Silvery lupine
Lupinus argenteus Pursh

Flora Americae Septentrionalis 2:468. 1814.

argente (Lat.): silvery, in reference to the color of the leaves.

GROWTH FORM: forb, flowers June to September, reproduces from seeds. LIFE-SPAN: perennial. ORIGIN: native. HEIGHT: 0.1–1 m. STEMS: herbaceous, erect or ascending, sparsely to densely strigulose, simple to shortly branched from a caudex surmounting a taproot. LEAVES: alternate, palmately compound; leaflets 5–9 (rarely 10), narrowly lanceolate to oblanceolate (1–5 cm long, 3–7 mm wide); apex acute to obtuse; base acute; upper surface dark green, glabrous to strigulose; lower surface silvery-green, strigulose to sericeous; petioles 2–10 cm long, 2 times or less the length of the blades; stipules adnate to the petioles for half their length. INFLORESCENCES: racemes (5–20 cm long) terminal, loose. FLOWERS: perfect, papilionaceous; calyx tube 1.5–2 mm long, upper side saccate or bulged at the base, not spurred, sericeous; bilabiate; upper lip 2–4.5 mm long, bidentate; lower lip 4.5–6 mm long, entire; corolla light to dark blue or white to pinkish-white (6–12 mm long, 7–11 mm wide); banner reflexed; keel falcate, often with a dark tip; stamens 10, monadelphous. FRUITS: legumes elliptic (1–3 cm long), linear, compressed, silky pubescent, brown, dehiscent; seeds usually 4–6. SEEDS: ovate (4–5 mm long), compressed, gray to light brown, smooth, indurate. 2n=48.

HABITAT: Silvery lupine is scattered to common in prairies, roadsides, rangelands, and open woodlands in both dry and moist soils. USES AND VALUES: Palatability is rated as only fair for horses, cattle, and sheep. Legumes and seeds of silvery lupine contain quinolizidine and piperidine alkaloids. These compounds are teratogenic and can cause lysosomal storage disease. Eating as little of 0.2 percent of animal's weight of seeds can cause death. Lower concentrations of the alkaloids are found in all plant parts. Clinical signs are excitement, frenzied actions, muscle twitching, frothing at the mouth, difficulty breathing, convulsions, and death. Calves born with skeletal defects and cleft palates may occur after cows eat silvery lupine from 40–70 days of gestation. Delaying grazing until after seeds drop reduces risk of poisoning. Children have been poisoned after eating the seeds. Silvery lupine flowers are visited by many kinds of bees and other insects. It is not used for erosion control, but it has been used in landscaping. ETHNOBOTANY: Members of some Great Plains tribes applied a poultice of crushed leaves to poison ivy blisters to obtain relief.

VARIETIES: This species is highly variable and has been separated into numerous varieties. SYNONYMS: *Lupinus abiesicola* C.P. Sm., *L. aduncus* Greene, *L. alpestris* A. Nelson, *L. alsophilus* Greene, *L. argentinus* Rydb., *L. candidissimus* Eastw., *L. cariciformis* C.P. Sm., *L. clarkensis* C.P. Sm., *L. corymbosus* A. Heller, *L. decumbens* Torr., *L. depressus* Rydb., *L. fremontensis* C.P. Sm., *L. helleri* Greene, *L. ingratus* Greene, *L. keckianus* C.P. Sm., *L. laxus* Rydb., *L. macounii* Rydb., *L. monticola* Rydb., *L. palmeri* S. Watson, *L. parviflorus* Nutt. *ex* Hook. & Arn., *L. pulcher* Eastw., *L. rosei* Eastw., *L. tenellus* Douglas *ex* G. Don (and about 40 additional synonyms). OTHER COMMON NAMES: blue bean, blue lupine, bluebead, silver lupine, perennial lupine, wild bean.

Tailcup lupine
Lupinus caudatus Kellogg

Proceedings of the California Academy of Sciences 2:197. 1863.
caudatus (Lat.): having a tail, in reference to the spurred calyx.

GROWTH FORM: forb, flowers June to August, reproduces from seeds. LIFE-SPAN: perennial. ORIGIN: native. HEIGHT: 20–60 cm. STEMS: herbaceous, erect to ascending or sometimes decumbent, 1 to several from a short-branching caudex surmounting a taproot, branched or simple, strigose. LEAVES: alternate, palmately compound leaflets 5–9, linear to oblanceolate (2–15 cm long, 5–15 mm wide), flat or folded; apex acute to rarely rounded, mucronate; base acute; margins entire; both surfaces densely sericeous; petioles 2–10 cm long; stipules lance-acuminate, persistent. INFLORESCENCES: racemes oblong (4–20 cm long), terminal; flowers many, scattered or somewhat whorled; peduncles 2–5 cm long. FLOWERS: perfect, papilionaceous; calyx tube asymmetrical (3–4 mm long); bilabiate, upper lip 5–9 mm long, lower lip 6–11 mm long, spurred at the base (0.4–1 mm long), pubescent; corolla light to dark blue or violet (8–14 mm long); banner reflexed above the midpoint, lighter blue or whitish in the middle, pubescent on the back; wings clawed; keel silky-ciliate; stamens 10, monadelphous. FRUITS: legumes oblong (2–3 cm long, 8–9 mm wide), linear, apex acuminate or long-acuminate, somewhat compressed, silky, may be slightly constricted between the seeds; seeds 3–10. SEEDS: ovate (3–6 mm long), compressed, brown, smooth, indurate. $2n=48$.

HABITAT: Tailcup lupine grows on dry prairies, foothills, mesas, and stream valleys. It is adapted to a broad range of soil textures but is most abundant on coarse-textured and well-drained soils. USES AND VALUES: It is grazed by livestock, especially when better forage is unavailable in sufficient quantities. Forage quality is poor for cattle and fair for sheep. Cattle may be attracted to the legumes and graze them selectively. However, it may be toxic to sheep, cattle, and horses, causing weakness and muscular trembling. Ewes and cows that eat these plants during pregnancy may give birth to lambs and calves with crooked legs. Alkaloids causing the poisoning are concentrated in the seeds and occasionally in the young plants. The toxic substances remain in the plants after drying. Deer, pronghorn, elk, and bighorn sheep often lightly graze the plants without being poisoned. Upland gamebirds and small mammals eat the seeds. It is pollinated by bumblebees and various other long-tongued bees and is visited by many different types of insects. Tailcup lupine is not used for erosion control, but its use in landscaping is increasing. ETHNOBOTANY: A drug has been extracted from tailcup lupine for management of cardiac arrhythmias.

VARIETIES: Several varieties have been described, including vars. *cutleri* (Eastw.) S.L. Welsh and *utahensis* (S. Watson) S.L. Welsh, which grow west of the Great Plains. Var. *caudatus* grows in the western part of the region, but most in the Great Plains are var. *argophyllus* (A. Gray) S.L. Welsh. SYNONYMS: *Lupinus argentinus* Rydb., *L. gayophytophilus* C.P. Sm., *L. holosericeus* var. *utahensis* S. Watson, *L. lupinus* Rydb., *L. meionanthus* var. *heteranthus* S. Watson, *L. montis-libertatis* C.P. Sm., *L. rosei* Eastw., *L. stinchfieldiae* C.P. Sm., *L. utahensis* Moldenke. OTHER COMMON NAMES: lupino (Spanish), spurred lupine.

Nebraska lupine
Lupinus plattensis S. Watson

Proceedings of the American Academy of Arts and Sciences 17:396. 1882.
plattensis: named after the Platte River region in Nebraska where the type specimen was collected.

GROWTH FORM: forb, flowers May to August, reproduces from rhizomes and seeds. LIFE-SPAN: perennial. ORIGIN: native. HEIGHT: 20–50 cm. STEMS: herbaceous, erect or ascending from rhizomes, patch-forming, simple or branching, pubescent; hairs of various lengths (to 2 mm long), some short and appressed to the surfaces, others longer and ascending. LEAVES: alternate, palmately compound; leaflets 5–11, narrowly to broadly oblanceolate or spatulate (2–5 cm long); apex acute to acuminate or rounded to retuse; base acute; margins entire, ciliate; upper surface glabrous to glabrate; lower surface appressed hairy; few lower leaves present at anthesis; petioles 2–6 cm long; stipules lance-acuminate. INFLORESCENCES: racemes 6–25 cm long, terminal, flowers 11–15; flowers scattered to partly whorled; peduncle short. FLOWERS: perfect, papilionaceous; calyx tube 2–3 mm long, more or less gibbous, asymmetrical, upper side saccate, but not projecting backward as a spur or sac, silky; bilabiate, upper lip 6–7 mm long, lower lip entire (7–9 mm long); corolla conspicuously bicolored (1.2–1.4 cm long); banner strongly reflexed, blue with a darker spot, glabrous or not conspicuously hairy; wings and keel white or suffused with blue; wings glabrous; keel ciliate; stamens 10, monadelphous. FRUITS: legumes oblong to linear (2–5 cm long, 8–9 mm wide), densely appressed-villous, constricted between the seeds; seeds usually 3–8. SEEDS: nearly circular (4–7 mm long, 3–6 mm wide), yellowish-brown to blackish, compressed, smooth, indurate. $2n=48$.

HABITAT: Nebraska lupine is infrequent to abundant in sandy soils of prairies, rangelands, stream valleys, and hillsides. It is less frequent in open woodlands. USES AND VALUES: Palatability is rated fair for cattle, sheep, and horses. Legumes and seeds contain quinolizidine and piperidine alkaloids. These compounds can cause lysosomal storage disease. Eating as little of 0.2 percent of an animal's weight of seeds can cause death of sheep within 24 hours of ingestion. Lower concentrations of the alkaloids are found in all plant parts. Clinical signs are excitement, frenzied actions, muscle twitching, frothing at the mouth, difficulty breathing, convulsions, and death. No antidote is available. Calves born with skeletal defects and cleft palates may occur when cows eat Nebraska lupine from 40–70 days of gestation. Children have been poisoned after eating the seeds. Nebraska lupine flowers are pollinated by bumblebees and various other long-tongued bees. The seeds are eaten by ground-foraging birds and small mammals without ill effect. It is used extensively in landscaping. ETHNOBOTANY: Members of some Great Plains Indian tribes applied a poultice of crushed leaves to poison ivy blisters to obtain relief from itching.

SYNONYMS: *Lupinus glabratus* (S. Watson) Rydb., *L. ornatus* Douglas *ex* Lindl. var. *glabratus* S. Watson, *L. perennis* subsp. *plattensis* (S. Watson) L.Ll. Phillips. OTHER COMMON NAMES: dune bluebonnet, plains bluebonnet, Platte lupine.

SIMILAR SPECIES: *Lupinus sericeus* Pursh, silky lupine, does not have rhizomes or a saccate calyx. At least two-thirds of the banner is hairy. It grows in the extreme western part of the region.

Rusty lupine
Lupinus pusillus Pursh

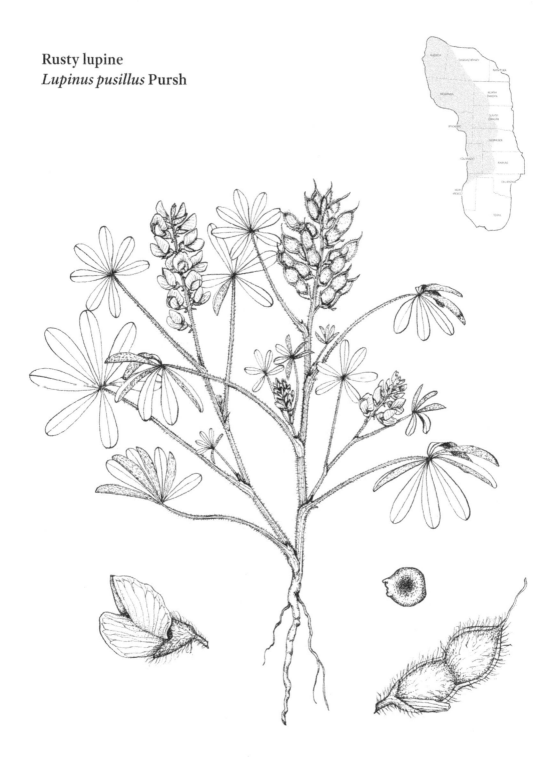

Flora Americae Septentrionalis 2:468. 1814.
pusillus (Lat.): small or insignificant, in reference to its low habit of growth.

GROWTH FORM: forb, flowers May to July, reproduces from seeds. LIFE-SPAN: winter annual. ORIGIN: native. HEIGHT: 5–20 cm. STEMS: herbaceous, erect to decumbent from a taproot, branching near the base from a winter rosette, densely pilose to hirsute. LEAVES: alternate, palmately compound; leaflets 5–9 (occasionally 3 on lowest leaves), elliptic-oblong to oblanceolate (1–4 cm long, 3–8 mm wide); apex acute to obtuse; base acute to cuneate; margins entire; upper surface glabrous or nearly so; lower surface sparsely pilose, sometimes ciliate; often folding along the midvein; petioles broadened and somewhat membranous at the base (2–5 cm long); stipules adnate to the petiole (6–7 mm long). INFLORESCENCES: racemes 3–7 cm long, usually equaling or exceeding the leaves, terminating branches; peduncles 1–3 cm long. FLOWERS: perfect, papilionaceous; calyx campanulate, tube 1.5–2.5 mm long, villous; bilabiate, upper lip 1.5–2 mm long, lower lip 5–6 mm long; corolla purple to blue (sometimes white), tinged with pink; banner nearly round (9–12 mm long), often with a white spot at the base; wings 8–10 mm long; keel curved upward, sometimes purple-spotted at the tip; stamens 10, monadelphous. FRUITS: legumes oblong (1–2.5 cm long, 5–7 mm wide), villous, constricted between the seeds; seeds usually 2. SEEDS: nearly circular to obliquely ovate (4–5 mm long, 3–4.5 mm wide), compressed (1.5 mm thick), light green to brown, mottled with darker brown, indurate. $2n=24$.

HABITAT: Rusty lupine is common in sandy soils of prairies, rangelands, pastures, waste places, badlands, and roadsides. USES AND VALUES: Forage quality and toxicity of rusty lupine is similar to the perennial lupines growing in the Great Plains. Palatability is rated fair for cattle, sheep, and horses. Legumes and seeds of rusty lupine contain quinolizidine and piperidine alkaloids. These compounds can cause lysosomal storage disease. Eating as little of 0.2 percent of animal's weight of seeds can cause death of sheep within 24 hours of ingestion. Lower concentrations of the alkaloids are found in all plant parts. Clinical signs are excitement, frenzied actions, muscle twitching, frothing at the mouth, difficulty breathing, convulsions, and death. No antidote is available. Calves born with skeletal defects and cleft palates may occur when cows eat rusty lupine from 40–70 days of gestation. Children have been poisoned after eating the seeds. Rusty lupine flowers are pollinated by bumblebees and various other long-tongued bees. It is grazed by deer, pronghorn, elk, and bighorn sheep without poisoning. The seeds are eaten by ground-foraging birds and small mammals without ill effect. It has no value for erosion control, but it is used occasionally in landscaping. ETHNOBOTANY: Members of some Great Plains Indian tribes used a decoction made from rusty lupine to treat ear and eye ailments.

VARIETIES: Vars. *intermontanus* (A. Heller) C.P. Sm. and *rubens* (Rydb.) S.L. Welsh grow west of the Great Plains. Var. *pusillus* is most common in this region. SYNONYM: *Lupinus intermontanus* A. Heller. OTHER COMMON NAMES: low lupine, small lupine.

Texas bluebonnet
Lupinus subcarnosus Hook.

Reprinted from *Texas Range Plants* (Stephan L. Hatch and Jennifer Pluhar) by permission of Texas A & M University Press.

Botanical Magazine 63: pl. 3467. 1836.

sub (Lat.): nearly; + *carnosus*: fleshy, in reference to the occasional thick, nearly fleshy, lower leaves.

GROWTH FORM: forb, flowers March to April, reproduces from seeds. LIFE-SPAN: winter annual. ORIGIN: native. HEIGHT: 15–45 cm. STEMS: herbaceous, erect to ascending, single to few from a taproot, mostly branched from the base, branches more or less decumbent, silky pubescent, hairs subappressed. LEAVES: alternate, palmately compound; leaflets usually 5 (sometimes 6 or 7), oblanceolate (3–5 cm long, 1.2–1.6 cm wide); apex rounded to truncate or obtuse; base acute to rounded; margins entire, silky; upper surface glabrous or nearly so; lower surface densely silky; petioles 4–6 cm long; stipules lanceolate (5–6 mm long), caducous; lower leaves sometimes with fleshy obovate leaflets. INFLORESCENCES: racemes (6–15 cm long) terminal; flowers 10–40. FLOWERS: perfect, papilionaceous, mildly fragrant; calyx 5–6 mm long; bilabiate, upper lip bifid (4–5 mm long), lower lip entire or with 3 teeth (5–6 mm long); corolla 1–1.3 cm long; banner suborbicular (1.1–1.3 mm long, 9–11 mm wide), bright blue (with a white spot), turning reddish-purple with age; wings blue (1–1.1 cm long, 9–13 mm wide); keel blue, slender point sharply upturned; pedicels 3–7 mm long; stamens 10, monadelphous. FRUITS: legumes (2–4 cm long, 8–10 mm wide), yellowish-gray to faintly brownish-gray, densely pubescent, constricted between the seeds; seeds usually 4–7. SEEDS: nearly circular to ovate (about 4 mm in diameter), gray to tawny unmarked or speckled with black, indurate. $2n=48$.

HABITAT: Texas bluebonnet is most abundant in sandy soils of roadsides, pastures, waste areas, and railroad rights-of-way. It does not tolerate poorly drained soils. Its range and frequency have been greatly increased by individuals hand-spreading seeds along roadsides. USES AND VALUES: Palatability is rated as only fair for cattle and horses. All plant parts, especially the seeds, contain quinolizidine and piperidine alkaloids and are poisonous to cattle, sheep, and horses. Deer graze these plants without apparent ill effect. Cats and dogs have been poisoned by eating the seeds and exhibit twitching, difficulty swallowing, respiratory problems, trembling, and loss of muscle control. A more detailed discussion of lupine poisoning may be found under *Lupinus plattensis*. It is pollinated by bumblebees and various other long-tongued bees. It is visited by many kinds of insects. This annual is not important for erosion control. Texas bluebonnet is used extensively in landscaping and roadsides. It is also grown in large pots, wooden barrels, and hanging baskets. ETHNOBOTANY: Members of some Great Plains tribes applied a poultice of crushed leaves to poison ivy blisters to obtain relief from itching.

OTHER COMMON NAMES: buffalo clover, el conejo (Spanish), sandyland bluebonnet, wolf flower.

SIMILAR SPECIES: *Lupinus texensis* Hook. is more showy and prolific than Texas bluebonnet. Both, along with species with similar appearance including *Lupinus argenteus* Pursh, silvery lupine; *L. concinnus* J. Agardh, bajada lupine; *L. havardii* S. Watson, Big Bend bluebonnet or Chisos bluebonnet; and *L. plattensis* S. Watson, Nebraska lupine, are considered collectively to be the state flower of Texas.

Medicago L.

medica (Lat.) named for a kind of clover (alfalfa) introduced from Media (Persia) into Greece about 2,500 years ago.

Annual or perennial herbs; stems erect to prostrate, angled or square, glabrous or pubescent; leaves pinnately trifoliate; leaflets with serrulate margins, terminal leaflet stalked; stipules shortly fused with the petiole, entire to lacerate, persistent; inflorescences comprised of axillary racemes, commonly umbellate or subcapitate; calyx tube campanulate; lobes 5, similar in length but slightly unequal; corolla papilionaceous, purple to blue, violet, or yellow; caducous after withering; banner obovate to oblong, longer than the wings; keel blunt, shorter than the wings; stamens 10, diadelphous; anthers all alike; fruits legumes, coiled 2–7 times, reniform or straight, exserted above the calyx, glabrous to spiny, usually indehiscent; seeds 1 to several. $x=7, 8$.

It is a genus of about 80 species. All are native to the Mediterranean region and extend to Europe, Africa, and Asia. None are native to North America. Seven grow in the Great Plains, and 3 are common.

1. Plants perennial; corollas blue to purple (rarely yellow), 8 mm long or more; legumes coiled in a loose spiral. .... *Medicago sativa*
1. Plants annual; corollas yellow, 6 mm long or less; legumes spirally coiled or not
  2. Corollas 4–6 mm long; legumes spirally coiled; seeds several. ....
  .... *Medicago polymorpha*
  2. Corollas 2–4 mm long or more; legumes reniform; seeds 1. ....
  .... *Medicago lupulina*

Black medic
Medicago lupulina L.

Species Plantarum 2:779. 1753.
lupulinus (Lat.): little hop, in reference to the hop-like clusters of legumes.

GROWTH FORM: forb, flowers April to November, reproduces from seeds. LIFE-SPAN: annual (or short-lived perennial). ORIGIN: introduced (from Eurasia). LENGTH: 10–80 cm. STEMS: herbaceous, prostrate to ascending (often sprawling) from a deep tap-root, branching, villous and becoming glabrate with age, light green to reddish-green. LEAVES: alternate, pinnately trifoliate; leaflets elliptic to obovate (1–2 cm long, 5–10 mm wide); apex spinose; base obtuse to acute; margins minutely serrate or sinuate in the apical half; surfaces glabrous to sparsely pubescent; petioles (to 3 cm long) pubescent; middle leaflet with a short petiolule; lateral leaflets sessile; stipules adnate with the petiole for one-fourth to one-half their length; stipules of upper leaves lanceolate (to 1.5 cm long, to 5 mm wide), long-acuminate, entire, upper surface glabrous, lower surface pilose; stipules of lower leaves winglike, deeply toothed. INFLORESCENCES: racemes, globose to short-cylindric (4–15 mm long), compact; flowers 10–50, each flower sub-tended by a linear bract; peduncles slender (to 8 cm long), exceeding the subtending leaves by 1–4 times. FLOWERS: perfect, papilionaceous; calyx less than half the length of the corolla, tube about 0.5 mm long; lobes 5, acuminate; corolla yellow 2–4 mm long; banner broad (to 3 mm wide), wings and keel smaller; stamens 10, diadelphous. FRUITS: legumes reniform (2–4 mm long), longitudinal veins conspicuous, nearly black at maturity, indehiscent; seeds 1. SEEDS: reniform (1.5–2 mm long, 1–1.5 mm wide), somewhat compressed, amber or olive to brown or black, indurate. $2n=16$.

HABITAT: Black medic is found throughout the Great Plains in well-drained soils of lawns, roadsides, old fields, waste places, meadows, prairies, and rangelands. It is often a contaminant in clover and alfalfa seed and has escaped to waste places. USES AND VALUES: Palatability to sheep is good, but it is only poor to fair for cattle. Black medic has minor value as a pasture plant and winter cover. However, forage production is low. Bloat is seldom a problem because animals are rarely able to consume enough of the forage to cause bloat. The foliage is eaten by deer, pronghorn, bighorn sheep, and elk. The seeds are eaten by upland birds, waterfowl, and small mammals. It attracts bees and is considered to be a good honey plant. This annual has little value for erosion control. It has no value for landscaping. It is considered to be a weed in lawns. It has been used for the remediation of soils containing heavy metals such as lead. ETHNOBOTANY: In Eastern Europe, the taproots are sometimes eaten raw. This species is recognized by many in Ireland as the "true shamrock."

SYNONYM: *Medicago lupulina* Scop. OTHER COMMON NAMES: black medick, black trefoil, blackseed hopclover, bur clover, hop clover, hop trefoil, horned clover, nonesuch, nonesuch clover, prostrate yellowclover, shamrock, yellow trefoil.

SIMILAR SPECIES: *Medicago minima* (L.) Bartal., prickly medic or small burclover is an annual with yellow corollas, but the foliage has uncinate prickles. It is much less common than *Medicago lupulina* and is found most frequently in the southeastern part of the region.

Burclover
Medicago polymorpha L.

Species Plantarum 2:779. 1753.

polys (Gk.): many; + *morpho* (Gk.): form or shape, in reference to the many forms within the species.

GROWTH FORM: forb, flowers March to June, reproduces from seeds. LIFE-SPAN: winter annual. ORIGIN: introduced (from the Mediterranean region). LENGTH: 10–50 cm. STEMS: herbaceous, procumbent to ascending, several from a shallow taproot, nearly glabrous to puberulent. LEAVES: alternate, pinnately trifoliate; leaflets cuneate-obovate to obcordate (8–20 mm long), mostly longer than wide; apex obcordate to emarginate; base obtuse to rounded; margins apically serrulate; upper surface sometimes with a whitish to purplish splotch; surfaces glabrous to puberulent; stipules leaflike (6–10 mm long), deeply lacerate to finely toothed, sinuses generally more than half of the length of the stipule. INFLORESCENCES: racemes oblong to oval (5–15 mm long), axillary; flowers 2–8; peduncles 1–3 cm long. FLOWERS: perfect, papilionaceous; calyx tubes campanulate (about 1 mm long), pubescent, lobes 5; corollas yellow (4–6 mm long); banner 2.5–4.5 mm long; stamens 10, diadelphous. FRUITS: legumes spheroid to shortly cylindric, spirally coiled 2–7 times (7–12 mm in diameter, including the spines), compressed, coriaceous, glabrous, brown or sometimes turning black; spines 2–3 mm long, in a double row, apices more or less hooked; seeds several. SEEDS: oblong to reniform (about 2 mm long), indurate. $2n=14$.

HABITAT: Burclover grows in valleys, plains, old fields, pastures, and waste places. It is a common weed in lawns and roadsides. It thrives in rich soils but can grow in poor soils where few other species can grow. USES AND VALUES: It is sometimes planted and may provide good to excellent forage for livestock and wildlife in the spring. Once established, it may reseed itself and not have to be replanted for several years. The legumes are highly nutritious to cattle and sheep and are eaten during the dry season. It is somewhat resistant to heavy grazing. However, it is not a dependable source of forage in dry years. It may cause bloat, but seldom is enough consumed to create the problem. The legumes are an important food for deer, birds, and small mammals. Hundreds of legumes may be produced on a single plant. It is pollinated by various kinds of bees. ETHNOBOTANY: Leaves and stems can be prepared as a pot vegetable. Young leaves can be used to garnish salads.

VARIETIES: This highly variable species has been divided into more than 20 varieties. Var. *vulgaris* (Benth.) Shinners grows in the southern Great Plains. SYNONYMS: *Medicago apiculata* Willd., *M. denticulata* Willd., *M. hispida* Gaertn., *M. lappacea* Desr., *M. nigra* (L.) Krock. OTHER COMMON NAMES: California burclover, carretilla (Spanish), medic, Shanghai trefoil, southern burclover, toothed burclover, toothed medic.

SIMILAR SPECIES: *Medicago arabica* (L.) Huds., burheart clover or California spotted-clover, leaves have a central dark spot, stipules are shallowly lacerate, and the fruits (9–14 mm in diameter) have spines. *Medicago orbicularis* (L.) Bartal., blackdisk medic, fruits (1–1.5 cm in diameter) do not have spines. Its leaves are about as long as broad. Its stipules are also deeply lacerate. Both grow in the southern Great Plains in Texas and Oklahoma.

Alfalfa
Medicago sativa L.

Species Plantarum 2:778–79. 1753.
sativus (Lat.): that which is sown, in reference to this being a cultivated species.

GROWTH FORM: forb, flowers May to September, reproduces from seeds. LIFE-SPAN: perennial. ORIGIN: introduced from Europe (originating in the Middle East). HEIGHT: 0.2–1.1 m. STEMS: herbaceous, erect to decumbent, freely branching, glabrous to finely pubescent, 5–25 from a woody caudex surmounting a deep, stout taproot. LEAVES: alternate, pinnately trifoliate; leaflets oblanceolate to narrowly obovate (1.5–4.5 cm long, 2–18 mm wide); apex acute to emarginate, apiculate; base obtuse to acute; margins toothed at the summit; surfaces glabrous to sparsely strigose; petioles of primary leaves 1–5 cm long; stipules ovate-lanceolate (5–20 mm long), adnate to the petioles for one-fourth to one-third of their length, slightly toothed. INFLORESCENCES: racemes subglobose to short-cylindric (1–3 cm long), axillary, dense; flowers 4–45; peduncles erect (1–4 cm long), about equaling the subtending leaves. FLOWERS: perfect, papilionaceous; calyx tube 1.5–3 mm long; lobes 5, lanceolate (2–4 mm long); corolla blue to blue-violet or dark purple (8–12 mm long), rarely yellow; banner 6–12 mm long; pedicels 2–3 mm long; stamens 10, diadelphous. FRUITS: legumes, coiled in a loose spiral of usually 1–3 complete turns (6–9 mm long, 3.5–10 mm in diameter), glabrous to pubescent or sparsely hairy, yellow to brown; seeds 2–12. SEEDS: reniform to ovate (2–3 mm long), yellowish- or orangish-brown to olive, smooth, indurate. $2n=32$.

HABITAT: Alfalfa is cultivated throughout the Great Plains. It sometimes escapes to roadsides, old fields, and waste places, but it seldom spreads. It grows best in well-drained soils with a pH of 6.5–7. It was introduced into the United States in the 1850s. USES AND VALUES: Alfalfa is primarily used for hay for all classes of livestock, and many commercial varieties of seed are available. It is sometimes grazed or dried and processed into pellets for animal feed. It is highly palatable and eaten by nearly all herbivores. Its seeds are eaten by small mammals, rabbits, upland birds, waterfowl, and songbirds. It is the primary honey plant in North America. It can be an important species for erosion control, but it has no horticultural value. Ruminant animals grazing fresh alfalfa or fed green-chopped alfalfa may be subjected to bloat. Bloat from dry hay or pellets is not a threat.

VARIETIES: Alfalfa is highly variable because of human manipulation for at least 2,000 years. More than 20 varieties and subspecies have been described. SYNONYMS: *Medicago afghanica* Vassilcz., *M. agropyretorum* Vassilcz., *M. beipinensis* Vassilcz., *M. grandiflora* Vassilcz., *M. ladak* Vassilcz., *M. mesopotamica* Vassilcz., *M. praesativa* Sinskaya, *M. tibetana* (Alef.) Vassilcz., and about 10 additional synonyms. OTHER COMMON NAMES: lucerne, burgundy clover, Dutch clover, great trefoil, holy hay, purple medic.

SIMILAR SPECIES: *Medicago falcata* L., yellow alfalfa, has small yellow flowers (5–9 mm long), and its legumes are straight rather than spiraled. It is sometimes classified as *Medicago sativa* subspecies *falcata* (L.) Arcang. It hybridizes freely with *Medicago sativa*. It is found in the central and northern Great Plains.

Melilotus L.

meli (Gk.): honey; + *lotus* (Gk.): unknown leguminous plant, in reference to being a source of nectar for honey.

Annual or biennial herbs, erect to ascending; leaves alternate, pinnately trifoliate, margins denticulate to subentire, terminal leaflet stalked; stipules partially adnate to the petiole, subulate to setiform, persistent; inflorescences elongated racemes, peduncled from upper leaf axils; calyx tube shortly campanulate; lobes nearly equal, subulate to lanceolate, acute to acuminate; corolla papilionaceous, small, corolla white or yellow; petals separate; banner oblong to obovate, usually longer than the wings and keel; stamens 10, diadelphous; ovary short, sessile to somewhat stipitate; fruits legumes, obovate to rotund, glabrous, coriaceous or reticulate, slightly compressed, usually indehiscent; seeds usually 1. $x=8$.

About 20 species have been described. All are originally from Eurasia, especially the Mediterranean region. None are native to North America. Two species, varying primarily in corolla color, are common in the Great Plains.

Yellow sweetclover
Melilotus officinalis (L.) Lam.

Flore Française 2:594. 1778.

officina (Lat.): of the shops or manufacturing laboratory, denoting a plant with uses in medicine and herbalism.

GROWTH FORM: forb, flowers May to October, reproduces from seeds. LIFE-SPAN: biennial (rarely annual). ORIGIN: introduced (from Eurasia). HEIGHT: 0.5–1.5 m. STEMS: herbaceous, erect or ascending, much-branched from a taproot, glabrous to sparsely pubescent. LEAVES: alternate, pinnately trifoliate; leaflets oblanceolate to obovate (1–3 cm long, 5–20 mm wide); apex acute to rounded; base cuneate to acute; margins serrate; surfaces glabrous; petiolule of center leaflet 3–7 mm long; petioles slender (1–3 cm long); stipules 6–10 mm long, persistent. INFLORESCENCES: racemes (5–15 cm long, including the peduncle), spikelike; flowers 30–70. FLOWERS: perfect, papilionaceous; calyx tube 1–2 mm long; lobes 5, deltoid to subulate (0.5–1 mm long); corolla yellow (4.5–7 mm long), fading with age; banner about equaling the wings; stamens 10, diadelphous; ovary stalked; pedicels recurved or decurved (1.5–2 mm long). FRUITS: legumes ovoid (2.5–5 mm long, 2–2.5 mm wide), cross-rugulose, brown to tan at maturity; seeds 1. SEEDS: ellipsoid to round (about 2 mm long), smooth, yellowish-green to olive or orangish-brown to tan, indurate. $2n=16$.

HABITAT: Yellow sweetclover grows in waste places, roadsides, old fields, and prairies. It was first reported in North America in 1739, and it is widely naturalized in the Great Plains. It is considered to be a weed by most landowners. USES AND VALUES: Yellow sweetclover was once planted and grown for grazing and hay. This practice is currently uncommon in the Great Plains. It is eaten by cattle after they become accustomed to its bitter taste. Most wild and domestic herbivores will graze the leaves and flowers of yellow sweetclover. Small mammals, gamebirds, and songbirds eat the seeds. Sweetclover poisoning occurs in cattle, and to a lesser extent in horses and sheep, after ingesting moldy hay following coumarin being converted into dicumarol. Dicumarol prevents coagulation of the blood, and animals may die from internal bleeding. A similar substance is used in many rodenticides. The nectar in the flowers attracts many kinds of insects including various bees, wasps, flies, butterflies, and skippers. It is an excellent honey plant. ETHNOBOTANY: Yellow sweetclover was recommended by Hippocrates (4th century BC) for external treatment of inflamed and swollen body parts and internal treatment of intestinal and stomach ulcers. In the Middle Ages, it was used in various forms for application to swollen joints, boils, and abscesses. In more recent times, it has been used in the formulation of anticoagulants.

SYNONYMS: *Brachylobus officinalis* (L.) Dulac, *Melilotus arvensis* Wallr., *M. graveolens* Bunge, *M. pallidus* Besser *ex* Ser., *M. suaveolens* Ledeb., *Sertula officinalis* (L.) Kuntze, *Trifolium officinale* L. OTHER COMMON NAMES: honey clover, least hoptrefoil, melilot, sweet lucerne, sweet yellowclover, yellow melilot.

SIMILAR SPECIES: *Melilotus albus* Medik., white sweetclover or Hubam sweetclover, is nearly identical except that it has white flowers. It may be taller (up to 2.5 m), and the flowers are 4–5 mm long. White sweetclover is found throughout the Great Plains.

Onobrychis Mill.

onos (Gk.): ass + *bruchis* (Gk.) a plant, referring to the grazing preference for this plant shown by donkeys.

Herbs, sometimes shrubby and spiny; leaves odd-pinnately compound with many leaflets; stipules amplexicaul, scarious; inflorescences long-peduncled axillary racemes, bracts and bracteoles present; calyx campanulate, nearly regular, lobes subulate-lanceolate; corolla papilionaceous, pink to purple (rarely white); banner obovate to obcordate, about equal to the keel, longer than the wings; stamens 10, monadelphous or diadelphous; fruits legumes, indehiscent, compressed, broadly oval to subround, blunt spines on the margins; seeds 1, occasionally 2. x=7, 8.

About 125 species occur in Eurasia, especially the Mediterranean region and western Asia. Only 1 has been successfully introduced into the Great Plains.

Sainfoin
Onobrychis viciifolia Scop.

Flora Carniolica, Editio Secunda 2:76. 1772.

vicia (Lat.): vetch; + *folium* (Lat.): leaf, in reference to the vetchlike leaves.

GROWTH FORM: forb, flowers June to July (may flower a second time in late summer or autumn), reproduces from seeds. LIFE-SPAN: perennial. ORIGIN: introduced (from western Asia through Europe). HEIGHT: 30–90 cm. STEMS: herbaceous, erect to ascending, sparingly branched, hollow, ribbed, glabrous to lightly pubescent, from a branching caudex surmounting a deep taproot. LEAVES: alternate, odd-pinnately compound (10–18 cm long); leaflets 11–29, linear-oblong to oval or narrowly elliptic (1–3 cm long, 4–10 mm wide); apex obtuse, often apiculate or mucronate; base obtuse; margins entire, sparsely hairy; upper surface glabrous, sometimes with reddish dots; lower surface appressed-pubescent, sparsely hairy on the midrib; petiole 2–5 cm long; stipules triangular-ovate to lanceolate (7–9 mm long), connate beyond the middle, scarious, midrib conspicuous. INFLORESCENCES: racemes spikelike (5–10 cm long), few to several, from the upper axils, compact; flowers 20–50; each flower subtended by a bract; bracts small, lanceolate, scarious; peduncles stout (10–30 cm long). FLOWERS: perfect, papilionaceous; calyx tube campanulate, irregular (2–3 mm long on lobed side); lobes 5, subulate-lanceolate (3–6 mm long), appressed-pilose; corolla pinkish-rose or white; banner obovate (8–12 mm long); wings much shorter (2–2.5 mm long); keel nearly equaling the banner (7.5–10 mm long); stamens 10 monadelphous. FRUITS: legumes broadly oval (5–8 mm long, 5–6 mm wide), exserted from the persistent calyx, compressed, coriaceous, strongly nerved, often with short, blunt spines on the dorsal suture, indehiscent; seeds 1. SEEDS: broadly reniform (4–7 mm long), compressed, dark olive to brown or black, indurate. $2n=28$.

HABITAT: Sainfoin is sparingly planted in the northern and northwestern Great Plains. It has occasionally escaped to fields, field margins, waste areas, and roadsides. It has a low tolerance of salt or wetter soils caused by a high water table. USES AND VALUES: It has been used for centuries for grazing and as a hay crop in Europe and western Asia. It is highly palatable to cattle and sheep. It does not cause bloat in ruminants because it contains condensed tannins that precipitate leaf proteins that could otherwise cause bloat. It is a preferred forage for elk and deer. It is unable to withstand heavy or continuous grazing. It is drought tolerant, but its yields are lower than those of alfalfa. Reseeding is usually necessary after five or six years. Seeds are eaten by upland birds, domestic fowl, and small mammals. It is visited by many types of bees, and the honey produced is of excellent quality. Sainfoin is said to be the nectar source for the finest European honey. Several cultivars are available commercially. Nodulation with rhizobial bacteria is not always effective, and nitrogen fertilizer may be required for maximum production. Sainfoin has been used for erosion control with mixed success, and it has few landscaping applications.

SYNONYMS: *Hedysarum onobrychis* L., *H. onobrychis* Lam., *Onobrychis sativa* Lam. The epithet *viciifolia* is sometimes spelled *viciaefolia*. OTHER COMMON NAMES: cockscomb, cockshead, esparcet, holy clover, lupinella, saintfoin.

Orophaca Britton

oros (Gk.): mountain; + *phakos* (Gk.): lentil (a legume), in reference to the habitat of an ancient, unknown legume.

Herbs, tufted or cushion-shaped, acaulescent or shortly caulescent, much-branched, prostrate and mat-forming, pubescence dolabriform, silvery; leaves palmately trifoliate; stipules connate, conspicuous and commonly hyaline; inflorescences racemes, flowers usually 1–5; bracts initially conspicuous, membranous or scarious; calyx tube cylindric or campanulate; flowers papilionaceous; corolla pinkish-purple or bicolored purple and white (often drying yellowish) or white or yellowish; stamens 10, diadelphous; fruits legumes, enclosed in the calyx or partially exserted, ascending, sessile, unilocular, conspicuously beaked; seeds usually 1 or 2. $x=12$.

Eight species grow in North America. Four grow in the Great Plains, and 3 are common. These Great Plains species have been moved from *Astragalus* to *Orophaca* in recent years.

1. Calyx tube less than 3.5 mm long; banner obovate-cuneate (5–7 mm long). ....
.... *Orophaca sericea*
1. Calyx tube more than 5.5 mm long; banner oblanceolate to narrowed in the middle, somewhat fiddle-shaped (1–2.9 cm long)
   2. Banner narrowed in the middle (1.1–1.9 cm long), not tapering from the tip to the base; flowers in summer (June to July). .... *Orophaca hyalina*
   2. Banner obovate-lanceolate to spatulate-oblanceolate (1.5–2.9 cm long), tapering from the tip to the base; flowering in spring (April to May). ....
.... *Orophaca caespitosa*

Plains orophaca
Orophaca caespitosa (Nutt.) Britton

An Illustrated Flora of the Northern United States 2:306. 1897.
caespitosus (Lat.): tuft, in reference to the growth form of the plant.

GROWTH FORM: forb, flowers April to May, reproduces from seeds. LIFE-SPAN: perennial. ORIGIN: native. HEIGHT: 2–15 cm. STEMS: herbaceous, acaulescent, growing from a cespitose caudex surmounting a deep taproot giving the plant a mounded cushion or densely tufted appearance, surfaces silvery or white strigose, hairs dolabriform. LEAVES: alternate, palmately trifoliate (2–12 cm long); leaflets narrowly oblanceolate to obovate-cuneate (5–20 mm long); apex acute or acuminate (rarely obtuse); base cuneate to attenuate; margins entire; surfaces white strigose, hairs dolabriform; petioles elongated, slender, wiry, strigose; stipules oblong-ovate to ovate-lanceolate (5–14 mm long), hyaline, connate, overlapping, glabrous, occasionally transversely corrugated. INFLORES-CENCES: racemes axillary, subsessile, shorter than the subtending leaves; flowers 1–3; peduncles short or obsolete. FLOWERS: perfect, papilionaceous; calyx tube cylindric (7–15 mm long); lobes 5, long-acuminate (1–5 mm long); corolla white or ochroleucous (rarely pinkish-purple), becoming yellow on drying, glabrous, keel tip purple (keel rarely all blue); banner obovate-lanceolate to spatulate-oblanceolate (1.5–2.9 cm long), tapering from tip to the base, recurved; wings 1.2–2.5 cm long; keel 1–2.2 cm long, claws 8–16 mm long; stamens 10, diadelphous; bracts linear-lanceolate; pedicels 0–1.5 mm long. FRUITS: legumes ovoid to ellipsoid (5–12 mm long, 3–6 mm wide), beaked, ascending, coriaceous, strigulose, tardily dehiscent; seeds usually 1. SEEDS: broadly reniform (1.5–2 mm wide), compressed, yellowish-brown to black, smooth, indurate. $2n=24$.

HABITAT: Plains orophaca is scattered on barren hilltops, slopes, and flats. It grows in a broad range of soils but is most common in gravelly to sandy soils. USES AND VAL-UES: Forage quality is only fair, but it is grazed by all classes of livestock as well as by pronghorn, elk, bighorn sheep, and deer. The amount of forage produced is limited. Its legumes and seeds are eaten by small mammals, and its seeds are eaten by birds. It is visited by various bees and other insects. Its tufted form and slow growth makes it unsuitable for erosion control. However, it is widely used in rock gardens and in other horticultural plantings. ETHNOBOTANY: Members of some Great Plains Indian tribes moistened the leaflets, rolled them into a ball, and placed them in the outer ear to relieve earache. Children used the legumes for rattles.

VARIETIES: Var. *gilviflorus* grows on the Great Plains. Var. *purpurea* (Dorn) Isely has a purple to blue corolla and is uncommon west of the region in Wyoming. SYNONYMS: *Astragalus gilviflorus* E. Sheld., *A. triphyllus* Pursh, *Phaca caespitosa* Nutt., *P. triphylla* (Pursh) Eaton & Wright, *Tragacantha triphylla* Kuntze. OTHER COMMON NAMES: cushion milkvetch, núŋǧoka yazáŋ pȟežúta (Lakota), plains milkvetch, sessileflowered milkvetch, threeleaf milkvetch.

SIMILAR SPECIES: *Orophaca barrii* (Barneby) Isely, Barr milkvetch, also has palmately trifoliate leaves. Its corollas are pinkish-purple, and its calyx tube is 5 mm long or less. It is uncommon and is found primarily in western South Dakota, western Nebraska, and eastern Wyoming.

Summer orophaca
Orophaca hyalina (M.E. Jones) Isely

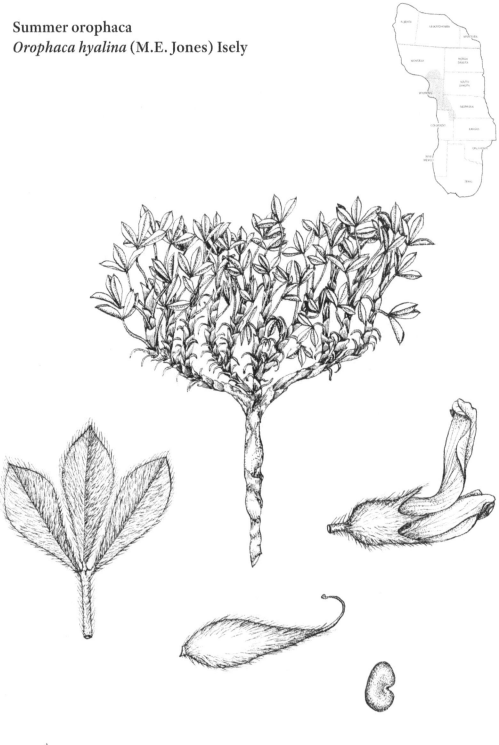

Systematic Botany 8(4):424. 1983.
hyalinus (Gk.): translucent, in reference to the stipules.

GROWTH FORM: forb, flowers in summer (June to July), reproduces from seeds. LIFE-SPAN: perennial. ORIGIN: native. HEIGHT: 10–70 cm. STEMS: herbaceous, initially cespitose, then pulvinate (may be reduced to a crown), growing from a repeatedly forking caudex surmounting a woody taproot, intricately branched and densely tufted giving the plant the appearance of a mounded cushion (10–50 cm in diameter), annual growth of the stipule-enclosed columns 0.5–1.5 cm. LEAVES: alternate, clustered, palmately trifoliate (5–40 mm long); leaflets oblanceolate to obovate or spatulate (3–15 mm long, 1–5 mm wide), often folded; apex acute or obtuse; base cuneate to attenuate; margins entire; surfaces silvery, villous-strigose; hairs dolabriform (up to about 2 mm long); petioles 2–20 mm long, stiff, recurving, persistent; stipules ovate to obovate (6–11 mm long), adnate behind the petiole, connate opposite to the petiole, hyaline-scarious, closely overlapping, ciliate, transversely wrinkled, persistent. INFLORESCENCES: racemes shorter than the subtending leaves; flowers 1–2 (rarely 3); peduncles short (up to 3.5 mm long) or obsolete, concealed by the stipules; bracts ovate or broadly lance-acuminate (3–5 mm long). FLOWERS: perfect, papilionaceous, erect; calyx tube cylindric (5.5–8 mm long), silvery-villous; lobes 5, triangular to acuminate (1.5–3.5 mm long), silvery-villous; corolla white during anthesis, becoming yellowish with age, drying yellow; tips of wings and keel sometimes light purple to lilac; banner erect to moderately reflexed (1.1–1.9 cm long), narrowed in the middle, not tapering from the tip to the base, claw oblanceolate (6–8 mm long); wings 1–1.8 cm long, claw 5–8 mm long; keel 1–1.4 cm long, claw 6–9 mm long; all petals dorsally villous; stamens 10, diadelphous; pedicels obsolete or to 0.5 mm long. FRUITS: legumes ellipsoid to ovoid (5–8 mm long), sessile, erect, hidden among the leaves, beak short-filiform, base obtuse, often rupturing the fragile calyx, densely strigulose; calyx apparently caducous; seeds 1–2. SEEDS: reniform to cordiform (2.5–3 mm long), olive to brown or black, smooth, indurate. $2n=24$.

HABITAT: Summer orophaca is usually uncommon and is considered to be a species of special concern in some states. It grows in rocky (usually shale or limestone), shallow soils of badlands, rangelands, prairie hilltops, and hillsides. USES AND VALUES: It is nearly unpalatable to cattle because of the relatively sharp stipules and low growth form. Sheep lightly graze it. It is sometimes grazed by pronghorn, elk, bighorn sheep, and deer. Summer orophaca is not an important forage because of its low palatability, low productivity, and uncommon occurrence. It is pollinated by various bees, and the flowers are visited by many kinds of insects. Seeds and legumes are eaten by small mammals, and the seeds are eaten by ground-foraging birds. It has little value for erosion control because of its tufted form and slow rate of growth. It is sometimes grown in rock gardens.

SYNONYMS: *Astragalus hyalinus* M.E. Jones, *Orophaca argophylla* (Nutt.) Rydb. OTHER COMMON NAMES: silver milkvetch, silvery milkvetch, summer milkvetch, transparent milkvetch.

Silky orophaca
Orophaca sericea (Nutt.) Britton

An Illustrated Flora of the Northern United States 2:307. 1897.
serikos (Gk.): silken, in reference to the pubescence on the herbage.

GROWTH FORM: forb, flowers May to June, reproduces from seeds. LIFE-SPAN: perennial. ORIGIN: native. HEIGHT: 1–8 cm. STEMS: herbaceous, caulescent, cespitose, prostrate to rarely ascending from a thatched caudex, creeping outward from the crown, forming mats (to 1 m in diameter), densely pubescent; hairs silvery-pilose, dolabriform; internodes often concealed by the stipules. LEAVES: alternate, dispersed or clustered, palmately trifoliate (1–5 cm long); leaflets broadly to narrowly oblanceolate or obovate (4–14 mm long, 2–7 mm wide), commonly folded; apex obtuse to acute; base cuneate to attenuate; margins entire; surfaces densely villous; hairs silvery, dolabriform (to 3 mm long); petioles 0.5–4 mm long; stipules hyaline (2–8 mm long), connate, sheathing, persistent, outer or lower stipules conspicuously pubescent, upper or inner stipules glabrous or ciliate. INFLORESCENCES: racemes, flowers 2–5, occasionally single-flowered; included in the stipular sheath or exserted; peduncles 5–25 mm long, generally exserted above the leaves. FLOWERS: perfect, papilionaceous; calyx tube campanulate (1.5–3 mm long), white strigulose; lobes 5, subulate to subsetaceous (0.5–1.5 mm long); corolla brightly pinkish-purple to violet (rarely white), drying yellow; banner recurved, obovate-cuneate to oblanceolate (5–7 mm long, 3–4 mm wide); wings 5–6 mm long, claw 2–2.5 mm long; keel obtuse to subacute (4–4.5 mm long), claw 2–2.4 mm long; stamens 10, diadelphous; pedicels ascending to erect (0.5–1.5 mm long); bracts lanceolate (1–2 mm long). FRUITS: legumes ovoid-ellipsoid (4–7 mm long), sessile, straight, slightly flattened, beaked, densely silky-strigose, tan, ascending or recurved, mostly enclosed in the persistent calyx, coriaceous, tardily dehiscent; seeds usually 1 or 2. SEEDS: reniform to cordiform or ellipsoid (1.2–1.8 mm long), greenish-brown to yellowish-brown, smooth, indurate. $2n=24$.

HABITAT: Silky orophaca is infrequent to locally abundant on exposed, eroded, shallow or gravelly soils of hilltops and rocky ridges of rangelands and badlands. USES AND VALUES: Palatability is poor for cattle and horses, but sheep may lightly graze the leaves, flowers, and legumes. Silky orophaca accumulates selenium and should be considered toxic to livestock. However, a sufficient quantity is rarely consumed by animals to cause poisoning. It is grazed by pronghorn, elk, and bighorn sheep. Use by deer is limited. The flowers are visited by various bees and many other kinds of insects, including butterflies and skippers. The legumes and seeds are eaten by small mammals, and ground-foraging birds eat the seeds. It is not used for erosion control. Silky orophaca is a popular rock garden plant. The mats are comprised of the thatch of the densely pubescent stipules but contain less dead material from year to year than do the other mat-forming *Orophaca* species.

SYNONYMS: *Astragalus sericoleucus* A. Gray, *Phaca sericea* Nutt., *Tragacantha sericea* (Nutt.) Kuntze. OTHER COMMON NAMES: hoary milkvetch, silk orophaca, silky milkvetch.

Oxytropis DC.

oxys (Gk.): sharp; + *tropis* (Gk.): keel, in reference to the characteristic distal appendage of the keel.

Perennial herbs, acaulescent or nearly so, commonly cespitose, pubescent throughout with hairs simple and basifixed or dolabriform; leaves and scapes crowded on the
woody caudex surmounting a taproot; leaves odd-pinnately compound, petiolate, often
dimorphic; leaflets numerous; stipules strongly or weakly adnate to the petioles; inflorescences racemes terminating scapes; flowers 2 to numerous; bracts ovate to lanceolate,
persistent; flowers perfect, papilionaceous; calyx cylindric or campanulate, somewhat
oblique at the base, generally with both white- and dark-colored hairs; lobes 5, narrow,
nearly equal; corolla pink to purple (drying purple to blue) or white to ochroleucous;
banner usually erect; wings commonly 2-lobed or emarginate at the apex, wings longer than the keel; keel abruptly narrowed to a distal beaklike appendage; stamens 10,
diadelphous; anthers all alike; fruits legumes, ascending to spreading, subspherical to
oblong-lanceolate, turgid to inflated, sessile or nearly so, membranous, coriaceous, or
woody in texture, partially bilocular by the intrusion of the ventral suture, dorsal suture
not intruded; seeds several to many; seeds reniform. x=8.

About 200–300 species grow in temperate to arctic Eurasia, some with circumboreal
distribution. They are most numerous at higher elevations. About 25 species grow in
North America. Of the 9 growing in the Great Plains, 4 are common.

1. Pubescent with dolabriform hairs. .... *Oxytropis lambertii*
1. Pubescent with basifixed hairs
    2. Racemes with fewer than 5 flowers; corolla pinkish-purple to purple; calyx inflated
       in fruit. .... *Oxytropis multiceps*
    2. Racemes with 5 or more flowers; corolla white to yellow (sometimes purple-tipped);
       calyx not inflated in fruit
        3. Legumes papery or membranous at maturity, not fleshy when immature, not
           rigid; banners usually 1.8 cm long or less. .... *Oxytropis campestris*
        3. Legumes woody to coriaceous at maturity, fleshy when immature, rigid; banners
           usually 1.8 cm long or more. .... *Oxytropis sericea*

Field locoweed
Oxytropis campestris (L.) DC.

Astragalogia, 74. 1802.
campestris (Lat.): flat or level ground, in reference to the habitat.

GROWTH FORM: forb, flowers May to July, reproduces from seeds. LIFE-SPAN: perennial. ORIGIN: native. HEIGHT: 20–50 cm. STEMS: herbaceous, acaulescent, scapes and leaves arising directly from a cespitose, much-branched caudex surmounting a stout taproot. LEAVES: alternate, odd-pinnately compound (2–25 cm long), dimorphic, primary leaves short with ovate leaflets, secondary leaves with 7–33 leaflets; leaflets oblong to lanceolate (1–2.5 cm long); apex acute; base obtuse; margins entire; surfaces pubescent; hairs appressed, basifixed; petioles pubescent; stipules lanceolate, membranous, adnate to the petioles (5–15 mm long), glabrate or pilose dorsally, persistent. INFLORESCENCES: racemes (4–40 cm long including the peduncle) ovoid to oblong, flowers 8–32, scapes ascending to erect; bracts conspicuous, pubescent. FLOWERS: perfect, papilionaceous; calyx tube cylindric to campanulate (5–7 mm long), not inflated in fruit, appressed-villous, hairs often black (to 1 mm long); lobes 5, lanceolate to acuminate (1.5–2.5 mm long), appressed-villous; corolla white to ochroleucous or yellow (sometimes pink, blue, or purple); banner ovoid to obovate (1.2–1.8 cm long, 6–10 mm wide, usually less than 1.5 cm long); wings oblanceolate to oblong (1–1.5 cm long, 2–6 mm wide); keel abruptly curved (1–1.5 cm long), may have purple blotches, distal appendage to 1 mm long; stamens 10, diadelphous. FRUITS: legumes oblong to ellipsoid (5–15 mm long, 3.5–5 mm wide), beak to 5 mm long, sessile, erect, pubescent, some hairs black; suture strongly intruded so that the legume is nearly bilocular, membranous or papery (rarely coriaceous), not fleshy, not rigid, dehiscing from the tip; seeds many. SEEDS: reniform (2–2.5 mm long), smooth, yellow to olive or brown to black, indurate. $2n=32$.

HABITAT: Field locoweed is scattered to locally abundant in prairies, rangelands, meadows, and open woodlands. It is most abundant in gravelly and rocky soils. USES AND VALUES: It is unpalatable to livestock, but it may be lightly grazed by pronghorn and other wildlife. The legumes and seeds are eaten by birds and small mammals.

VARIETIES: This species is circumboreal in distribution and has been divided into about 20 varieties. Vars. *cusickii* (Greenm.) Barneby, *dispar* (A. Nelson) Barneby, and *spicata* Hook. are most common in the Great Plains. SYNONYMS: *Aragallus alpicola* Rydb., *Astragalus campestris* L., *Oxytropis cusickii* Greenm., *O. paysoniana* A. Nelson. OTHER COMMON NAMES: field oxytropis, plains locoweed, yellow locoweed.

SIMILAR SPECIES: Three similar species of *Oxytropis* are less common on the Great Plains. *Oxytropis splendens* Douglas, showy locoweed, has reddish-purple corollas, and some of the leaflets are in fascicles on one side of the rachis. *Oxytropis besseyi* (Rydb.) Blank., Bessey locoweed, has pink to reddish-purple corollas, bracts are rhombic-lanceolate, and the calyx surface is not obscured by the pubescence. *Oxytropis lagopus* Nutt., haresfoot locoweed, also has reddish-purple corollas. Its bracts are lanceolate, and the surface of the calyx is obscured by the pubescence. All three of these species grow in the northwestern Great Plains.

Purple locoweed
Oxytropis lambertii Pursh

Flora Americae Septentrionalis 2:740. 1814.

lambertii (Lat.): named after English botanist Aylmer Bourke Lambert (1761–1842).

GROWTH FORM: forb, flowers April to June (occasionally later in September), reproduces from seeds. LIFE-SPAN: perennial. ORIGIN: native. HEIGHT: 10–40 cm. STEMS: herbaceous, acaulescent, scapes and leaves arising directly from a cespitose caudex surmounting a stout taproot, 1 to several in clusters, pubescent; hairs dolabriform. LEAVES: alternate, odd-pinnately compound, erect to ascending, dimorphic, principal leaves 4–20 cm long; leaflets 7–19, linear to narrowly oblong or sometimes ovate (5–40 mm long, 2–7 mm wide); apex acute or acuminate; base acute; margins entire; both surfaces densely pilose, hairs dolabriform; petioles of principal leaves shorter than the rachis; stipules adnate to the petioles. INFLORESCENCES: racemes (4–12 cm long), flowers 6–28; bracts lanceolate to narrowly ovate (2–10 mm long); scapes erect to ascending. FLOWERS: perfect, papilionaceous; calyx tube campanulate (5–9 mm long), ruptured by the expanding fruit, densely villous; lobes 5, triangular-subulate (2–4 mm long), densely villous; corolla purple or rose to pink or blue (white not uncommon); banner somewhat reflexed (1.5–2.5 cm long, 6–12 mm wide); wings (1.2–2 cm long) enveloping the keel; keel 1–2 cm long, distal appendage 5–25 mm long; stamens 10, diadelphous; pedicels 1–3 mm wide. FRUITS: legumes lanceolate to ellipsoid (1.5–3 cm long, 5–6 mm wide), straight or divergent, beak prominent (3–7 mm long), surface strigose-silky or strigose, soon glabrate to glabrescent, suture intruded about halfway across, dehiscing from the top; seeds many. SEEDS: reniform to nearly circular (about 2 mm wide), compressed, brown, smooth, indurate. $2n=32$.

HABITAT: Purple locoweed is common on dry prairie uplands and rangelands. USES AND VALUES: The forage is nearly worthless to livestock, but they will eat it if other plants are unavailable. They may become habituated to purple locoweed and eat it in preference to other plants. All plant parts are toxic, including the flowers and mature seeds. The plants remain toxic when dry. It contains swainsonine, an indolizidine alkaloid, which disrupts cellular metabolism. There is no antidote, and death can result. Horses may become unmanageable when excited, causing them to become dangerous or "locoed" (a Spanish word meaning *crazy*). ETHNOBOTANY: Plains Indians recognized that purple locoweed was toxic to horses and managed their herds to avoid these plants.

VARIETIES: Var. *lambertii* is most common in the Great Plains. Var. *bigelovii* A. Gray grows in the southwestern part of the region, and var. *articulata* (Greene) Barneby grows in the southcentral part of the Great Plains. SYNONYMS: *Aragallus angustatus* Rydb., *A. aven-nelsonii* Lunell, *A. falcatus* Greene, *A. formosus* Greene, *A. involutus* A. Nelson, *Astragalus lambertii* (Pursh) Greene, *Oxytropis angustata* (Rydb.) A. Nelson, *O. aven-nelsonii* (Lunell) A. Nelson, *O. bushii* Gand., *O. falcata* (Greene) A. Nelson, *O. hookeriana* Nutt., *O. involuta* (A. Nelson) K. Schum., *O. plattensis* Nutt., *Spiesia lambertii* (Pursh) Kuntze. OTHER COMMON NAMES: Lambert locoweed, stemless crazyweed, stemless locoweed, śuŋktá pejútia (Lakota), whitepoint locoweed.

Dwarf locoweed
Oxytropis multiceps Nutt.

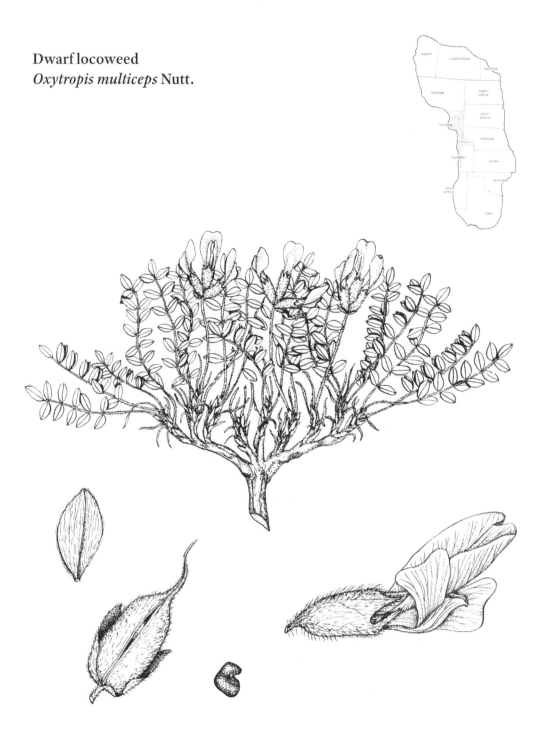

Flora of North America 1(2):341. 1838.

multi (Lat.): many; + *ceps* (Lat.): head, in reference to the many inflorescences.

GROWTH FORM: forb, flowers April to June, reproduces from seeds. LIFE-SPAN: perennial. ORIGIN: native. HEIGHT: 2–10 cm. STEMS: herbaceous, acaulescent, densely cespitose, scape and leaves arising from a caudex surmounting a taproot, forming a silvery mound. LEAVES: alternate, odd-pinnately compound (1–7 cm long); leaflets 5–11, lanceolate to oblanceolate or oblong (3–15 mm long, 2–6 mm wide), often folded or involute: apex obtuse; base obtuse to acute; margins entire; both surfaces densely silky, hairs appressed, antrorse, basifixed; petioles 5–30 mm long, usually equaling or longer than the rachis; stipules adnate to the petioles, clasping, not connate. INFLO-RESCENCES: racemes (1.5–3 cm long), subcapitate scapes ascending; flowers 1–4; axis scarcely elongating; bracts ovate to ovate-lanceolate, thinly pilose. FLOWERS: perfect, papilionaceous; calyx cylindric or campanulate (5–10 mm long) becoming inflated in fruit (8–18 mm long, 5–9 mm wide), silky-villous, reddish; lobes 5, unequal (2–3 mm long); corolla pinkish-purple to purple (drying blue); banner oblong-ovate (1.7–2.5 cm long, 7–9 mm wide), emarginate; wings 4.5–5.5 mm wide near apex; keel 1.3–1.8 cm long, distal appendage straight or curved (0.5–1.5 mm long); stamens 10, diadelphous. FRUITS: legumes ovoid to ellipsoid (6–11 mm long, 3–5 mm wide), wholly or partially included in the inflated calyx, papery, not rigid, short-villous, apex contracted into a beak, stipe 0.5–1.5 mm long; seeds several. SEEDS: reniform, reddish-brown, often mottled with purple, smooth, hard. $2n=32$.

HABITAT: Dwarf locoweed is infrequent to locally common on prairie uplands, ridges, rangelands, sagebrush flats, and hillsides of open woodlands. USES AND VALUES: Dwarf locoweed is usually not palatable to livestock, but they will graze it if other plants are not available. It is toxic to livestock because it contains the indolizidine alkaloid swainsonine which may cause death. There is no antidote for this type of poisoning. Wildlife lightly graze it without being poisoned. It is pollinated by bees and many other types of insects. The legumes are eaten by small mammals, and the seeds are eaten by small mammals and ground-foraging birds. It has no application in erosion control, but it is used in borders in landscapes and in rock gardens.

VARIETIES: This species is often divided into var. *multiceps* and var. *minor* A. Gray. SYN-ONYMS: *Aragallus minor* (A. Gray) Cockerell, *A. multiceps* (Nutt.) A. Heller, *Astragalus bisontum* Tidestr., *Oxytropis minor* (A. Gray) Cockerell, *Physocalyx multiceps* Nutt. *ex* A. Gray, *Spiesia multiceps* (Nutt.) Kuntze. OTHER COMMON NAMES: dwarf crazyweed, manyhead crazyweed, Nuttall oxytrope, tufted oxytropis.

SIMILAR SPECIES: *Oxytropis nana* Nutt., Wyoming locoweed, is another dwarf species. It has rigid and coriaceous legumes that are canescent rather than short-villous. The flowers are purple or white and tinged with pink or with purple spots. It is found only in central and east-central Wyoming. It is somewhat similar in appearance to *Oxytropis besseyi* (Rydb.) Blank., Bessey locoweed, which has pink to reddish-purple corollas, but its bracts are rhombic-lanceolate.

White locoweed
Oxytropis sericea Nutt.

A Flora of North America 1(2):339. 1838.
serikos (Gk.): silken, in reference to the pubescence.

GROWTH FORM: forb, flowers April to June, reproduces from seeds. LIFE-SPAN: perennial. ORIGIN: native. HEIGHT: 20–35 cm. STEMS: herbaceous, acaulescent, scapes and leaves arising from a woody caudex surmounting a taproot, cespitose, much-branched, silky-pilose, hairs basifixed. LEAVES: alternate, odd-pinnately compound (4–30 cm long), ascending or erect, dimorphic; leaflets 7–25 on primary leaves, elliptic to oblong (5–35 mm long, 2–10 mm wide); apex acute or obtuse; base acute; margins entire; surfaces gray or silvery, hairs appressed and basifixed; petiole 1–15 cm long; stipules 2–25 mm long. INFLORESCENCES: racemes (5–10 cm long) elevated above the leaves on erect to ascending scapes, flowers 6–30, bracts lanceolate (4–16 mm long). FLOWERS: perfect, papilionaceous; calyx tube cylindric (8–12 mm long), not inflated in fruit, appressed silky, some black hairs; lobes 5, acuminate (2–5 mm long), unequal; corolla white (fading to yellow), purple-tipped keel; banner oblong-ovate (1.8–2.5 cm long, 8–10 mm wide), emarginate or deeply lobed, claw broad; wings 1.5–2 cm long, widening distally, 5–9 mm wide near the emarginate apex; keel 1.2–1.8 cm long, distal appendage curved or straight (1–2 mm long); stamens 10, diadelphous. FRUITS: legumes oblong or ovoid-oblong (1–2.5 cm long, 5–8 mm wide), beak short, erect, short-pilose to strigose, woody to coriaceous, fleshy when immature, rigid; seeds many. SEEDS: reniform (2–2.5 mm long), brown, smooth, indurate. 2*n*=32.

HABITAT: White locoweed grows on prairie uplands, rangelands, stream banks, valleys, and open woodlands. It is most common on sandy, gravelly, or rocky soils. USES AND VALUES: It is generally unpalatable to livestock, but they will graze it if other forage is unavailable. Animals may become habituated to it and eat it in preference to other plants. Horses are especially susceptible. It contains the toxic substance swainsonine. There is no antidote for the poison, and it can cause death. Ground-foraging birds and small mammals eat the seeds without poisoning. It is pollinated by bees and other insects. This low-growing plant has attractive corollas and silvery-foliage, making it a favorite for rock gardens. ETHNOBOTANY: Blackfoot Indians applied an infusion of leaves to open sores to accelerate healing.

VARIETIES: This taxon has been divided into vars. *sericea, speciosa* (Torr. & A. Gray) S.L. Welsh, and *spicata* (Hook.) Barneby. Var. *sericea* grows in the Great Plains. SYNONYMS: *Aragallus aboriginum* Greene, *A. albiflorus* A. Nelson, *A. invenustus* Greene, *A. majusculus* Greene, *A. pinetorum* A. Heller, *A. saximontanus* A. Nelson, *A. sericeus* (Nutt.) Greene, *A. veganus* (Cockerell) Wooton & Standl., *Astragalus albiflorus* (A. Nelson) Tidestr., *A. saximontanus* (A. Nelson) Tidestr., *Oxytropis albiflora* (A. Nelson) K. Schum., *O. condensata* (A. Nelson) A. Nelson, *O. pinetorum* (A. Heller) K. Schum., *O. saximontana* (A. Nelson) A. Nelson, *O. vegana* (Cockerell) Wooton & Standl. OTHER COMMON NAMES: silky crazyweed, white crazyweed, whitepoint crazyweed.

SIMILAR SPECIES: *Oxytropis deflexa* (Pall.) DC., nodding locoweed, is shortly caulescent and has blue to purplish corollas. Its stipules are adnate to the petioles for only 1–3 mm.

Pediomelum Rydb.

pedio (Gk.): field or plain; + *malium* (Gk.): fruit or apple, in reference to the habitat and tuberlike roots of some of the species.

Perennial herbs, occasionally shrubs, caulescent or acaulescent or nearly so; from rootstalks or tuberlike roots; stems erect to ascending, simple or branched, hirsute or glabrate; leaves alternate, palmately compound (sometimes appearing to be pinnately compound); leaflets 3–5 (occasionally 7), often glandular-punctate, margins entire; petioles well-developed; stipules ovate or lanceolate to linear, usually persistent; inflorescences racemes or spikelike racemes, from upper axils, bracts persistent; flowers prefect, papilionaceous; calyx campanulate, enlarging in fruit, often gibbous, sometimes oblique, persistent; lobes 5, equal or lowest lobe longer; corolla blue to purple or rarely white; banner obovate, usually clawed; wings oblong to oblanceolate, usually equal to the banner; keel oblong to oblanceolate, shorter than the banner and wings, incurved; pedicels 0–6 mm long; stamens 10, diadelphous; ovary sessile or nearly so; fruits legumes ovate to elliptic, short, slightly compressed or turgid, usually indehiscent, sometimes terminated by an indurate and persistent style, smooth or glandular-punctate, stipe short; seeds 1; seeds ellipsoid to ovoid or reniform. x=11.

About 21 species grow in North America. They are most numerous in the plains and Pacific states. Of the 7 in the Great Plains, 4 are common.

1. Plants low, stemless or stems to 25 cm tall; from tuberous or thickened roots; petioles longer than the leaflets. .... *Pediomelum esculentum*
1. Plants taller than 25 cm, stems leafy and much-branched; lacking tuberous roots; petioles shorter than or equaling the length of the leaflets
    2. Flowers greater than 1.5 cm long in dense spikelike racemes; continuous; leaflets mostly 5, rarely 7. .... *Pediomelum cuspidatum*
    2. Flowers less than 1.5 cm long in spikelike racemes; interrupted; leaflets 3–5
        3. Plants densely white-silky throughout; both leaf surfaces densely sericeous or white-silky. .... *Pediomelum argophyllum*
        3. Plants greenish to gray; leaflets glabrate to sparingly pubescent on the upper midvein, lower surface densely strigose to sericeous. .... *Pediomelum digitatum*

Silverleaf scurfpea
Pediomelum argophyllum (Pursh) J.W. Grimes

Memoirs of the New York Botanical Garden 61:69. 1990.
argos (Gk.): shining; + *phyllon* (Gk.): leaf, in reference to the silvery hairs on the leaves.

GROWTH FORM: forb, flowers June to September, reproduces from rhizomes and seeds, forming colonies. LIFE-SPAN: perennial. ORIGIN: native. HEIGHT: 30–80 cm. STEMS: herbaceous, erect or ascending from adventitious buds on woody roots and creeping rhizomes, much-branched, densely white-silky. LEAVES: alternate, odd-palmately compound, often appearing pinnately compound; leaflets 3 on branches, 5 (sometimes 4) on main stem, elliptic or oblanceolate to narrowly obovate (1–5 cm long, 6–18 mm wide); apex obtuse to acute, usually with a short mucro; base rounded to acute; margins entire; both surfaces densely sericeous or white-silky; upper surface less pubescent, darker green, sparingly glandular-punctate; petioles shorter than or equaling the leaflets (1–3 cm long); stipules lanceolate to linear, 1–2 cm long on lower leaves, 5–10 mm long above. INFLORESCENCES: racemes (2–8 cm long, 1–1.5 cm wide) spikelike, axillary, comprised of 1–5 or more whorls; whorls separated, interrupted, each whorl with 2–8 flowers, flowers in lowest whorls blooming first; bracts linear to lanceolate (3–9 mm long); peduncles 2–9 cm long. FLOWERS: perfect, papilionaceous, sessile; calyx tube campanulate (2–5 mm long), densely white-silky; lobes 5, lowest lobe attenuate (8–10 mm long), nearly twice as long as the upper 4 lobes; corolla dark blue to purple, fading to yellow or brown to tan; banner obovate (5–8 mm long); wings 4–6 mm long; keel 4–5 mm long; stamens 10, diadelphous. FRUITS: legumes ovoid to oblong-lanceolate (5–9 mm long), mostly enclosed in the persistent calyx, beak 1–3 mm long; silky, not glandular-punctate, pericarp rigid; seeds 1. SEEDS: reniform to orbicular (4–5 mm long), compressed, olive to reddish-brown or black, smooth, indurate. $2n=22$.

HABITAT: Silverleaf scurfpea is scattered to locally abundant on dry prairies, hills, rangelands, and open woodlands. It is most common in sandy or rocky soils. USES AND VALUES: Low palatability makes it of little value to livestock. Pronghorn, elk, and deer lightly graze the foliage, flowers, and legumes. Pocket gophers eat the roots. Other small mammals eat the legumes and seeds, and ground-foraging birds eat the seeds. Pollination is by bees, and many small insects visit the flowers. Ingested seeds may cause photosensitization in animals. One case of severe poisoning was reported after a child ate a relatively large quantity of seeds. It is not used for erosion control because the stems break off near the soil surface in autumn. They scatter seeds as they tumble in the wind. It has little potential for landscaping. However, it will attract butterflies and bees. ETHNOBOTANY: Some Great Plains Indians used and infusion of silverleaf scurfpea leaves to treat wounds and made a tea to drink to reduce fevers. A mild stimulant was made from the roots, and the roots were fed to tired horses to restore energy. Lakotas made baskets from the green stems.

SYNONYMS: *Lotodes argophylla* (Pursh) Kuntze, *Psoralea argophylla* Pursh, *P. collina* Rydb., *Psoralidium argophyllum* (Pursh) Rydb. OTHER COMMON NAMES: gi'ziso'bûgons' (Chippewa), Nebraska psoralea, silverleaf indianbreadroot, silverleaf psoralea, silvery scurfpea, tičaničahu xloxota (Lakota).

Tallbread scurfpea
Pediomelum cuspidatum (Pursh) Rydb.

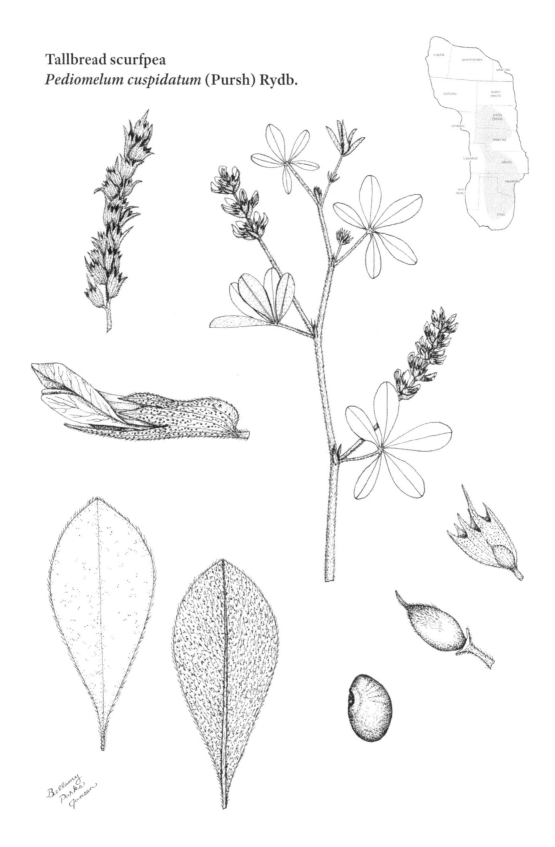

North American Flora 24(1):19. 1919.
cuspidatus (Lat.): pointed, in reference to the calyx lobes.

GROWTH FORM: forb, flowers May to July, reproduces from seeds. LIFE-SPAN: perennial. ORIGIN: native. HEIGHT: 30–90 cm. STEMS: herbaceous, prostrate to ascending, rarely erect, stout, from an elongated fusiform or ellipsoidal taproot (not tuberous), much-branched, thinly strigose to glabrate. LEAVES: alternate, odd-palmately compound; leaflets generally 5 (rarely 7, sometimes 3 on upper nodes), elliptic to narrowly obovate or oblong-obovate (1.5–6 cm long, 5–30 mm wide), center leaflet longest; apex acute to obtuse or rounded, sometimes with a mucro; base acute to attenuate; margins entire, pubescent; upper surface glabrate to glabrous, glandular-punctate; lower surface strigose; petioles 1–5 cm long (shorter than or equaling the length of the leaflets); stipules of lower nodes 1–2 cm long, upper stipules lanceolate to ovate (about 1 cm long, 2–3 mm wide). INFLORESCENCES: racemes spikelike (2–9 cm long, 2–4 cm wide), cylindric, axillary, dense, not interrupted; peduncles (1–15 cm long) longer than the petioles; bracts lanceolate or ovate to elliptic (1–1.5 cm long). FLOWERS: perfect, papilionaceous; calyx tube campanulate to tubular (8–15 mm long), gibbous at the base, strigose, densely glandular-punctate; lobes 5, acuminate, lowest lobe 7–12 mm long, upper 4 lobes 3–6 mm long; corollas blue or purple to violet (rarely white); banner obovate (1.5–2 cm long, claw 4–6 mm long); wings oblong to oblanceolate (about 1 cm long, claws 5–7 mm long); keel oblong to oblanceolate (7–9 mm long); pedicels 4–5 mm long; stamens 10, diadelphous. FRUITS: legumes obovate to ovoid (6–8 mm long), hidden by the persistent calyx; walls thin, papery, glandular-punctate; style persistent (about 2 mm long); seeds 1. SEEDS: round to broadly elliptic (4–5 mm long, 3–3.5 mm wide), slightly compressed, grayish-brown to brown, sometimes mottled, shiny, indurate. $2n=22$.

HABITAT: Tallbread scurfpea is infrequent to common on prairie uplands and rangelands, especially in dry, rocky soils. It usually does not grow in sandy soils. USES AND VALUES: Palatability is relatively low, and usually only the flowers and legumes are eaten by cattle and horses, but the plants still decrease in abundance with grazing. Pronghorn, elk, and deer eat the foliage, as well as the legumes and seeds. The flowers are pollinated by bees, and it is visited by many types of other insects and butterflies. It has little value for erosion control or landscaping. ETHNOBOTANY: Members of some tribes of Great Plains Indians peeled and ate the roots either raw or cooked. The roots were also dried and ground or pounded into flour and used for cooking.

SYNONYMS: *Lotodes cuspidata* (Pursh) Kuntze, *Pediomelum caudatum* Rydb., *P. parksii* Tharp & F.A. Barkley, *Psoralea aromatica* Payson, *P. caudata* (Rydb.) Cory, *P. cryptocarpa* Torr. & A. Gray, *P. cuspidata* Pursh, *P. macrorhiza* Fraser, *P. roemeriana* Scheele. OTHER COMMON NAMES: largebract scurfpea, largebract psoralea.

SIMILAR SPECIES: *Pediomelum latestipulatum* (Shinners) Mahler, plains indianbreadroot, is restricted to Texas in the southern Great Plains. Its stipules are wider and more conspicuous than those of *Pediomelum cuspidatum*, and its stems are not branched.

Palmleaf scurfpea
Pediomelum digitatum (Nutt. *ex* Torr. & A. Gray) Isely

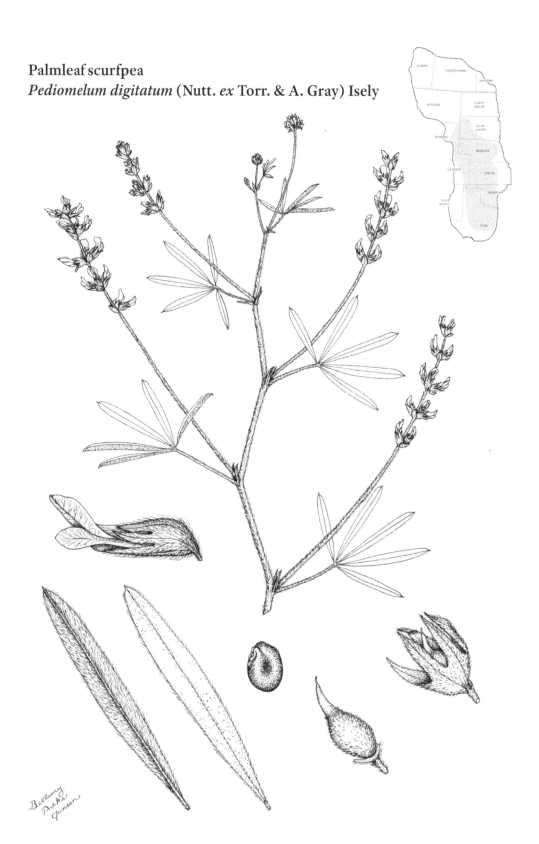

Sida 11(4):430. 1986.

digitatus (Lat.): having fingers, in reference to the palmately compound leaves.

GROWTH FORM: forb, flowers May to July, reproduces from rhizomes and seeds. LIFE-SPAN: perennial. ORIGIN: native. HEIGHT: 0.3–1 m. STEMS: herbaceous, erect to ascending from rhizomes or thick (not fleshy) taproots, divaricately branched above, appressed canescent, greenish-gray. LEAVES: alternate, odd-palmately compound; leaflets usually 5, but sometimes 3 on the uppermost leaves or 7 on the lowermost leaves, lanceolate to oblanceolate (1.5–6 cm long, 3–7 mm wide), middle leaflet longest; apex acute to obtuse, mucronate; base acute to attenuate; margins entire; upper surface glandular-punctate, glabrate to sparingly pubescent on the midvein; lower surface densely strigose to sericeous; petioles 2–6 cm long; stipules lanceolate (3–10 mm long), reflexed, involute on drying, sericeous. INFLORESCENCES: racemes (2–7 cm long) spikelike, axillary; flowers in 3–10 whorls, whorls separated or interrupted; peduncles (5–22 cm long) much longer than the leaves; bracts spatulate to obovate (2–7 mm long). FLOWERS: perfect, papilionaceous, nearly sessile; calyx tube campanulate (2.5–3.5 mm long) at anthesis, exceeding the fruits when mature, densely sericeous, glandular-punctate; lobes 5, upper 4 lobes 3–4 mm long, lower lobe slightly longer, acute to acuminate, sericeous; corolla blue or lavender to purple (rarely white), fading to yellowish-tan; banner obovate (8–10 mm long); wings 6–7 mm long; stamens 10, diadelphous. FRUITS: legumes obovate (5–8 mm long), tapering to a compressed beak (2–4 mm long), nearly entirely enclosed in the persistent calyx, walls thin, papery, glandular-punctate, pubescent; seeds 1. SEEDS: broadly elliptic (4–5 mm long, 3 mm wide), slightly compressed, olive to gray or brown, shiny, indurate. $2n=22$.

HABITAT: Palmleaf scurfpea is infrequent to locally common in prairies, rangelands, and open woodlands. It is most common in sandy or sandy clay soils. USES AND VALUES: Palmleaf scurfpea is relatively unpalatable to cattle and horses. Sheep occasionally graze the leaves and eat the flowers and legumes. Pronghorn, elk, and deer lightly graze the foliage and eat the flowers and legumes. Ground-foraging birds and small mammals eat the legumes and seeds. Palmleaf scurfpea is pollinated by various bees and attracts butterflies and many insects. It has little value for erosion control and has limited applications in landscaping. ETHNOBOTANY: Members of some Great Plains Indian tribes gathered the roots and ate them raw or cooked.

VARIETIES: Var. *digitatum* is most common in the Great Plains. Var. *parvifolium* (Shinners) Gandhi & L.E. Br. has narrow leaflets and grows only in Texas. SYNONYMS: *Lotodes campestre* (Nutt.) Kuntze, *L. digitata* (Nutt. *ex* Torr. & A. Gray) Kuntze, *Psoralea campestris* Nutt., *P. digitata* Nutt. *ex* Torr. & A. Gray, *Psoralidium digitatum* (Nutt. *ex* Torr. & A. Gray) Rydb. OTHER COMMON NAMES: finger scurfpea, palmleaf breadroot.

SIMILAR SPECIES: *Pediomelum linearifolium* (Torr. & A. Gray) J.W. Grimes, slimleaf scurfpea, plants are taller (0.5–1.5 m tall), and the leaflets are generally narrower (2–4 mm wide). It grows in the western and southern Great Plains.

Breadroot scurfpea
Pediomelum esculentum (Pursh) Rydb.

North American Flora 24(1):20. 1919.

esculentus (Lat.): edible, in reference to the root.

GROWTH FORM: forb, flowers May to July, reproduces from seeds. LIFE-SPAN: perennial. ORIGIN: native. HEIGHT: 10–25 cm. STEMS: herbaceous, erect to ascending, usually solitary, with axillary peduncles giving the appearance of lateral branches, densely villous-hirsute; from a tuberous fusiform to subglobose taproot, also serving as a storage organ (3–6 cm long, 2–5 cm wide), surface brown; inside white, starchy. LEAVES: alternate, odd-palmately compound; leaflets usually 5, sometimes 3 above, elliptic or oblanceolate to obovate (2–6 cm long, 6–16 mm wide), middle leaflet longest; apex obtuse to acute, sometimes mucronate; base acute to obtuse; margins entire, ciliate; upper surface glabrous to glabrate; lower surface densely sericeous, sparingly or not glandular-punctate; petioles 2–15 cm long (longer than the leaflets), villous-hirsute; stipules oblong to lanceolate (1.2–1.8 cm long). INFLORESCENCES: racemes (2–8 cm long, 2–2.5 cm wide), axillary, dense, few; peduncles 1–12 cm long; bracts ovate to lanceolate (8–16 mm long). FLOWERS: perfect, papilionaceous (1.3–2 cm long); calyx about 1.2 mm long in flower, elongating to 1.5 mm long in fruit; tube campanulate (5–7 mm long), persistent or not, villous to hirsute; lobes 5, acuminate (5–8 mm long), nearly equal or lowest lobe slightly longer; corolla blue to purple (rarely tinged with white) at anthesis, fading to yellow, drying to brown; banner oblong (9–12 mm long); wings 9–11 mm long; keel 4–5 mm long; stamens 10, diadelphous. FRUITS: legumes broadly elliptic or narrowly ovate (2 cm long), beak 1–2 cm long, thin, papery; seeds 1 or 2. SEEDS: Nearly round to broadly elliptic or narrowly ovate (4–6 mm long, 3–4.5 mm wide), plump to slightly compressed, olive to grayish-brown, sometimes mottled with purple, indurate. $2n=22$.

HABITAT: Breadroot scurfpea is infrequent to common in dry prairies, rangelands, bluffs, valleys, and open woodlands. It does not grow in dense stands. USES AND VALUES: It has little forage value for cattle and horses, but it is fair for sheep, pronghorn, elk, and deer. It increases with light grazing and rapidly disappears with moderate to heavy use. It has little value for erosion control because the plants break off near the soil surface in late summer and tumble in the wind. ETHNOBOTANY: Breadroot scurfpea was one of the most important foods for Great Plains Indians who ate both raw and cooked roots. Fresh roots were sometimes dried and stored for long periods. Dried roots were pulverized, mixed with water, and baked over coals.

SYNONYMS: *Lotodes esculenta* (Pursh) Kuntze, *Psoralea brachiata* Douglas *ex* Hook., *P. esculenta* Pursh. OTHER COMMON NAMES: aha (Crow), breadroot, esharusha (Crow), ground potato, Indian breadroot, Indian potato, Indian turnip, mu-gar-re (Omaha-Ponca), patsuroka (Pawnee), prairie turnip, tdokewihi (Winnebago), tiŋpsila (Lakota), tipsinah (Sioux).

SIMILAR SPECIES: *Pediomelum hypogaeum* (Nutt.) Rydb., little breadroot, is stemless or has short stems (1–2 cm long). Leaves are comprised of 5–7 linear-lanceolate to linear-oblong leaflets. Its corolla is less than 1.5 cm long in fruit. It is found primarily in the southwestern Great Plains.

Pisum L.

pisum (Lat.): pea.

Annual herbs; stems prostrate; sprawling or twining and climbing, glabrous, glaucous; leaves even-pinnately compound, glabrous, glaucous; leaflets 2–8, oval to oblong or suborbicular, upper pair modified into branched tendrils; stipules as large or larger than the leaflets, attached laterally, lobe downwardly directed, foliaceous; racemes axillary, flowers few; calyx tube generally shorter than the lobes; flowers perfect, papilionaceous; stamens 10, diadelphous; style longitudinally grooved; fruits legumes, oblong, laterally compressed, dehiscent, seeds several. $x=7$.

A single species, originally from the Mediterranean region, is commonly planted in gardens and occasionally for forage in the Great Plains.

Austrian winterpea
Pisum sativum L.

Reprinted from *Texas Range Plants* (Stephan L. Hatch and Jennifer Pluhar) by permission of Texas A & M University Press.

Species Plantarum 2:727. 1753.

sativum (Lat.): cultivated, in reference to it being a planted crop.

GROWTH FORM: forb, flowers April to June, reproduces from seeds. LIFE-SPAN: annual. ORIGIN: introduced (from Europe, but native to the Mediterranean region). LENGTH: 1–3 m long. STEMS: herbaceous, prostrate, branching above, twining and climbing, round, hollow, glabrous to glaucous, bluish-green. LEAVES: alternate, even-pinnately compound; leaflets 2–8, oval to oblong or suborbicular (2–7 cm long, 1–4 cm wide); apex acuminate to aristate with a mucro; base obtuse to cuneate; margins entire or dentate; surfaces glabrous to glaucous; upper pair of leaflets modified into branched tendrils; leaflets sessile or nearly so; midrib of leaf rachis sometimes slightly winged; petioles 3–7 cm long; stipules foliaceous (5–10 cm long, 3–6 cm wide), glabrous, glaucous. INFLORESCENCES: racemes axillary, flowers 1–3, bracts minute. FLOWERS: perfect, papilionaceous; calyx campanulate (8–15 mm long), foliaceous; lobes 5, exceeding the tube in length; corolla usually white or bicolored white and purple, sometimes pink or blue and purple (1.5–3.5 cm long); banner obovate (1.6–3 cm long); wings often purple; keel shorter than the wings, often white to greenish-white; style longitudinally grooved, or folded, bearded on the inside, curved, compressed; ovary nearly sessile; stamens 10, diadelphous. FRUITS: legumes oblong (3–12 cm long, 1–2.5 cm wide), laterally compressed; seeds usually 2–10. SEEDS: globose to angled (4–9 mm in diameter), smooth to wrinkled, yellow to green or brown, sometimes mottled. $2n=14$.

HABITAT: Austrian winterpea is cultivated for winter forage and grown in home gardens. It grows best in well-drained silty or sandy soils. It does not tolerate acidic or waterlogged soils. USES AND VALUES: This is the commonly planted garden pea, and numerous cultivars have been developed to ensure vegetable production in all parts of the region. However, production of grain is not the only goal of growing Austrian winterpea in the Great Plains. Its forage is high in protein and is grown for hay, pasture, and silage. It provides fair forage for grazing cattle and wildlife. After grazing, young plants will regrow multiple times furnishing additional forage. Hay quality is good, and it is often grown with cereal grasses to fortify protein content of the hay. It is grown as green manure and a cover crop because it grows rapidly and contributes nitrogen to the soil from nitrogen-fixing bacteria. Also, the vines break down quickly following tillage, adding additional nitrogen to the soil. It fits well into many crop rotations. It is pollinated by bees and visited by many other kinds of insects. ETHNOBOTANY: It has been a valuable food source for humans for millennia. It has been cultivated for as long as 8,000 years. Seeds are high in protein and contain important vitamins and minerals. The young leaves are sometimes boiled and eaten.

VARIETIES: This species has been divided into several varieties including *arvense* (L.) Poir., *pumilio* Meikle, *quadratum* L., *sativum*, and *umbellatum* L. Var. *arvense* is grown in the Great Plains for forage and *sativum* is grown in gardens. SYNONYM: *Pisum vulgare* Judz. OTHER COMMON NAMES: common pea, English pea, field pea, garden pea, spring pea.

Psoralidium Rydb.

psoraleos (Gk.): scurfy, referring to the many glands.

Perennial herbs from rhizomes or woody roots, stems erect to ascending, branching, pubescent or glabrate; leaves alternate, odd-palmately compound; leaflets usually 5 (often 3 on the upper leaves), often glandular-punctate, margins entire; petioles well-developed; stipules ovate or lanceolate to linear, usually persistent; inflorescences racemes, axillary; bracts persistent to caducous; calyx campanulate, not enlarging in fruit; lobes 5, equal or lower lobe longer; corolla papilionaceous, blue to purple or white to bicolored; banner orbicular to obtuse-obovate, usually clawed; stamens 10, diadelphous; ovary sessile or nearly so; fruits legumes, globose to oblong, short, slightly flattened or turgid, usually indehiscent, sometimes terminated by an indurate and persistent style, smooth or glandular-punctate, stipe short; seeds 1; seeds nearly round to reniform. x=11.

Three species occur in North America and the Great Plains. Two are common. These species are sometimes placed in *Pediomelum* or *Psoralea*.

1. Plants rhizomatous, forming large colonies; glands on leaves of various sizes; bracts caducous; legumes globose (3–6 mm long), pubescent. .... ***Psoralidium lanceolatum***
1. Plants from a woody taproot, seldom rhizomatous, not forming large colonies; glands on leaves uniform in size; bracts persistent; legumes elliptic or oblong (6–9 mm long), glabrous. .... ***Psoralidium tenuiflorum***

Lemon scurfpea
Psoralidium lanceolatum (Pursh) Rydb.

North American Flora 24(1):13. 1919.
lanceolatus (Lat.): spearlike, in reference to the shape of the leaflets.

GROWTH FORM: forb, flowers May to August, reproduces from rhizomes and seeds. LIFE-SPAN: perennial. ORIGIN: native. HEIGHT: 10–80 cm. STEMS: herbaceous, erect or ascending, from long rhizomes (up to 10 m), forming large colonies, much-branched, glabrate to sparingly strigose, glandular-punctate. LEAVES: alternate, odd-palmately compound; leaflets 5 (often 3 on upper leaves), emitting a lemonlike fragrance when bruised, variable, linear to narrowly lanceolate (1.5–5 cm long, 2–13 mm wide); apex acute, often mucronate; base attenuate; margins entire; upper surface glabrate, glandular-punctate, gland size variable; lower surface glabrate to sparingly strigose; petioles (1–2 cm long), about equaling the leaflets; stipules linear-lanceolate (3–10 mm long). INFLORESCENCES: racemes (1–3 cm long), axillary; peduncles 2–5 cm long, scarcely projecting above the foliage; bracts caducous. FLOWERS: perfect, papilionaceous; calyx tube campanulate (2–3 mm long), glandular, sparsely strigose; lobes 5, broadly triangular (less than half as long as the tube), nearly equal, 2 united and shortly bifid, glandular-punctate; corolla white, yellowish, or tinged with violet; banner orbicular (4–6 mm long, claw about 1 mm long); wings oblong to oblanceolate (3–4 mm long, claw about 1 mm long). FRUITS: legumes globose (3–6 mm long), abruptly short-beaked, glabrate to densely villous, glandular-punctate, usually indehiscent; seeds 1. SEEDS: nearly round (3–6 mm long, 3–4.5 mm wide), slightly compressed, reddish-brown, indurate. $2n=22$.

HABITAT: Lemon scurfpea grows in sandy soils of dry prairies and rangelands and is especially common in sandy blowouts. USES AND VALUES: Lemon-scented glands make it unpalatable to livestock. It is visited by various bees and butterflies. Small mammals and ground-foraging birds eat the seeds. Its rhizomatous growth and low palatability make it excellent for binding sand, and it is a pioneer species in blowouts. ETHNOBOTANY: Some Arapahos applied an infusion of leaves to the head for headache. Cheyennes used the plants in ceremonies and ate the flowers as a gastrointestinal aid.

VARIETIES: Var. *lanceolatum* is most common in the Great Plains, while vars. *stenophyllum* (Rydb.) S.L. Welsh and *stenostachys* (Rydb.) S.L. Welsh are found west of the region. SYNONYMS: *Lotodes ellipticum* (Pursh) Kuntze, *L. micranthum* (A. Gray) Kuntze, *Psoralea arenaria* Nutt., *P. elliptica* Pursh, *P. lanceolata* Pursh, *P. micrantha* A. Gray, *P. purshii* Vail, *P. scabra* Nutt., *P. stenophylla* Rydb., *P. stenostachys* Rydb., *Psoralidium micranthum* (A. Gray) Rydb., *P. purshii* (Vail) Rydb., *P. stenophyllum* (Rydb.) Rydb., *P. stenostachys* (Rydb.) Rydb. OTHER COMMON NAMES: lanceleaf psoralea, lanceleaf scurfpea, lemonweed, shoestring plant.

SIMILAR SPECIES: *Psoralidium linearifolium* (Torr. & A. Gray) Rydb., narrowleaf scurfpea, is not as common as lemon scurfpea and is found primarily in the southwestern portion of the region. Narrowleaf scurfpea racemes (2–8 cm) are much longer than the leaves. The lower leaves have 3 linear leaflets, and the upper leaves have single or 2 leaflets. These plants are 30–70 cm tall.

Slimflower scurfpea
Psoralidium tenuiflorum (Pursh) Rydb.

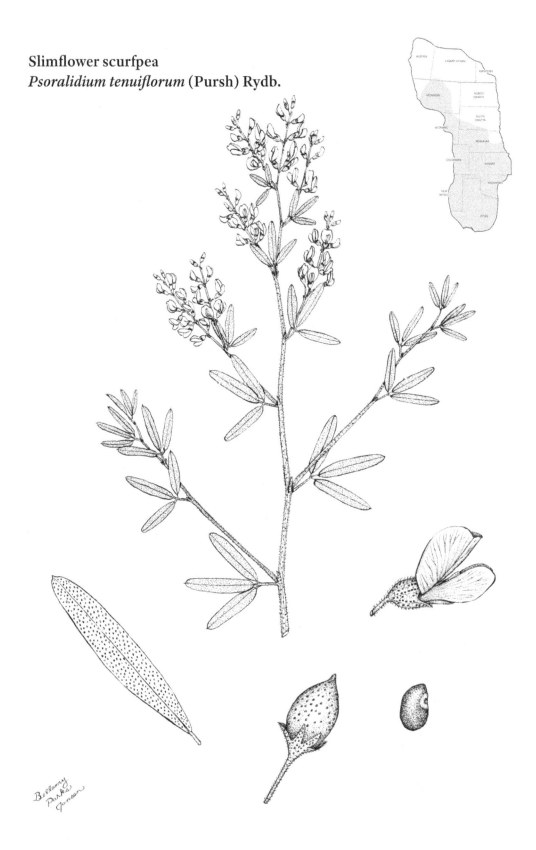

North American Flora 24(1):15. 1919.
tenuis (Lat.): thin; + *floris* (Lat.): flower, referring to the slender flowers.

GROWTH FORM: forb, flowers May to July, reproduces from seeds, seldom rhizomatous. LIFE-SPAN: perennial. ORIGIN: native. HEIGHT: 0.3–1.3 m. STEMS: herbaceous, erect to ascending from a woody taproot, not forming large colonies, much-branched, initially hoary pubescent, strigose later, glandular-punctate, breaking at the ground level in late summer and tumbling in the wind, scattering seeds. LEAVES: alternate, odd-palmately compound; leaflets 5 (often 3 on upper leaves), linear-oblanceolate to obovate (1–5 cm long, 4–12 mm wide); apex obtuse to rounded, mucronate; base acute; margins entire; upper surface glabrate, glandular-punctate, glands uniform in size; lower surface gray-strigose, scattered glandular-punctate; tending to be caducous; petioles (4–20 mm long) mostly shorter than the leaflets; lower stipules deltoid (6–9 mm long), upper stipules lanceolate (2–3 mm long). INFLORESCENCES: racemes 2–10 cm long, axillary, dense to loose; flowers 3–4 at each node; peduncles (2–9 cm long) longer than the leaves; bracts ovate to lanceolate (2–6 mm long), persistent. FLOWERS: perfect, papilionaceous; calyx tube campanulate (1.5–2.5 mm long or more), glandular-punctate, strigose or loosely villous-hirsute; lobes 5, acuminate (lower lobe 1.2–2.5 mm long, upper lobes 1–1.5 mm long); corolla blue to purple, rarely white; banner obtuse-obovate (4–8 mm long); wings 3–6 mm long; pedicels 1.5–4 mm long. FRUITS: legumes elliptic or oblong (6–9 mm long), beak short, glabrous, densely glandular-punctate, often asymmetric, usually indehiscent; seeds 1. SEEDS: reniform to nearly round (4–5.5 mm long), slightly compressed, orangish-brown, sometimes spotted with purple, shiny, indurate. 2n=22.

HABITAT: Slimflower scurfpea is common on dry prairies, rangelands, pastures, rocky banks, and openings in woodlands. USES AND VALUES: Palatability is low, and it is generally not grazed by cattle or horses. After curing in hay, it is readily eaten. It is pollinated by various bees and visited by many kinds of insects. It has little value for erosion control or landscaping. ETHNOBOTANY: Some Lakotas made tea from the roots for headache, burned the plants to repel mosquitoes, and made garlands from the tops to be worn for protection from the sun. Other tribes reportedly used slimflower scurfpea for fish poison.

VARIETIES: Var. *floribundum* (Nutt. *ex* Torr. & A. Gray) Rydb. plants have crowded, showy flowers and grow in the eastern and southeastern Great Plains. Var. *bigelovii* (Rydb.) J.F. Macbr. grows in the southern part of the region. SYNONYMS: *Lotodes floribundum* (Nutt. *ex* Torr. & A. Gray) Kuntze, *L. tenuiflora* (Pursh) Kuntze, *Psoralea bigelovii* (Rydb.) Tidestr., *P. floribunda* Nutt. *ex* Torr. & A. Gray, *P. obtusiloba* Torr. & A. Gray, *P. tenuiflora* Pursh, *Psoralidium batesii* Rydb., *P. bigelovii* Rydb., *P. floribundum* (Nutt. *ex* Torr. & A. Gray) Rydb., *P. obtusilobum* (Torr. & A. Gray) Rydb., *P. youngiae* Tharp & F.A. Barkley. OTHER COMMON NAMES: fewflowered scurfpea, scurvy pea, slender-flowered scurfpea, tičaničahu taŋka-hu (Lakota), wild alfalfa.

Pueraria DC.

pueraria: named after Swiss botanist Marc Nicolas Puerari (1766–1845).

Perennial vines, woody or herbaceous, stems prostrate, trailing or climbing; pubescent; leaves alternate, pinnately compound, leaflets 3; leaflets orbiculate to ovate, lobes on 1 or both sides, lobes asymmetric; upper surface glabrate to pubescent; lower surface pubescent; petioles well-developed; stipules generally with retrorse lobes; inflorescences pseudoracemes, sometimes branched; bracts small, caducous; flowers perfect, papilionaceous, many; calyx lobes 4, shorter than the tube; flowers regular, papilionaceous; stamens submonadelphous to diadelphous; ovary pubescent; style glabrous; fruits legumes, oblong, stipitate, tardily dehiscent, seeds several. x=10, 11.

About 15 species grow in Asia. One has been introduced to North America as a forage. It is presently regarded as a serious weed and grows in the southeastern Great Plains.

Kudzu
Pueraria montana (Lour.) Merr.

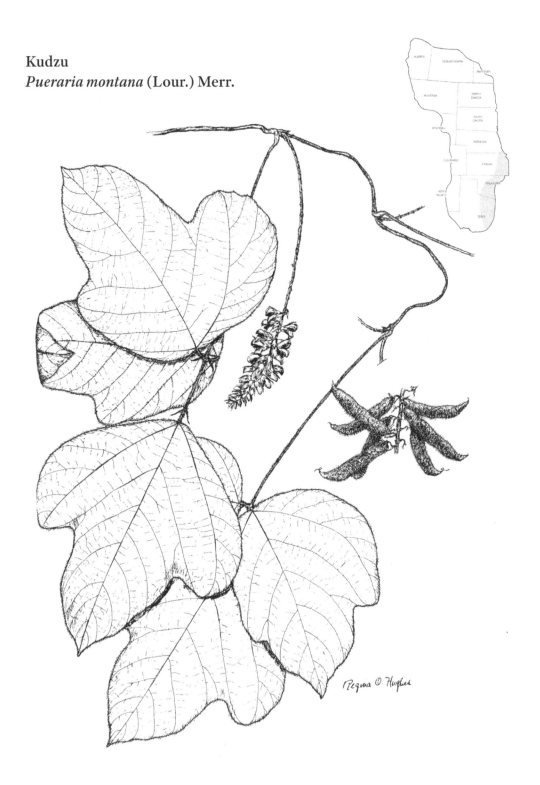

Reprinted from *Selected Weeds of the United States*
(Agricultural Research Service Handbook 366).

Transactions of the American Philosophical Society, New Series 24(2):210. 1935.
montana: of or from the mountains.

GROWTH FORM: forb, flowers July to September, reproduces from seeds and creeping stems. LIFE-SPAN: perennial. ORIGIN: introduced (from Asia). LENGTH: 5–30 m. STEMS: herbaceous, becoming woody, from a large taproot (a single root may weigh more than 40 kg), fleshy, prostrate and climbing or trailing, climbing vines may survive over winter, surfaces hispid. LEAVES: alternate, odd-pinnately compound; leaflets 3, rachis not extending into a tendril; central leaflets orbiculate to ovate or ovate-rhombic (5–25 cm long, 5–20 cm wide); apex acute to acuminate; base cuneate; margins entire to sinuate or lobed on 1 or both margins, lateral lobes asymmetric; surfaces appressed-pubescent, not glandular-punctate; petioles 8–20 cm long; stipules lanceolate to ovate-lanceolate (1.4–2.1 cm long); basal lobes incised or entire, 1 often retrorse. INFLORESCENCES: racemes (10–30 cm long), axillary and terminal; flowers 12–40, crowded; peduncle short. FLOWERS: perfect, papilionaceous, seldom produced in the northern portion of its range in the Great Plains; calyx 1–1.2 cm long, tube campanulate; lobes 4, upper lobes united more than half their length, shorter than the tube, pubescent; corollas purple or reddish-purple, showy; banner obovate (1–2.5 cm long) with a distinct yellow or green patch at its base; wings oblong to ovate, united and slightly longer than the keel; keel oblong with a pointed tip, yellowish; stamens 10, monadelphous to diadelphous. FRUITS: legumes linear-oblong (3–10 cm long, 5–10 mm wide), straight, laterally compressed, densely pubescent (hairs 2–3 mm long), stipe absent; reddish-brown to brown; seeds 10–20. SEEDS: reniform to oblong (3–5 mm long), compressed, reddish-brown, sometimes mottled with black, smooth, indurate. $2n=20, 22$.

HABITAT: Kudzu grows in well-drained soils in disturbed areas, abandoned fields, fence rows, roadsides, and forest edges. It can grow in nutrient deficient soils and requires relatively mild winters to survive. USES AND VALUES: It produces good quality forage for cattle, horses, and sheep. It is nutritionally similar to alfalfa. Its seeds are eaten by quail, other ground-foraging birds, and small mammals. It is pollinated by various bees and is visited by butterflies and many types of insects. Honey is made from the nectar. The flowers are visited by numerous butterflies and skippers. It is not recommended for protection from soil erosion because it has the ability to quickly spread and overwhelm the site. For the same reason, it has no application in landscaping. It may cause bloat. ETHNOBOTANY: In Asia the roots are sometimes boiled and eaten.

VARIETIES: Plants in the Great Plains are var. *lobata* (Willd.) Maesen & S.M. Almeida *ex* Sanjappa & Predeep. Vars. *chinensis* (Benth.) Maesen & S.M. Almeida *ex* Sanjappa & Predeep, *montana*, and *thomsonii* (Benth.) M.R. Almeida grow in eastern and southeast parts of the United States. SYNONYMS: *Dolichos montanus* Lour., *Pachyrhizus montanus* (Lour.) DC., *Pueraria thunbergiana* var. *formosana* Hosok., *P. tonkinensis* Gagnep. OTHER COMMON NAMES: kudzu vine, foot-a-night vine.

Robinia L.

Robinia: named after Jean Robin (1550–1629) and his son, Vespasian Robin (1579–1662), herbalists and botanists for the king of France.

Trees or shrubs; armed or unarmed; leaves odd-pinnately compound, petioled; leaflets several to many; stipules setaceous or modified into spines; inflorescences racemes, often pendant, axillary, bracts caducous; flowers perfect, papilionaceous, large; calyx tube hemispheric or broadly campanulate; lobes 5, bilabiate, lower 3 about equal, upper 2 connate for one-third or more of their length; corolla usually white, sometimes rose or purple; banner suborbicular, more or less reflexed; wings obliquely obovate, claws long; keel strongly curved upwards, claws long; stamens 10, diadelphous; ovary elongate; legumes oblong to elongate (5 cm long or more), laterally compressed, papery to subcoriaceous, tardily dehiscent; seeds several to many. x=10, 11.

Four species grow in North America. Three are found in the Great Plains. Two are shrubs. One is common in eastern North America, and another is often planted on the western edge of the region. One is a common tree and is described here.

Black locust
Robinia pseudoacacia L.

Species Plantarum 2:722. 1753.

pseudo (Gk..): false; + *akakia* (Gk.): acacia, in reference to the similarity this species and some in the genus *Acacia*.

GROWTH FORM: tree, crown oblong to rounded, flowers May to June, reproduces from seeds and forming colonies from root sprouts. LIFE-SPAN: perennial. ORIGIN: native. HEIGHT: to 18 m. TWIGS: woody (1.8–2.1 mm in diameter), gray or reddish-brown; spines scattered, short, in pairs, extending at right angles from the twigs. TRUNKS: to 1.2 m in diameter. LEAVES: deciduous, alternate, odd-pinnately compound (10–35 cm long); leaflets 7–29, elliptic to oval (2–5 cm long, 1–1.2 cm wide); apex rounded to obtuse, mucronate; base obtuse; margins entire; upper surface glabrous; lower surface lighter green, lightly pubescent to glabrate; petiole 1–5 cm long, glabrous or pubescent, base enlarged; petiolules short; stipules linear-subulate (3–25 mm long), membranous first, then developing into spines (3–25 mm long). INFLORESCENCES: racemes 5–20 cm long, axillary; flowers 10–40. FLOWERS: perfect, papilionaceous, fragrant; calyx tube campanulate (4–5 mm long), pubescent, sometimes purple; lobes 5, bilabiate, lower 3 acuminate (1–2 mm long), upper 2 connate (2–3.5 mm long); corolla white, banner with a small yellow patch; banner suborbicular (1.5–2.5 cm long), claw 4–6 mm long; wings 2–2.2 cm long, lobed, claws 6–7 mm long; keel semicircular (1.9–2 cm long), apex acute; stamens 10, diadelphous; pedicels 5–10 mm long, pubescent, sometimes purple. FRUITS: legumes straight (5–10 cm long, 1–1.5 cm wide), compressed; apex apiculate, short-stipitate; valves rigid, glabrous; seeds 2–12. SEEDS: reniform to oval (4–6 mm long, 2–3.5 mm wide), slightly beaked, compressed, brown, often mottled with darker brown or purple, smooth, indurate. $2n=20$.

HABITAT: Black locust is common in moist but well-drained soils in pastures, rangelands, roadsides, valleys, and thickets. It has been planted over a wide geographic area and has commonly escaped. USES AND VALUES: Cattle and deer browse the foliage. It has a high tannin content, resulting in low digestibility. Ground-foraging birds, squirrels, and other small mammals eat the seeds. It is pollinated by various bees, and some use the nectar in making honey. The bark contains the toxin lectin. Cattle have been poisoned by eating shoots sprouting from stumps. Sheep have been poisoned by eating the seeds. Horses have been poisoned by stripping and eating the bark, often while being tied to a black locust tree. It is often considered to be a weed. Black locust was once used for fence posts and railroad ties. It is used in erosion control and landscaping. ETHNOBOT-ANY: Flowers can be eaten fresh or after frying in batter. A tea is made from the flowers.

SYNONYMS: *Robinia pringlei* Rose, *R. pyramidalis* Pépin. OTHER COMMON NAMES: American locust, false acacia, peaflower locust, post locust, shipmast locust, sweet locust, yellow locust.

SIMILAR SPECIES: *Robinia hispida* L., bristly locust, is a shrub (to 3 m tall) and grows in the southeastern part of the region. *Robinia neomexicana* A. Gray, New Mexico locust, is a shrub (to 3 m tall) with pink or pinkish-purple corollas, and its legumes are bristly to glandular-pubescent. It grows on the western edge of the region as far north as Wyoming.

Securigera DC.

securigera (Lat.): armed with an ax; in reference to the shape of the loment joints of the type specimen.

Herbs or shrubs, annual or perennial, glabrous throughout, stems angular or sulcate, leaves odd-pinnately compound, leaflets few to many; inflorescences umbels, axillary, long-peduncled; flowers 8 or more, perfect, papilionaceous; calyx campanulate to hemispheric; lobes 5, bilabiate, broader lower lip with 3 short triangular teeth, upper lip narrow and shallowly cleft; corolla pink to white or yellow, petals about equal in length, clawed; keel curved upwards; stamens 10, diadelphous; fruits loments, linear or lanceolate, terete or 4-angled, straight or curved, transversely jointed; seeds few to several. $x=6$.

A genus of about 12 species. All are native to Eurasia and northern Africa. Several have been introduced into the United States. Only 1, described here, grows in the Great Plains.

Crownvetch
Securigera varia (L.) Lassen

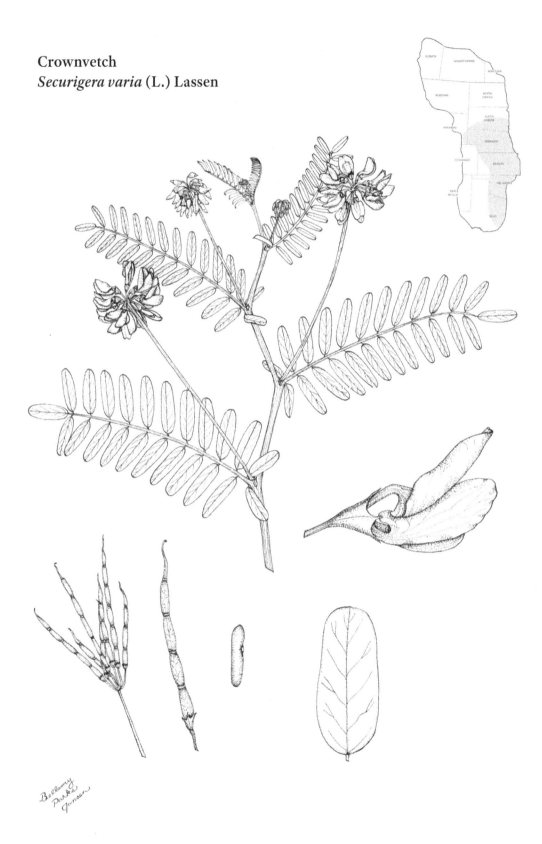

Svensk Botanisk Tidskrift 83:86. 1989.
varia (Lat.): variable, referring to the color of the corollas.

GROWTH FORM: forb, flowers May to September, reproduces from rhizomes and seeds. LIFE-SPAN: perennial. ORIGIN: Introduced through Europe (from the Mediterranean region). LENGTH: 0.3–2 m. STEMS: herbaceous, trailing, ascending, slender, glabrate, from rhizomes and a caudex and taproot. LEAVES: alternate, odd-pinnately compound (4–13 cm long), rachis not extended into a tendril; leaflets 9–25, ovate to obovate or ovate-lanceolate (1–3 cm long, 2–10 mm wide); apex rounded to retuse or acute, mucronate; base broadly obtuse to cuneate; margins entire; surfaces glabrous; short-petiolate below to nearly sessile above; stipules narrowly lanceolate to ovate-lanceolate (2–3 mm long), persistent or caducous. INFLORESCENCES: umbels globose; flowers 8–20; peduncles stout (4–12 cm long), often surpassing the subtending leaves. FLOWERS: perfect, papilionaceous, fragrant; calyx tube campanulate to hemispheric (1–1.9 mm long); lobes 5, upper 2 lobes nearly united, lower 3 lobes distinct; corolla pink or bicolored; banner pink, wings and keel white; banner orbiculate (1–1.3 cm long, claw 1–2 mm long); wings ovate-oblong (1.1–1.5 cm long, claw 2–4 mm long); keel oblanceolate (1–1.3 cm long, claw about 3 mm long), curved upward, tipped with purple; stamens 10, diadelphous. FRUITS: loments linear (2–5 cm long, 2–3 mm wide), borne in crownlike clusters, 4-angled, slightly constricted between the seeds, coriaceous, breaking apart at maturity, stipe 2–9 mm long; pedicels 3–7 mm long; joints and seeds 3–12. SEEDS: oblong (3–4 mm long, 1 mm wide), slightly compressed, smooth, light to dark brown, indurate. $2n=24$.

HABITAT: Crownvetch grows on roadsides, embankments, stream banks, fields, and gardens. It will grow on acidic soils, but it is not tolerant of saline or alkaline soils. USES AND VALUES: Crownvetch produces good-quality forage for cattle, sheep, and goats. Grazing reduces its persistence more than does frequent mowing. Crownvetch does not cause bloat, but it contains the toxic nitroglycoside coronarian. Its toxicity was discovered nearly 200 years ago. However, it is not poisonous to ruminants because coronarian is detoxified in the rumen. Poisoned nonruminants, such as horses, exhibit slowed growth, ataxia, and posterior paralysis. Poisoning only occurs after consumption of large amounts of herbage, and it may result in death. It is grazed lightly by deer, and birds and small mammals eat the loments and seeds. It is often used for cover by ground-nesting birds and small mammals. One of its most valuable uses is as a bank stabilizer along roadways and waterways. However, on steep slopes, the canopy of crownvetch may only conceal erosion occurring under it. Crownvetch grows well in poor soils. It has been used as an understory in orchards and Christmas tree farms. It is somewhat difficult to start from seeds and is often propagated by transplanting crowns. It may slowly spread, and can become a problem in natural areas such as native prairies. Once established, it is difficult to control.

SYNONYM: *Coronilla varia* L. OTHER COMMON NAMES: axseed, hatchet vetch, hivevine, purple crownvetch, trailing crownvetch.

Sophora L.

sufayra (Ar.): referring to a yellow dye made from dry, immature flower buds of some members of this genus.

Deciduous or evergreen trees, shrubs, or perennial herbs; unarmed; leaves alternate, odd-pinnately compound, leaflets several to numerous, not glandular-punctate; stipules minute on herbaceous species, caducous; inflorescences racemes (occasionally panicles), terminal, occasionally axillary; calyx campanulate, base gibbous; lobes generally 4 through fusion of upper 2 lobes, shorter than the tube; flowers perfect, papilionaceous; corollas yellow, white, or purple; banner obtuse, tapered to the base; wings straight, clawed; keel nearly straight, clawed; stamens 10, distinct; fruits legumes, oblong, turgid to laterally compressed, more or less constricted between seeds, stipitate, woody or fleshy, terete, indehiscent or tardily dehiscent; seeds many. x=9.

A genus of about 40 species scattered in the temperate to tropic regions of the world. Eight are native to the United States, and others are horticultural introductions. Of the 3 species growing in the Great Plains, only 1 is relatively common. The second extends only into the southern Great Plains, and the third is an introduced ornamental.

Silky sophora
Sophora nuttalliana B.L. Turner

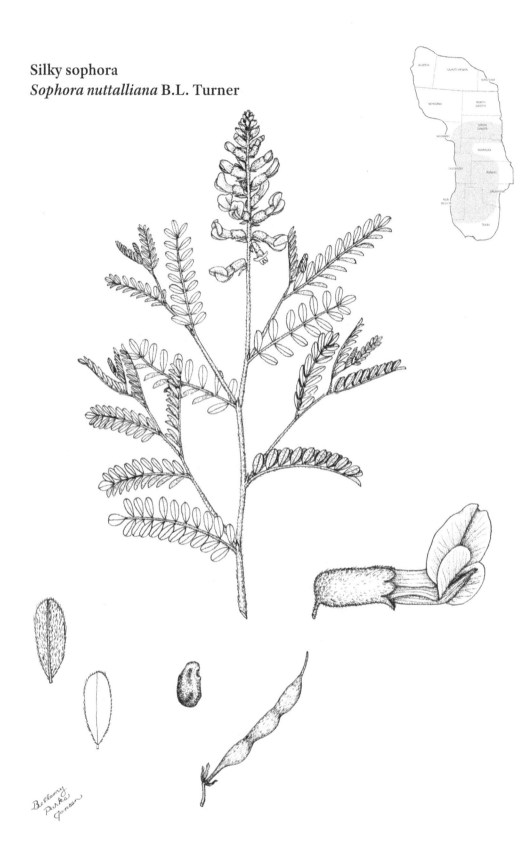

Field & Laboratory 24(2):42. 1956.

nuttalliana: named for Thomas Nuttall (1787–1859), explorer and naturalist, who was the first to collect this species.

GROWTH FORM: forb, flowers May to June, reproduces from seeds and lateral shoots. LIFE-SPAN: perennial. ORIGIN: native. HEIGHT: 10–70 cm. STEMS: herbaceous, erect to ascending, simple from the base or much-branched above, arising from taproots and forming colonies by shoots from adventitious buds on lateral roots, silky-canescent, not glandular-punctate. LEAVES: alternate, odd-pinnately compound; leaflets 7–31, obovate or narrowly oblong to elliptic (7–18 mm long, 1.5–9 mm wide), rachis not extended into a tendril; apex acute to obtuse or rarely retuse; base acute; margins entire; lower surface silky-canescent; upper surface glabrous or sparsely sericeous; petioles 1–9 mm long; stipules bristlelike, sheathing on the lowermost nodes, obsolete on the uppermost nodes. INFLORESCENCES: racemes (7–11 cm long) terminal, occasionally axillary, lax to densely flowered; flowers 8–80; peduncles 1–4 cm long; bracts lanceolate (4–11 mm long), semipersistent. FLOWERS: perfect, papilionaceous; calyx tube campanulate (5–8 mm long), gibbous, strigulose; lobes 4, lower 3 lobes 1–2.5 mm long, upper lobe double, short-acuminate to triangular (1–2 mm long), connivent for half their length; corolla white to ochroleucous or rarely purple; banner oblanceolate (1.2–1.6 cm long), widened above the middle, strongly recurved, claws 5–7 mm long; wings oblong (8–11 mm long), asymmetric, claws 5–6 mm long; blades of keel 7–8 mm long, claws 5–6 mm long, 2 apical beaks (1–3 mm long); stamens 10, distinct. FRUITS: legumes linear to oblong (2–7 cm long), straight or curved, turgid, beaked, short-stipitate, constricted between the seeds, densely sericeous, indehiscent or tardily dehiscent; seeds usually 1–7. SEEDS: oblong (4.5–5 mm long, 2.5–3 mm wide), somewhat compressed, smooth, olive to brown, indurate. $2n=18$.

HABITAT: Silky sophora is infrequent to common in dry prairies, rangelands, badlands, roadsides, abandoned fields, and stream valleys. It is most abundant on sandy and gravelly soils. USES AND VALUES: It has little forage value for livestock. It has been reported to be poisonous to horses, but toxicity was not proven in feeding trials. Alkaloids have been isolated from the seeds. The seeds are eaten by ground-foraging birds and small mammals without apparent poisoning. The flowers are pollinated by various bees. It is not used for erosion control or in landscaping. ETHNOBOTANY: Some Great Plains Indians chewed its sweet-tasting roots for the pleasant flavor.

SYNONYMS: *Patrinia sericea* Raf., *Pseudosophora sericea* (Raf.) Sweet, *Radiusia sericea* (Nutt.) Heynh., *Sophora carnosa* (Pursh) Yakovlev, *S. sericea* J. St.-Hil., *S. sericea* Nutt., *Vexibia sericea* (Raf.) Raf. OTHER COMMON NAMES: Nuttall sophora, white loco.

SIMILAR SPECIES: *Sophora affinis* Torr. & A. Gray, Texas sophora, leaves have 9–15 leaflets, rosy-white corollas, and black legumes. This woody plant forms thickets and grows in southern Oklahoma and northern and central Texas. *Sophora japonica* L., pagodatree, is an introduced ornamental. This small tree has a rounded crown and short trunk. It has large leaves and racemes to 40 cm long.

Strophostyles Elliott

strophe (Gk.): turning; + *stylos* (Gk.): pillar, referring to the recurved style.

Annual or perennial, twining or trailing herbs, retrorsely pilose or glabrate; leaves usually alternate, pinnately trifoliate, appressed-pubescent or glabrous; margins entire, may be lobed; stipules triangular to ovate, persistent; inflorescences racemes, axillary, subcapitate, long-peduncled; flowers perfect, papilionaceous, few; calyx campanulate; lobes 4 through fusion of upper 2 lobes, lowest lobe longest, exceeding the length of the tube; corollas purplish-pink to white; banner orbicular to broadly obtuse-ovate; wings shorter than the keel, oblong, curved upward; keel falcate; stamens 10, diadelphous; style bearded; fruits legumes, linear, subterete, strigose or glabrous, coiled after dehiscence; seeds few to several. x=11.

A genus of three species in North America. One is perennial and grows in the southeastern United States. The other two are annuals and grow in the Great Plains.

1. Leaflets narrowly ovate to oblong-lanceolate, at least 2 times longer than wide; margins entire; calyx tube densely hirsute; banner 5–8 mm long; legume 2–4 cm long. .... ***Strophostyles leiosperma***
1. Leaflets ovate-oblong or ovate to rhombic, less than 2 times longer than wide; margins entire to lobed; calyx tube glabrous to sparsely pubescent; banner 1–1.4 cm long; legume 4 cm or longer. .... ***Strophostyles helvola***

Trailing wildbean
Strophostyles helvola (L.) Elliott

A Sketch of the Botany of South-Carolina and Georgia 2(3):230. 1823.
helvus (Lat.): light yellow, in reference to the color of the dried corollas.

GROWTH FORM: forb, flowers June to September, reproduces from seeds. LIFE-SPAN: annual. ORIGIN: native. LENGTH: 0.3–2.5 m. STEMS: herbaceous, initially erect then becoming twining, decumbent, may be reddish-purple at maturity, branched at the base and occasionally above, from taproots. LEAVES: principal leaves alternate (leaves on lower 1–3 nodes opposite), pinnately trifoliate (8–12 mm long); leaflets variable in shape, usually ovate-oblong or ovate to rhombic (2–7 cm long, 1.5–5 cm wide), length usually less than 2 times the width, terminal leaflet usually largest; apex acute, usually mucronate; base rounded; margins entire to deeply lobed on 1 or both sides, ciliate; upper surface sparsely appressed-pubescent to glabrate; lower surface lighter in color, appressed-pubescent; petioles 1–9 cm long; stipules lanceolate (4–6 mm long), persistent. INFLORESCENCES: racemes axillary; flowers 2–8; peduncles 4–22 cm long; bracteoles lanceolate (2–3 mm long), acute, extending beyond the sinuses of the calyx. FLOWERS: perfect, papilionaceous (8–14 mm long); calyx tube campanulate (1.5–3 mm long); lobes 4, upper lobe with 2 united teeth, lower lobe longest (4–6 mm long), glabrous or with a few appressed hairs; corolla purplish-rose to greenish purple or pinkish-yellow, fading to light yellow or pinkish-yellow; banner orbicular or rounded-ovate (1–1.4 cm long, 1–1.2 cm wide), claw 1–2 mm long; wings slender (6–9 mm long), claw 1.5–2 mm long; keel maculate (1.2–1.4 cm long), claw 2–3 mm long; stamens 10, diadelphous. FRUITS: legumes linear (4–9 cm long, 5–8 mm wide), subterete, initially strigose, becoming glabrate with maturity, style bent, persistent; valves coiled after dehiscence; seeds several. SEEDS: rectangular to oblong with truncated ends (5–10 mm long, 2–5 mm wide), sometimes scurfy or woolly, but with the scurfy coating readily detaching, brown to brownish-black, indurate. $2n=22$.

HABITAT: Trailing wildbean is infrequent to locally common in prairie ravines, open woodlands, damp thickets, disturbed areas, roadsides, and old fields, along streams, and in sand and gravel bars in streams. USES AND VALUES: It is palatable to all classes of livestock. It is not highly valued because it is seldom abundant. It is lightly grazed by deer. Its seeds are eaten by many kinds of birds and small mammals. Trailing wildbean is pollinated by bees, especially leaf-cutting bees and bumblebees. Butterfly and skipper larvae feed on the foliage. It has been used for erosion control in moist soils, but it has little potential for landscaping. ETHNOBOTANY: Seeds were gathered by members of several tribes of Great Plains Indians for food, and seeds have been found in many archaeological sites. Roots were boiled and mashed for food. Some treated poison ivy and warts by rubbing the leaves on the affected areas. A decoction made from the seeds was used to treat typhoid.

SYNONYMS: *Cajanus helvolus* (L.) Spreng., *Glycine helvola* (L.) Elliott, *Phaseolus helvolus* L., *Strophostyles helvula* (L.) Elliott. OTHER COMMON NAMES: amberique bean, makatominiča (Lakota), pink wildbean, sand bean, trailing fuzzybean, wild bean.

Smoothseed wildbean
Strophostyles leiosperma (Torr. & A. Gray) Piper

Contributions from the United States National Herbarium 22(9):668. 1926.
leio (Gk.): smooth to the touch; + *sperma* (Gk.): seed, referring to the smooth seed.

GROWTH FORM: forb, flowers May to September, reproduces from seeds. LIFE-SPAN: perennial. ORIGIN: native. LENGTH: 0.2–2 m. STEMS: herbaceous, twining, prostrate, from a taproot, 1 to several branches at the base, generally villosulous with spreading or obliquely retrorse hairs (less than 1 mm long). LEAVES: leaves on lower 1–4 nodes opposite, simple (8–10 cm long); principal leaves alternate, pinnately trifoliate; leaflets narrowly ovate or oblong-lanceolate (2–6 cm long, 2–18 mm wide), length is usually greater than 2 times the width; middle leaflet usually short-stalked; apex obtuse to acute, sometimes mucronate; base rounded to acute; margins entire, not lobed, ciliate; upper surface pilose; lower surface more densely pilose; petioles 2–5 cm long; stipules about 2 mm long. INFLORESCENCES: racemes; flowers 1–6; usually only 1 or 2 blooming at a time; peduncles slender (1–12 cm long); bracteoles lanceolate (0.7–1.5 mm long), hirsute. FLOWERS: perfect, papilionaceous; calyx tube campanulate (1.5–2 mm long), densely hirsute; lobes 4; 2 teeth fused to form the upper lobe (1.5–2 mm long), middle tooth of lower lobe longest (2–4 mm long); corolla light rose to purple or rarely green, fading to yellow or yellowish-green; banner 5–8 mm long, claw about 1 mm long, dark spot at the base; wings 5–6 mm long, claw 1.5–2 mm long; keel abruptly bent to a slightly twisted, narrowed, and maculate beak; stamens 10, diadelphous. FRUITS: legumes elongate (2–4 cm long), subterete, sessile, appressed-pubescent or spreading pubescent to rarely glabrate, valves coiled after dehiscence; seeds usually few to several. SEEDS: round to reniform (2.5–5 mm long, 2–3 mm thick), scurfy pubescence easily detached, gray to brown, sometimes mottled with purple or black, smooth. $2n=22$.

HABITAT: Smoothseed wildbean is scattered to uncommon in dry or moist soils of prairies, rangelands, meadows, old fields, open woodlands, dunes, and shores. It is most abundant in sandy soils. USES AND VALUES: Forage quality is fair, but it is not valued for forage because it is seldom abundant. It is palatable to cattle, horses, and sheep and is lightly grazed. Deer, pronghorn, and elk eat the foliage. Mourning doves, quail, sharp-tailed grouse, and prairie chickens eat the seeds. Small mammals eat the legumes and the seeds. Smoothseed wildbean is pollinated by various bees, and it attracts many kinds of insects. It has little value for erosion control. It is used occasionally as a bedding plant in areas with well-drained soils and full sun. It is relatively intolerant of competition from other plants. ETHNOBOTANY: Some members of the Lakota and other tribes ate the beans after cooking.

SYNONYMS: *Phaseolus leiospermus* Torr. & A. Gray, *P. pauciflorus* Benth., *Strophostyles pauciflora* S. Watson. OTHER COMMON NAMES: omníča hú (Lakota), slickseed fuzzybean, slickseed wildbean, small wildbean, trailing fuzzybean, trailing pea, wild beanvine.

SIMILAR SPECIES: The third species in this genus is *Strophstyles umbellata* (Muhl. *ex* Willd.) Britton, perennial fuzzybean. It is a wiry-stemmed perennial that grows in the eastern and southeastern United States.

Stylosanthes Sw.

stylos (Gk.): pillar or column; + *anthos* (Gk.): flower, in reference to the stalklike hypanthium and calyx tube.

Herbaceous perennials or annuals, low or prostrate, commonly hispid; leaves alternate, trifoliate; petioles fused with the stipule sheath; flowers axillary, solitary or clustered, perfect, papilionaceous, small, usually only 1–2 flowering at a time; calyx lobes 4 or 5, lowest lobe longest; corolla orangish-yellow; banner obcordate to broadly obovate or obtuse, longer than the wings and keel, narrowed at the base; wings oblong to obovate, clawed; keel curved upward, nearly as long as the banner; stamens 10, monadelphous, anthers alternately subglobose and oblong; fruits loments, segments 1 or 2; lowest segment usually sterile, indehiscent; seeds irregular, 1 per loment segment. x=10.

This genus is represented by about 25 species found primarily in the tropics. They are most numerous in tropical America. Four grow in the United States, and only 1 grows in the Great Plains.

Pencilflower
Stylosanthes biflora (L.) Britton, Sterns & Poggenb.

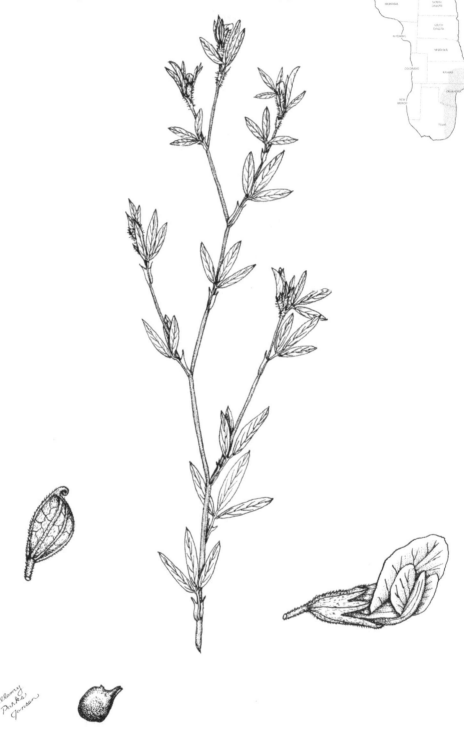

Preliminary Catalogue of Anthophyta and Pteridophyta Reported as Growing Spontaneously within One Hundred Miles of New York 13. 1888.

bis (Lat.): two; + *flos* (Lat.); flower, referring to the commonly occurring inflorescence with only 2 flowers.

GROWTH FORM: forb, flowers May to September, reproduces from seeds. LIFE-SPAN: perennial. ORIGIN: native. HEIGHT: 15–60 cm. STEMS: herbaceous, ascending to erect, 1 to several, from a woody caudex, more or less branched, glabrous or minutely pubescent to hispid. LEAVES: alternate, trifoliate, rachis not extending into tendrils; leaflets elliptic to oblanceolate or lowermost sometimes obovate (1–4 cm long, 4–8 mm wide); apex acute to acuminate, occasionally mucronate; base cuneate to acute; margins entire, sometimes sparsely hispid, sometimes with short spiny teeth on the upper margin; upper surface glabrous to glabrate; lower surface lighter green, glabrous, veins conspicuous, sometimes glandular-punctate; petiolule of terminal leaflet 1–3 mm long; lateral leaflets sessile or nearly so; petioles 1–3 mm long, fused with stipule sheaths; stipules 4–12 mm long, adnate to the base of the petiole, forming a tube around the stem, distinct portions linear or subulate. INFLORESCENCES: racemes subcapitate or spikelike (5–10 mm long), usually from axils near the ends of the branches; flowers commonly 1–2, as many as 6. FLOWERS: perfect, papilionaceous; calyx tube campanulate (4–6 mm long); lobes 4 or 5, lower and lateral lobes acute, nearly as long or longer than the tube, upper 2 lobes rounded and united near the apex; corolla orangish-yellow (drying pink to cream); banner obcordate to obtuse (4–7 mm long, 3–5 mm wide), claw cuneate (about 1.5 mm long); wings oblong to obovate (3–5 mm long), claws linear (about 1.5 mm long); keel oblong (3–6 mm long), curved upward; stamens 10 (about 5.5 mm long), monadelphous; anthers alternating subglobose and oblong; style glabrous or pubescent. FRUITS: loments obliquely ovate (3–6 mm long, 2–3 mm wide), longer than the persistent calyx, slightly compressed, terminated by a short curved or coiled style, minutely pubescent, veins reticulate; segments 1 or 2, indehiscent, lowest segment usually sterile, stalklike; upper segment plump, puberulent; seeds usually 1. SEEDS: variable, sometimes triangular (2–2.5 mm long, about 1.5 mm wide), slightly compressed, smooth, yellow to tan, indurate. $2n=20$.

HABITAT: Pencilflower is scattered to locally common in dry, rocky or sandy soils of prairies, open woodlands, glades, clearings, disturbed areas, and roadsides. It is most common in sandy or rocky soils. It grows well in acidic soils. USES AND VALUES: It is palatable and is grazed by cattle, horses, and sheep. It is not an important forage species in the Great Plains because the plants are small and produce only a small amount of forage. The loments and seeds are eaten by ground-foraging birds and small mammals. Bees are the primary pollinators. It has little value for erosion control or landscaping. ETHNOBOTANY: Some Cherokees used an infusion of the roots to promote menstruation.

SYNONYMS: *Arachis aprica* Walter, *Stylosanthes elatior* Sw., *S. floridana* S.F. Blake, *S. hispida* Michx., *S. riparia* Kearney, *Trifolium biflorum* L. OTHER COMMON NAMES: decumbent pencilflower, sidebeak pencilflower.

Tephrosia Pers.

tephros (Gk.): ash-colored, referring to the gray pubescence.

Herbaceous perennials, pubescent, more or less woody at the base; leaves odd-pinnately compound, leaflets several to many, not glandular-punctate; inflorescences racemes or pseudoracemes, usually terminal; flowers several, papilionaceous; calyx tube campanulate to hemispheric, slightly oblique; lobes 4, lanceolate, lower 3 longer than the tube; upper lobe consisting of a partially fused pair, slightly shorter than lower lobes; corollas yellow to purplish-rose; petals all clawed; banner subround, sericeous on back; wings and keel connivent, broadly auriculate on upper side above the short claw; wings ovate-oblong; keels semicircular; stamens 10, upper 1 at least partially free; fruits legumes, oblong to linear, laterally compressed, sessile, dehiscent, valves papery, generally impressed between the seeds; seeds few to several. $x=11$.

About 400 species occur in tropical and subtropical regions in both hemispheres. Species are most abundant in Africa, and several grow in the southern and southeastern United States. Two grow in the Great Plains, and 1 is common.

Goats rue
Tephrosia virginiana (L.) Pers.

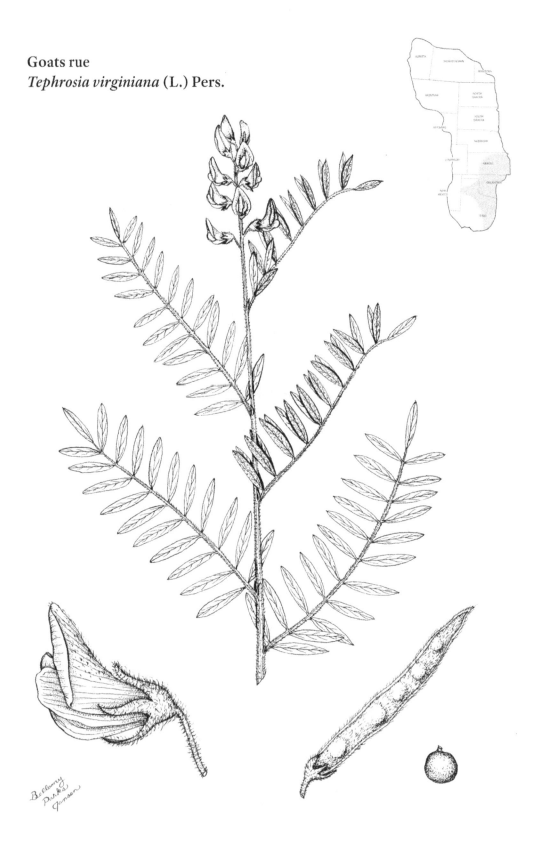

272

Synopsis Plantarum 2(2):329. 1807.
virginiana: of or from Virginia.

GROWTH FORM: forb, flowers May to July, reproduces from seeds. LIFE-SPAN: perennial. ORIGIN: native. HEIGHT: 20–70 cm. STEMS: herbaceous, ascending to erect; 1 to several from a woody caudex, villous to glabrous, not glandular-punctate. LEAVES: alternate, odd-pinnately compound (6–15 cm long), nearly sessile, not extending into tendrils; leaflets 9–31, elliptic to linear-oblong (1–3 cm long, 4–10 mm wide); apex acute to obtuse, terminal leaflet mucronate; base acute or obtuse; margins entire; upper surface sparsely appressed-pubescent to glabrate; lower surface densely silky, gray; petiole usually shorter than the lowermost leaflets; stipules linear or lanceolate to narrowly elliptic (8–10 mm long), caducous. INFLORESCENCES: racemes (4–10 cm long) terminal, compact, simple (rarely branched), villous; flowers 8–40; peduncle short (to 4 cm). FLOWERS: perfect, papilionaceous; calyx tube campanulate (3–5 mm long), densely strigose to villous; lobes 5, triangular or lanceolate-acuminate, longer than the tube; corolla bicolored; banner broadly ovate (1.4–2.1 cm long), lemon-yellow on the outside, cream inside, sericeous on back, claw 2–3 mm long; wings oblong to obovate (1.5–2 cm long), rose, claw 2–3 mm long; keel apex yellow and rose-striped (1.4–1.5 cm long), claw 2–3 mm long; stamens 10, at least upper stamen partially free; pedicels 4–18 mm long. FRUITS: legumes narrowly oblong to linear (2–6 cm long), straight or slightly curved, laterally compressed, sparsely to densely strigose; seeds usually 5–11. SEEDS: round to reniform (3–5 mm long, 4–5 mm wide), somewhat compressed, yellowish-brown to green or olive, mottled with black, white membrane readily flaking loose from the smooth surface, indurate. $2n=22$.

HABITAT: Goats rue is infrequent to locally abundant on dry prairies, rangelands, fields, open woodlands, and dunes. It is most frequent on sandy soils on sites that are periodically burned. USES AND VALUES: Goats rue is palatable to livestock. It decreases in continuously grazed areas. Deer eat the foliage, and upland birds and small mammals eat the seeds. It has limited potential for erosion control on sandy soils. It has been planted in borders and native plant gardens. ETHNOBOTANY: Goats rue contains rotenone. Some Plains Indians placed crushed plants in streams or ponds to harvest fish. Rotenone decreases oxygen exchange at the gills, and the fish float to the surface, where they are easily caught. Members of some tribes used crushed roots as an insecticide. It was also used to treat pulmonary problems, rheumatism, and hair loss.

SYNONYMS: *Cracca holosericea* (Nutt.) Britten & Baker f., *C. latidens* Small, *C. leucosericea* Rydb., *C. mohrii* Rydb., *C. virginiana* L., *Galega virginica* (L.) J.F. Gmel., *Tephrosia holosericea* Nutt., *T. latidens* (Small) Standl., *T. leucosericea* (Rydb.) Cory, *T. virginica* Bigelow. OTHER COMMON NAMES: catgut, devil's shoestring, distai'yĭ (Cherokee), hoary pea, rabbit pea, turkey pea, Virginia tephrosia, wild sweetpea.

SIMILAR SPECIES: *Tephrosia onobrychoides* Nutt., multibloom hairypea, is a rank and vigorous species that is often decumbent. Its corollas are white when they open and then darken to red or purple. It grows in the southeastern Great Plains.

Thermopsis R. Br.

thermos (Gk.): lupine; + *opsis* (Gk.): appearance, in reference to the similar appearance of the plants to those of the genus *Lupinus*.

Herbaceous perennials from rhizomes or a caudex topping a taproot; leaves pinnately or palmately trifoliate; leaflets broadly ovate to oblanceolate, margins entire; stipules foliaceous, persistent; inflorescences racemes, terminal; flowers perfect, papilionaceous; calyx campanulate; lobes 4, bilabiate, upper lip shallowly lobed or merely notched, lower lip deeply 3-lobed; corolla yellow; petals nearly equal in length; banner suborbicular; wings and keel oblong; stamens 10, distinct; ovary sessile or short-stipitate; fruits legumes, linear, curved or straight, laterally compressed, sessile or short-stipitate; seeds few to numerous. $x=9$.

About 10 species have been described in temperate North America and eastern Asia. Two grow in the eastern states, another is restricted to the western states, and only 1 is common in the Great Plains.

Goldenpea

Thermopsis rhombifolia (Nutt. *ex* Pursh) Richardson

Narrative of a Journey to the Shores of the Polar Sea 737. 1823.
rhombos (Gk.): 4-sided figure whose opposite sides and angles are equal; +*folium* (Lat.): leaf, in reference to the shape of the leaflets.

GROWTH FORM: forb, flowers April to July, reproduces from rhizomes and seeds. LIFE-SPAN: perennial. ORIGIN: native. HEIGHT: 10–70 cm. STEMS: herbaceous, erect to ascending, singly or in clusters from rhizomes, grooved, appressed-pubescent to glabrate. LEAVES: alternate, palmately trifoliate (or appearing palmately trifoliate); leaflets obovate to elliptic (2–5 cm long, 2–5 mm wide); apex acute to acuminate; base acute to attenuate; margins entire, ciliate; upper surface usually glabrous; lower surface usually appressed-pubescent; petioles 1.1–3.2 cm long; stipules foliaceous, broadly ovate to ovate-lanceolate, persistent. INFLORESCENCES: racemes subterminal (3–10 cm long); loosely flowered; flowers 10–20. FLOWERS: perfect, papilionaceous, sweet-pea-scented; calyx tube campanulate (4–5 mm long), somewhat gibbous, pubescent, lobes generally shorter than the tube; bilabiate, upper lobe with 2 united teeth; corolla yellow; petals nearly equal in length; banner suborbicular (1.6–2.1 cm long), may have purple dots; wings and keel oblong; stamens 10, distinct; pedicels 4–10 mm long. FRUITS: legumes linear (5–10 cm long, 5–8 mm wide), arcuate to falcate or curved in a half-circle, laterally compressed, coriaceous, pubescent to glabrous at maturity, short-stipitate; seeds 3–15. SEEDS: reniform (3–5.5 mm long), green or olive to black, smooth, indurate. $2n=18$.

HABITAT: Goldenpea is infrequent to common on dry plains, prairies, rangelands, open woodlands, and badlands. It often grows in sandy soils on roadsides and railroad rights-of-way. USES AND VALUES: Goldenpea has little to no forage value for livestock. It contains toxic quinolizidine alkaloids. Livestock losses are rarely reported because animals rarely eat the plants. Sheep have been nonfatally poisoned by the seeds. Pronghorn, elk, and deer lightly graze the plants without being poisoned. Circumstantial evidence links goldenpea with poisoning children. The legumes and seeds are eaten by small mammals, and the seeds are eaten by ground-foraging birds. It is pollinated by various bees. Its deep roots and rhizomes make it potentially valuable for erosion control. It is used in landscaping because of its showy flowers and attractive foliage. ETHNOBOTANY: Some Lakotas used the flowers to treat inflammatory rheumatism. Flowers were mixed with hair and burned producing smoke that was caught under a robe or blanket and inhaled.

VARIETIES: This species has been divided into several varieties. Var. *rhombifolia* grows in the Great Plains. Most of the others grow west of the region. SYNONYMS: *Cytisus rhombifolius* Nutt. *ex* Pursh, *Drepilia rhombifolia* (Nutt. *ex* Pursh) Raf., *Scolobus rhombifolius* (Nutt. *ex* Pursh) Raf., *Thermia rhombifolia* (Nutt. *ex* Pursh) Nutt., *Thermopsis angustata* Greene, *T. annulocarpa* A. Nelson, *T. arenosa* A. Nelson, *T. divaricarpa* A. Nelson, *T. montana* Nutt., *T. pinetorum* Greene, *T. stricta* Greene, *T. xylorhiza* A. Nelson. OTHER COMMON NAMES: bush pea, false lupine, golden banner, prairie buckbean, prairie goldenpea, yellow pea, yellow wildpea.

Trifolium L.

tres (Lat.): three; +*folium* (Lat.): leaf, in reference to the compound leaves with 3 leaflets.

Annual, biennial, or perennial herbs with weak stems, erect to decumbent, sometimes stoloniferous; leaves palmately or pinnately trifoliate, or appearing so; leaflets rarely more than 3; leaflets elliptic to oblong or ovate to obovate; terminal leaflet sessile or stalked; margins serrulate or denticulate; surfaces pubescent or glabrous; petioles well-developed; stipules adnate to the petioles; inflorescences globose heads, racemes, spikes, or umbels; capitate or spicate, terminal or axillary, sessile or peduncled; flowers perfect, papilionaceous; calyx campanulate to tubular, sometimes oblique, usually bilabiate; lobes triangular to setaceous, often unequal; corollas yellow, white, pink, rose, red, or purple; petals separate to more or less united into a tube; banner ovate to oblong or obovate, folded about the wings in bud, reflexed in flower; wings more or less hooked over the keel; stamens 10, diadelphous; fruits legumes, short, straight, membranous, often enclosed in the persistent calyx, indehiscent; seeds 1–6. x=8 (5, 6, 7).

About 300 species have been described. Most are in the northern temperate zone. Twenty grow in the Great Plains, but only 7 are common.

1. Corollas yellow. . . . . *Trifolium campestre*
1. Corollas white, purple, red, or pink
  2. Flowers sessile or with pedicels 1 mm long or less
    3. Plants perennial (occasionally biennial)
      4. Plants with stolons, rooting at the nodes; heads not subtended by a pair of reduced leaves. . . . . *Trifolium fragiferum*
      4. Plants without stolons; erect to decumbent; heads subtended by a pair of reduced leaves. . . . . *Trifolium pratense*
    3. Plants annual. . . . . *Trifolium incarnatum*
  2. Flowers with pedicels 2 mm long or more
    5. Plants ascending to decumbent; with stolons, rooting at the nodes. . . . .
. . . . *Trifolium repens*
    5. Plants erect to ascending; stems not stoloniferous nor rooting at the nodes
      6. Calyx lobes more than 2 times longer than the tube; tube nerves 10. . . . .
. . . . *Trifolium reflexum*
      6. Calyx lobes less than 2 times longer than the tube; tube nerves 5. . . . .
. . . . *Trifolium hybridum*

Hop clover
Trifolium campestre Schreb.

Deutschlands Flora oder Botanisches Taschenbuch 1(4):16, pl. 253. 1804. *campestre* (Lat.): of fields, in reference to its habitat.

GROWTH FORM: forb, flowers May to September, reproduces from seeds. LIFE-SPAN: annual or winter annual. ORIGIN: introduced (through Europe from the Mediterranean region). HEIGHT: 10–40 cm. STEMS: herbaceous, ascending to decumbent from a taproot, not rooting at the nodes, simple to much-branched, glabrate to finely appressed-pubescent, green to reddish-green. LEAVES: alternate, appearing as pinnately trifoliate; leaflets obovate to oblanceolate or elliptic (5–15 mm long, 3–8 mm wide); apex obtuse to retuse; base acute; margins denticulate above the middle; surfaces glabrous to sparsely pubescent; terminal petiolule longer (1–3 mm) than the lateral ones (0.2–0.5 mm long); petioles 1–3 cm long below, only 1–2 mm above, glabrous; stipules ovate-lanceolate (5–8 mm long), adnate to the petiole for half their length. INFLORESCENCES: racemes globose to short-cylindric (6–15 mm long, 6–10 mm wide), headlike, axillary; flowers 18–36; peduncles (5–90 mm long) longer than the subtending leaves. FLOWERS: perfect, papilionaceous; calyx tube campanulate (0.5–1 mm long), membranous, glabrate or with a few villosulous hairs, nerves 5; lobes 5, long-acuminate or bristlelike, unequal, lower lobe 1–1.5 mm long, upper lobes reduced; corolla bright to pale yellow (rarely purple-tinged), drying light brown; banner prominently veined (3.5–4.5 mm long, 2–4 mm wide), longer than the wings and keel; stamens 10, diadelphous; pedicels 0.2–0.8 mm long. FRUITS: legumes oblong to ovate (2–3 mm long), exserted from the calyx tube, stipe 1 mm long; seeds 1. SEEDS: ovate (about 1 mm long), slightly compressed, smooth, orangish-yellow to brown, indurate. $2n=14$.

HABITAT: Hop clover is infrequent to locally common where it has escaped from planted areas to prairies, savannas, pastures, roadsides, lawns, disturbed areas, abandoned fields, and open woodlands. USES AND VALUES: Hop clover is planted for forage and as a cover crop to improve soil properties. It is palatable to all classes of livestock. Deer frequently graze the foliage and flowers. It is pollinated by various kinds of bees. Small butterflies, skippers, and bee flies visit the flowers as well. Caterpillars of numerous moths and sulphur butterflies (*Colias* spp.) feed on the foliage. Upland gamebirds feed on the foliage, inflorescences, and fruits. Since it is an annual, it has limited application for erosion control. It is not used in landscaping. ETHNOBOTANY: In Europe leaves were eaten in salads, and the leaves and flowers were used to make tea. Seeds were collected in autumn, eaten raw or roasted, or ground into flour.

VARIETIES: This species has been divided into 4 varieties. SYNONYMS: *Chrysaspis campestris* (Schreb.) Desv., *Trifolium agrarium* L., *T. procumbens* L. OTHER COMMON NAMES: dwarf hopclover, low hopclover, field clover, plains clover, small hopclover, small hoptrefoil.

SIMILAR SPECIES: *Trifolium dubium* Sibth., least hopclover or yellow sucklingclover, also is an annual with yellow corollas. It usually has only 5–18 flowers per inflorescence. Its petioles are shorter than the leaflets. It grows in the southeastern Great Plains.

Strawberry clover
Trifolium fragiferum L.

Species Plantarum 2:772. 1753.

fragum (Lat.): strawberry; + *fer* (Lat.): bearing, in reference to the strawberry-like appearance of the inflorescences.

GROWTH FORM: forb, flowers May to September, reproduces from stolons and seeds. LIFE-SPAN: perennial. ORIGIN: introduced (through Europe from the Middle East). LENGTH: 10–40 cm. STEMS: herbaceous, decumbent, creeping and rooting at the nodes, glabrate to pubescent. LEAVES: alternate, palmately trifoliate; leaflets broadly elliptical to obovate (1–2.5 cm long); apex retuse; base obtuse; margins serrulate; surfaces glabrous; veins are distinctive as they meet the leaf margin at nearly right angles; petioles 3–20 cm long; stipules lanceolate to subulate (1.5–2 cm long), papery at the base, adnate to the petioles, overlapping, sheathing the stem. INFLORESCENCES: spikes globose to cylindric (1–1.5 cm wide at anthesis, 1.5–3 cm wide in fruit), axillary; surrounded by whorled bracts (3–5 mm long), lowest bracts united to form an involucre; peduncles 4–20 cm long, longer than the leaves, usually pubescent. FLOWERS: perfect, papilionaceous, subsessile; calyx tube reticulate-veined (1.5–2.2 mm long at anthesis), dorsally pilose; lobes 5, bilabiate, subequal, equaling the tube in fruit, green, nerves striate, upper lip inflated in fruit; corolla pink to pinkish-white (rarely red), persistent; banner 6–8 mm long; stamens 10, diadelphous; usually sessile. FRUITS: legumes ovoid (1.8–2.5 mm long), inflated; seeds 1–3. SEEDS: smooth (about 2 mm long), dull, green to dark brown, indurate. $2n=16$.

HABITAT: Strawberry clover is infrequently planted on wet saline and alkaline soils in the western Great Plains and has escaped to disturbed areas, roadsides, lawns, irrigation ditches, wet meadows, and pastures. USES AND VALUES: It produces excellent forage for all classes of livestock, but it can cause bloat. It is planted mainly for pastures and in mixtures with grasses for hay. It will withstand flooding for extended periods. It is grazed by deer, elk, bighorn sheep, and pronghorn. Its seeds are eaten by ground-foraging birds and small mammals. It is pollinated by various bees, and the flowers are visited by numerous insects.

VARIETIES: This species has been divided into several varieties. Var. *fragiferum* is most common in the Great Plains. SYNONYMS: *Galearia fragifera* (L.) C. Presl, *G. fragifera* Bobrov. OTHER COMMON NAMES: strawberry-head clover.

SIMILAR SPECIES: *Trifolium dasyphyllum* Torr. & A. Gray, alpine clover, has ascending flowers on short pedicels. The calyx is 5–7 mm long. The banner is white to ochroleucous, and the other petals are violet or tipped with violet. Generally growing at higher altitudes, it has moved into the extreme western part of the Great Plains. *Trifolium subterraneum* L., subterranean clover, has been planted for forage in the western part of the region. It has two kinds of yellowish-white to purple or purple-striate flowers. The outer flowers are petaliferous and fertile, and the inner flowers are usually sterile. *Trifolium wormskioldii* Lehm., cows clover, is rhizomatous and has an involucre with fused bracts. The petals are purplish-pink with white tips. It grows on the southwestern edge of the Great Plains.

Alsike clover
Trifolium hybridum L.

Species Plantarum 2:766–67. 1753.

hybrida (Lat.): mongrel or crossbreed, because it was once thought to be a cross between two species of clover.

GROWTH FORM: forb, flowers May to October, reproduces from seeds. LIFE-SPAN: perennial. ORIGIN: introduced (from Europe). HEIGHT: 20–80 cm. STEMS: herbaceous, erect to ascending, numerous from taproots, somewhat cespitose, glabrous to sparsely pubescent. LEAVES: alternate, palmately trifoliate; leaflets oval to elliptic (1.5–3.5 cm long, 8–25 mm wide); apex obtuse to retuse; base obtuse; margins denticulate above, serrate below; surfaces glabrous; petiolules about 1 mm long; petioles slender (3–15 cm long below, only 1–3 cm long above); stipules ovate-lanceolate (1–3 cm long), tapering to an attenuate apex, adnate to the petiole for up to half their length, veins 3. INFLORESCENCES: racemes globose to ovoid (1.5–3.5 cm in diameter), numerous, not involucrate; flowers 30–80; peduncles 2–12 cm long, equaling or exceeding the subtending leaf. FLOWERS: perfect, papilionaceous; calyx tube campanulate (1.5–2.5 mm long), nerves 5; lobes 5, linear-subulate, exceeding the length of the tube, less than 2 times longer than the tube, slightly unequal, glabrous to glabrate; corolla white to pinkish-white (drying brown to tan); banner obovate-oblong (5–11 mm long), 2–3 mm longer than the wings and keel; stamens 10, diadelphous; pedicels about 2 mm long at anthesis, expanding to 5–7 mm long in fruit. FRUITS: legumes linear-oblong to reniform (3–4 mm long), walls thin, exserted from the calyx tube, indehiscent; seeds usually 2–4. SEEDS: cordiform (about 1.5 mm long), smooth, brown and mottled with red or yellow to greenish-black, compressed, indurate. 2*n*=16.

HABITAT: Alsike clover has escaped from plantings, usually in wet meadows, to roadsides, waste ground, fields, pastures, lawns, and stream valleys. It grows well on poorly drained acidic soils. USES AND VALUES: It is used for hay, pasture, and soil improvement and provides good forage for all classes of livestock. However, it may cause bloat. Large quantities in the diets have been reported to cause photosensitivity in horses, cattle, and sheep. The condition is called trifoliosis and is most prevalent in horses. Deer, elk, pronghorn, and rabbits eat the foliage. Ground-foraging birds and small mammals eat the seeds. It is pollinated by various bees and is a good honey plant. Caterpillars of some butterflies and moths feed on the foliage. It has little potential for erosion control and landscaping.

VARIETIES: Several varieties have been described. SYNONYM: *Amoria hybrida* (L.) C. Presl. OTHER COMMON NAMES: Alsatian clover, honey clover, hybrid clover, Swedish clover.

SIMILAR SPECIES: In *Trifolium bejariense* Moric., Bejar clover, calyces are reticulately veined, and the corollas are white to greenish-white. It grows in Texas in the extreme southeastern Great Plains. *Trifolium longipes* var. *reflexum* A. Nelson, longstalk clover, is rhizomatous and has whitish to ochroleucous corollas. It grows on the western edge of the region in Colorado and Wyoming. *Trifolium nanum* Torr., dwarf clover, is diminutive and mat-forming. Its calyx has 10 strong nerves. Its corollas are lavender to purple, and the wings and keel are sometimes pink.

Crimson clover
Trifolium incarnatum L.

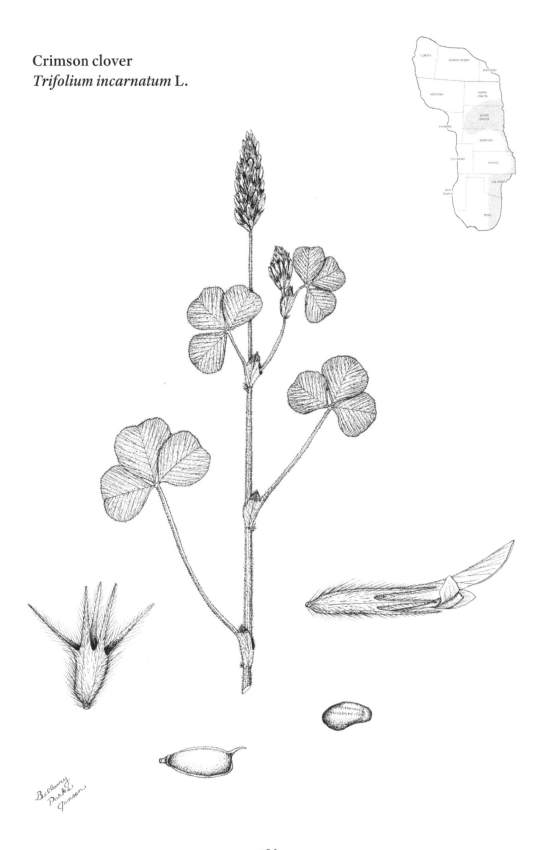

Species Plantarum 2:769. 1753.
incarnatus (Lat.): fleshlike, in reference to the color of the corollas.

GROWTH FORM: forb, flowers May to August, reproduces from seeds. LIFE-SPAN: annual or winter annual. ORIGIN: introduced (from southern Europe). HEIGHT: 20–80 cm. STEMS: herbaceous, erect, branching or rarely simple from a taproot, appressed-pubescent. LEAVES: alternate, palmately trifoliate; leaflets broadly obovate (1–4 cm long, 8–35 mm wide); apex rounded or truncate; base obtuse to acute; upper half of the margins erose to denticulate; surfaces pubescent; lower petioles 5–20 cm long, upper petioles 5–30 mm long, densely pubescent; stipules irregularly shaped (1–2 cm long), mostly adnate to the petioles, erose, may be red to purple near the summit. INFLORESCENCES: spikes ovoid to cylindric (2–7 cm long, 1–2.5 cm in diameter), solitary, terminal, not involucrate; flowers 75–125, subsessile; peduncles erect (4–12 cm long). FLOWERS: perfect, papilionaceous; calyx tube cylindrical to campanulate (2.6–3.3 mm long), densely villous, nerves 10; lobes 5, linear-subulate, about equal, equaling or longer than the tube; corolla crimson (rarely white); banner oblanceolate, obtuse, 2–5 mm longer than the wings and keel; stamens 10, diadelphous; usually sessile. FRUITS: legumes ovoid (3–4 mm long), sessile, enclosed in the calyx tube; seeds 1. SEEDS: ovoid (2–2.5 mm long), shiny, smooth, yellow to reddish-brown, indurate. $2n=14$.

HABITAT: Crimson clover is planted for pasture and hay. It has escaped locally to old fields, disturbed areas, waste areas, and roadsides. USES AND VALUES: Crimson clover is used for forage and provides good-quality pasture, hay, and silage. It is palatable and nutritious to all classes of livestock. It is one of the earliest forage legumes available in spring. It causes minor bloat or no bloat in ruminants. Hay from mature crimson clover may be dangerous to horses. The stiff, wiry hairs become feltlike balls that may block the intestines. It is grazed by deer and eaten by rabbits and wild turkeys. The legumes are eaten by songbirds, gamebirds, and small mammals. It is pollinated by bees and is a good honey plant. It is also used as a cover crop and green manure crop. This annual plant is not used for erosion control. It is occasionally used in landscaping. ETHNOBOTANY: The seeds can be sprouted and eaten in salads or dried and ground into flour.

OTHER COMMON NAMES: carnation clover, French clover, German clover, great haresfoot, Italian clover, Napoleon clover, scarlet clover.

SIMILAR SPECIES: *Trifolium arvense* L., rabbitfoot clover, is an annual species with a glabrous or uniformly pubescent calyx tube. Its corollas are rose to pink or white and is found in a few scattered location in the eastern Great Plains. *Trifolium resupinatum* L., Persian clover, is an annual species with pink to purple corollas. The lower lip of the calyx is glabrous, and the upper lip is densely pubescent. It is rare in scattered locations and is occasionally a contaminant in lawn seed. *Trifolium vesiculosum* Savi, arrowleaf clover, is another annual, but its corollas are white to ochroleucous (turning pink). Its calyx tubes are multistriate and glabrate to sparsely villous. It is found in the southeastern Great Plains.

Red clover
Trifolium pratense L.

Species Plantarum 2:768. 1753.

pratensis (Lat.): growing in a meadow, in reference to its habitat.

GROWTH FORM: forb, flowers May to September, reproduces from seeds. LIFE-SPAN: biennial or short-lived perennial. ORIGIN: introduced (from southern Europe). HEIGHT: 0.2–1 m. STEMS: herbaceous, erect to decumbent, 1 to several from a long taproot, cespitose, simple or short-branched from the base, thinly pubescent. LEAVES: alternate, palmately trifoliate; leaflets ovate or broadly elliptic (2–5 cm long, 1–3.5 cm wide); apex obtuse to retuse; base rounded to obtuse; margins finely serrate on youngest leaflets, older leaflets entire, ciliate; upper surface with a light green crescent or an inverted V (fading on drying); lower surface thinly pubescent to pilose; petiolules 1–1.5 mm long; lower leaves with long petioles (to 9 cm), upper leaves sessile or with short petioles; stipules ovate to ovate-lanceolate (1–3 cm long), adnate to the petiole for one-half or three-fourths of their length, veins conspicuous, membranous, distinct portion abruptly narrowed to a short bristle. INFLORESCENCES: heads globose to round-ovoid (1.5–3.5 cm long, 2–4 cm wide), mostly terminal, subtended by a pair of reduced globose to ovoid leaves; flowers 40–120. FLOWERS: perfect, papilionaceous; calyx tube campanulate (3–4 mm long), membranous, pilose to glabrous, nerves 10; lobes 5, long-acuminate from a triangular base (longest 4–8 mm long, others 2–5 mm long); corollas reddish-purple to purplish-pink (1.2–2 cm long); banner obovate-oblong, not prominently nerved, equaling or slightly exceeding and enveloping the wings and keel; stamens 10, diadelphous; usually sessile. FRUITS: legumes ovoid to oblong (2–3 mm long, 1.5–2 mm wide), thickened above; seeds usually 1. SEEDS: reniform to ellipsoid with a slight lateral lobe (1.5–2 mm long), smooth, yellowish-brown or green mottled with purple, indurate. 2n=14.

HABITAT: Red clover has spread from seeded fields to lawns, roadsides, pastures, disturbed areas, and waste areas. USES AND VALUES: It produces high quality hay that is palatable to all classes of livestock. Grazing green foliage may cause bloat. Second cutting or late season hay may produce a syndrome characterized by slobbering in cattle, horses, and sheep. It may progress into bloating, stiffness of gait, decreased milk flow, and diarrhea. Symptoms stop soon after the animals eat a different kind of hay. Red clover has been reported to cause photosensitization. It is grazed by several species of wildlife. Legumes are eaten by songbirds, gamebirds, and small mammals. It has little value for erosion control or landscaping. ETHNOBOTANY: Dried flowers were used in past centuries in Europe to prepare medicine used to treat whooping cough and ulcers.

VARIETIES: This species has been divided into many varieties and subspecies. Var. *sativum* Schreb. is most common. OTHER COMMON NAMES: beebread, broadleaf clover, early redclover, giant clover, honeysuckle clover, mammoth clover, meadow clover, zigzag clover.

SIMILAR SPECIES: *Trifolium beckwithii* Brewer *ex* S. Watson, Beckwith clover, is a perennial with reddish to light purple corollas. It is glabrous and is found in the Great Plains only in a few counties in South Dakota and Montana.

Buffalo clover
Trifolium reflexum L.

Species Plantarum 2:766. 1753.

flexus (Lat.): bend, in reference to the deflexed pedicels.

GROWTH FORM: forb, flowers May to August, reproduces from seeds. LIFE-SPAN: biennial or short-lived perennial. ORIGIN: native. HEIGHT: 20–50 cm. STEMS: herbaceous, erect to ascending; 1 to several from a taproot, not stoloniferous, branched from the base, glabrous to finely villous. LEAVES: alternate, palmately trifoliate; leaflets obovate to oblanceolate or broadly elliptic (1–4 cm long, 8–15 mm wide); apex obtuse or occasionally retuse; base acute or short-acuminate; margins finely serrate; surfaces pubescent to glabrate; upper surface marked with a purple band or blotch; petioles 2–14 cm long; stipules lanceolate to ovate (1–3.5 cm long), foliaceous, adnate to the petiole for one-fourth their length. INFLORESCENCES: umbels globose (2–4 cm in diameter at anthesis), headlike, terminal and axillary; flowers 10–60; peduncle 2–10 cm long, pilose to glabrous. FLOWERS: perfect, papilionaceous; calyx tube broadly campanulate (1.2–1.7 mm long), membranous, glabrous to short-pubescent; nerves 10, strong; lobes 5, linear to long-acuminate (4–7 mm long), 4 equal and the lowest slightly shorter; corollas bicolored; banner broadly elliptic-ovate (8–12 mm long, 5–6 mm wide), pink to reddish-pink or reddish-purple, slightly exceeding the wings and keel, prominently veined, glabrous, slightly erose; wings pink or creamy-white, connate to the keel to about half their length; keel pink or creamy-white; stamens 10, diadelphous; pedicels slender (4–9 mm long in flower, 6–13 mm long in fruit), deflexed or arcuate. FRUITS: legumes oblong to obovate (body 3–5 mm long, stipe 1–1.5 mm long), inflated; seeds usually 1–3. SEEDS: reniform to nearly round (1–1.5 mm long, about 1 mm wide), smooth, yellowish-brown, indurate. $2n=16$.

HABITAT: Buffalo clover is scattered and not common in upland prairies, open woodlands, and gravelly stream valleys. It usually grows in acidic soils and grows best in drier rather than moist soils. Its abundance and distribution has decreased in the last century. Fire suppression is suggested as a significant cause for the decline. It is on the endangered/threatened species lists in some states east of the Great Plains. USES AND VALUES: It is palatable and nutritious to livestock and wildlife. It is not considered to be an important constituent of their diets because it is not abundant. Its legumes and seeds are eaten by ground-foraging birds and small mammals. Buffalo clover is pollinated by long-tongued bees, and butterflies visit the flowers to drink the nectar. It has little potential for erosion control. It has been used as a border or specimen plant in landscaping.

VARIETIES: Based largely on the form of the calyx and pubescence, this species has been divided into four varieties. SYNONYMS: *Amoria reflexa* (L.) C. Presl, *Trifolium adscendens* Hornem., *T. platycephalum* Bisch.

SIMILAR SPECIES: *Trifolium carolinianum* Michx., Carolina clover, is similar in appearance, except that the corollas are yellowish-white or purple and the inflorescences are only 1–1.5 cm in diameter. It grows in the southeastern Great Plains, and its range partially overlaps with that of *Trifolium reflexum*.

White clover
Trifolium repens L.

Species Plantarum 2:767. 1753.
repens (Lat.): creeping, in reference to the stems.

GROWTH FORM: forb, flowers May to October, reproduces from stolons and seeds. LIFE-SPAN: perennial. ORIGIN: introduced from Europe. LENGTH: 5–40 cm. STEMS: herbaceous, ascending to decumbent, from fibrous roots, stoloniferous, rooting at the nodes, mat-forming, glabrous to slightly pubescent. LEAVES: alternate, palmately trifoliate; leaflets broadly elliptic to obovate (1–3.5 cm long, 5–20 mm wide); apex rounded or obtuse to retuse; base obtuse to acute; margins serrulate to denticulate; upper surface glabrous; lower surface puberulent along the veins; petioles 3–20 cm long; stipules lanceolate (3–10 mm long), white, membranous, adnate to the petiole for nearly their whole length, bases united into a membranous sheath. INFLORESCENCES: racemes globose (1–3 cm in diameter), axillary, without involucres; flowers 20–90; peduncles 5–20 cm long, longer than the leaves. FLOWERS: perfect, papilionaceous; calyx campanulate to cylindric, tube 2–4 mm long, whitish, glabrous, nerves 10; lobes 5, narrowly triangular (1–2 mm long), slightly unequal, acuminate, glabrous, sometimes purplish; corolla 7–12 mm long, initially white or tinged with pink, becoming pinkish; banner elliptic-obovate, obtuse at the summit, not conspicuously veined, exceeding the wings; wings obtuse; stamens 10, diadelphous; pedicels about 2 mm long at anthesis, elongating to 5–7 mm long in fruit. FRUITS: legumes oblong to linear (3–6 mm long), constricted between the seeds; seeds usually 2–4. SEEDS: cordiform (1–1.4 mm long), smooth, yellow or tan to brown, indurate. $2n=32$.

HABITAT: White clover is common in lawns, pastures, fields, roadsides, disturbed areas, and waste areas. It grows in all soil textures but is most common in silt and clay. USES AND VALUES: It has excellent palatability and forage quality for all classes of livestock, but yields are generally low in the Great Plains. Bloat may occur in cattle grazing lush white clover. It may contain cyanogenic glycosides, but loss of life is seldom attributed to hydrocyanic poisoning. Deer, elk, pronghorn, rabbits, and small mammals eat the foliage. The legumes and seeds are eaten by ground-foraging birds and small mammals. White clover is pollinated by various bees, and it is an excellent honey plant. It is valuable for erosion control and improvement of soil properties. Several cultivars are available for seeding. It is sometimes used for ground cover in landscaping. ETHNOBOTANY: An eyewash was made in Europe by steeping the flowers, and an infusion of dried leaves was taken for cold and coughs.

VARIETIES: It has been divided into many varieties. SYNONYMS: *Amoria repens* C. Presl, *Lotodes repens* (L.) Kuntze, *Trifolium limonium* Phil., *T. stipitatum* Clos. OTHER COMMON NAMES: Dutch clover, honeysuckle clover, ladino clover, wild whiteclover, white shamrock, white trefoil.

SIMILAR SPECIES: *Trifolium stoloniferum* Muhl., running buffaloclover, is also a creeping plant that roots at the nodes. Its calyx lobes are generally more than 2 times as long as the tube. The corolla is white and tinged with purple. It is rare in the Great Plains and can only be found scattered in the eastern part of the region.

Vicia L.

vicia (Lat.): ancient name of vetch.

Annual, winter annual, biennial, or short-lived perennial herbs; stems usually decumbent, commonly climbing or trailing; leaves even-pinnately compound, leaflets 2–24; terminal leaflets in most species modified into tendrils; petioles obsolete or to 3 mm long; stipules herbaceous, small, persistent, often semisagittate; inflorescences racemes, axillary; pedicels short; calyx regular or irregular, campanulate, often gibbous at the base, lobes 5; flowers perfect, papilionaceous; corolla purple, blue, white, or yellow; banner obovate to subround, apically notched, claw broad and overlapping the wings; wings oblong or narrowly obovate, coherent with the keel petals, usually exceeding the keel; stamens 10, diadelphous; ovary sessile or short-stipitate; style filiform, bearded with a tuft or ring of hairs only at the apex; fruits legumes, flat to terete, dehiscent, valves 2, sometimes transversely separate between the seeds; seeds 2 to several. x=7 (5, 6).

A genus of about 150 species mostly in Eurasia and temperate North America. Nine are found in the Great Plains, and 3 are common.

1. Peduncles obsolete; stipules not dentate or hastate, with a basal gland. ..... *Vicia sativa*
1. Peduncles 2–9 cm long; stipules dentate or hastate, without a basal gland
   2. Flowers 10–60; calyx strongly gibbous; stipules with a single hastate lobe; leaflets 16–24. .... *Vicia villosa*
   2. Flowers 10 or fewer; calyx slightly gibbous; stipules coarsely dentate; leaflets 8–16.
   .... *Vicia americana*

American vetch
Vicia americana Muhl. *ex* Willd.

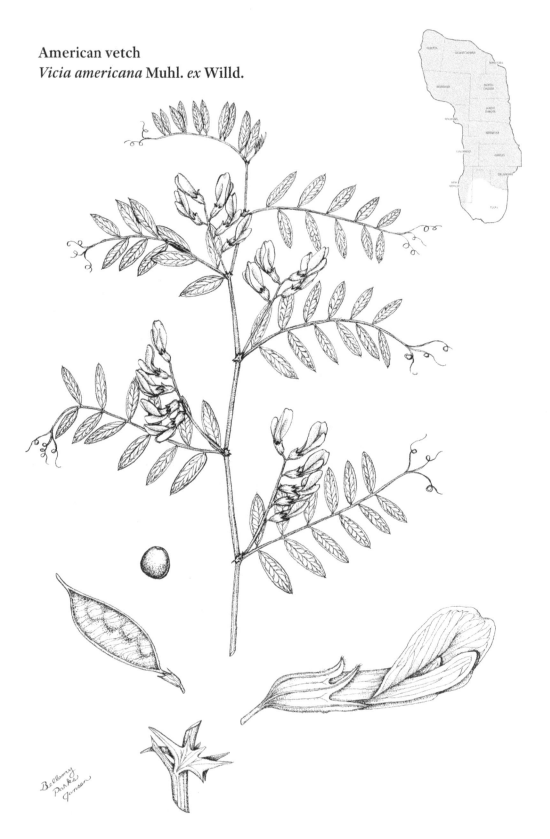

Species Plantarum. Editio quarta 3(2):1096. 1802.
americana (Lat.): of or from America.

GROWTH FORM: forb, flowers May to August, reproduces from rhizomes and seeds. LIFE-SPAN: perennial. ORIGIN: native. LENGTH: 0.2–1.1 m. STEMS: herbaceous, sprawling to scrambling or climbing, glabrous to sparsely pubescent. LEAVES: alternate, even-pinnately compound, terminating in tendrils, leaflets 8–16; leaflets narrowly oblong or elliptic to linear (1–4 cm long, 3–11 mm wide), coriaceous; veins conspicuous, not branched; tendrils branched, unbranched, or reduced to bristles; apex obtuse or truncate, mucronate; base obtuse to acute; margins entire (rarely toothed); surfaces glabrous or glabrate; stipules coarsely dentate (8–10 mm long), often appearing starlike. INFLORESCENCES: racemes (1–5 cm long), usually shorter than the subtending leaves, axillary; flowers 2–10; peduncle 2–6 cm long. FLOWERS: perfect, papilionaceous; calyx tube slightly gibbous at the base (3.5–6 mm long); lobes 5, variable, unequal; lowest lobe lance-attenuate (1–4 mm long), longest; upper pair of lobes short, broad; corolla bluish-purple (rarely white); banner 1.2–2.5 cm long, longer than the wings and keel. FRUITS: legumes obliquely elliptic (2.5–4 cm long), laterally compressed, glabrous; seeds 2–12. SEEDS: nearly round to oblong (2.5–3.5 mm long), slightly compressed, dark brown, indurate. $2n=14$.

HABITAT: American vetch is infrequent to locally common in upland prairies, rangelands, badlands, roadsides, and waste places. USES AND VALUES: It provides excellent forage and is palatable to all classes of livestock. It decreases and may disappear with continued heavy use. It has infrequently been linked to cases of photosensitization. Deer, pronghorn, elk, and rabbits eat the herbage. While occasionally planted for ground cover, it has little potential for landscaping. ETHNOBOTANY: Some Lakotas boiled and ate the young shoots and leaves.

VARIETIES: This species has been divided into 10–15 varieties and subspecies. Two varieties grow in the Great Plains. Var. *americana* stems have branched tendrils while var. *minor* Hook. has shorter stems and unbranched or weakly branched tendrils. SYNONYMS: *Abacosa americana* Alef., *Lathyrus linearis* Nutt., *Orobus diffusus* Nutt., *Vicia acicularis* Greene, *V. caespitosa* A. Nelson, *V. californica* Greene, *V. copelandii* Eastw., *V. oregana* Nutt., *V. perangusta* Greene, *V. pumila* A. Heller, *V. vexillaris* Greene, and about 15 additional synonyms. OTHER COMMON NAMES: buffalo pea, narrowleaf vetch, peavine, stiffleaf vetch, tasúsu (Lakota), wild sweetpea, wild vetch.

SIMILAR SPECIES: *Vicia caroliniana* Walter, Carolina vetch, is a rhizomatous perennial with white to lavender corollas. Its legumes are 1.5–3.5 cm long. It grows in the southern Great Plains. *Vicia cracca* L., bird vetch, is a perennial and is infrequent in the northern Great Plains. Its bluish-purple flowers are only 9–12 mm long, and the stipules are entire. *Vicia ludoviciana* Nutt. *ex* Torr. & A. Gray, Louisiana vetch, is an annual and grows in the southern Great Plains. Its legumes (1.5–3.5 cm) are asymmetrically acute-tipped, and the racemes contain 2–12 flowers. *Vicia minutiflora* D. Dietr., pygmyflower vetch, leaves have only 2–4 leaflets with small, entire stipules. It grows in the southeastern Great Plains.

Common vetch
Vicia sativa L.

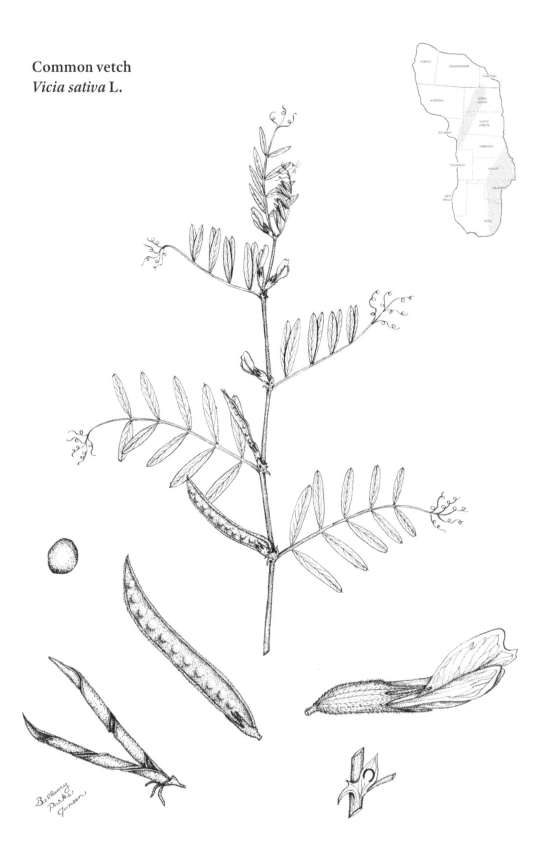

Species Plantarum 2:736. 1753.

sativus (Lat.): that which is sown, in reference to the cultivation of this species.

GROWTH FORM: forb, flowers June to September, reproduces from seeds. LIFE-SPAN: annual. ORIGIN: introduced (from Eurasia). LENGTH: 30–90 cm. STEMS: herbaceous, erect to decumbent, climbing, branching occasionally, ribbed, glabrous to sparsely pubescent or villous. LEAVES: alternate, even-pinnately compound (2–12 cm long), terminal leaflets modified into a branched tendril; leaflets 8–16, oblong to elliptic or obovate to linear (1.4–3.2 cm long, 2–8 mm wide); apex rounded to truncate or emarginate, often apiculate; base obtuse to cuneate; margins entire; upper surface glabrous; lower surface glabrescent to strigose along the midvein; petiole short; stipules variable (5–7 mm long); basal gland small, dark. INFLORESCENCES: racemes, loose, axillary; flowers often solitary or in pairs; peduncles obsolete. FLOWERS: perfect, papilionaceous; calyx 7–12 mm long, tube weakly gibbous at base; corolla bright violet to purple or rarely white (1.2–1.8 cm long); banner often with a patch of white at the base; stamens 10, diadelphous. FRUITS: legumes (2.5–7 cm long, 4–12 mm wide), glabrous or glabrescent, constricted between the seeds, brown or black when mature, valves coiling at dehiscence; seeds usually 4–12. SEEDS: oblong to round (2.5–3.5 mm long), not strongly compressed, brown and often mottled, indurate. $2n=14$ (10, 12).

HABITAT: Common vetch is planted in fields but has escaped to other cropland, fallow fields, roadsides, weedy meadows, disturbed sites, and waste areas. It grows in a broad range of soils. USES AND VALUES: It is planted as a source of forage for livestock. Fresh herbage is palatable and readily eaten by cattle, horses, sheep, deer, rabbits, ground hogs, small mammals, and gamebirds. Common vetch hay quality is good to excellent. The seeds are potentially toxic but are eaten without apparent harm by gamebirds, songbirds, and small mammals. The exact lethal dose is unknown, but possibly as little as a handful of seed could kill a cow. If eaten in sufficient quantities, the herbage can cause bloat in ruminants. It is often planted in mixtures with grasses or cereal grains and has been used as a cover crop and a green manure crop. Since it is an annual, its value for erosion control is limited. It has few applications in landscaping. The flowers are pollinated by bumblebees and other long-tongued bees. Caterpillars and skipper larvae feed on the foliage as do leafhoppers and thrips. Ants are attracted to nectar in the gland at the base of the stipules. ETHNOBOTANY: Fresh seeds can be cooked and eaten. However, they are not very palatable, and caution is advised because of the potential toxicity.

VARIETIES: This species has been divided into about 30 varieties and subspecies. SYNONYMS: *Vicia alba* Moench, *V. bacla* Moench, *V. bobartii* E. Forst., *V. canadensis* Zuccagni, *V. communis* Rouy, *V. cuneata* Gren. & Godr., *V. glabra* Schleich., *V. incisa* M. Bieb., *V. maculata* C. Presl, *V. nemoralis* Boreau, *V. subterranea* Gérard *ex* Dorthes, *V. vulgaris* Uspensky, and about 15 additional synonyms. OTHER COMMON NAMES: garden vetch, narrowleaved vetch, spring vetch, subterranean vetch, tare.

Hairy vetch
Vicia villosa Roth

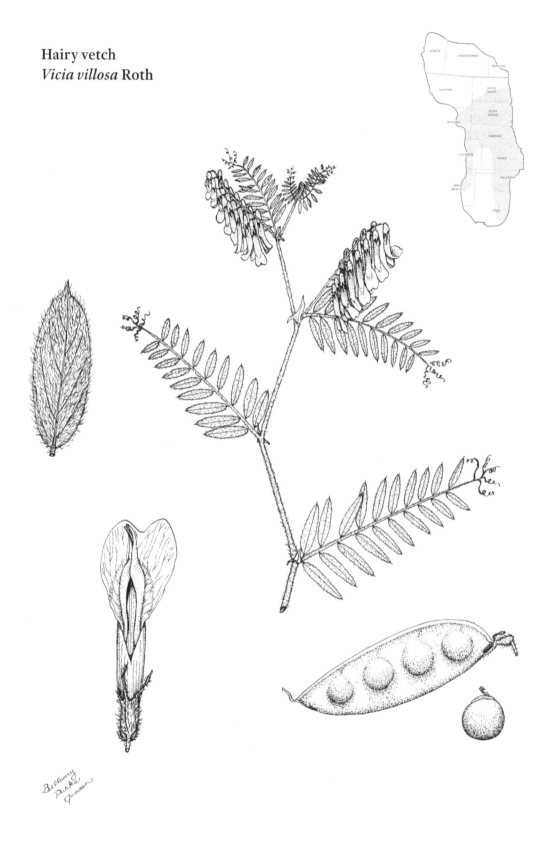

Tentamen Florae Germanicae 2(2):182–83. 1793.
villosus (Lat.): hairy, in reference to the pubescence on the stems, leaves, and peduncles.

GROWTH FORM: forb, flowers April to September, reproduces from seeds. LIFE-SPAN: annual (sometimes biennial). ORIGIN: introduced (from Europe). LENGTH: 0.5–2 m. STEMS: herbaceous, decumbent to ascending, twining or climbing from a taproot, villous. LEAVES: alternate, even-pinnately compound (6–15 cm long); leaflets opposite or alternate along the rachis; leaflets 16–24, narrowly oblong to linear-lanceolate (1–2.5 cm long, 3–7 mm wide), final pair modified into tendrils; tendrils much-branched; apex obtuse to acute, mucronate; base rounded to cuneate; margins entire; surfaces usually softly villous, hairs 1–2 mm long; petioles short; stipules lanceolate to ovate (5–10 mm long), each with a single hastate basal lobe. INFLORESCENCES: racemes axillary, flowers 10–60, dense, secund; peduncle 3–9 cm long, softly villous. FLOW-ERS: perfect, papilionaceous; calyx tube irregular (2–4 mm long), strongly gibbous at the base, villous; lobes 5, upper lobes linear-triangular (0.8–1.5 mm long), lateral and lower lobes linear above triangular bases (1–5 mm long); corolla bluish-purple to violet (rarely white), drying blue (1.2–1.8 cm long); banner oblong (7 mm wide, less than half as long as the claw), spreading; pedicel appears to be laterally inserted. FRUITS: legumes oblong (2–4 cm long, 7–10 mm wide), compressed, glabrous to pubescent; seeds usually 3–5. SEEDS: globose (4–5 mm in diameter), reddish-brown to black. $2n=14$.

HABITAT: Hairy vetch is planted in fields and on roadsides. It has escaped to other fields, waste places, disturbed sites, meadows, and valleys. It is most common on sandy soils and can tolerate moderate alkalinity. USES AND VALUES: It is planted for hay or pasture and is usually sown with a support crop of a cereal grain. Hay quality is good to excellent. Numerous species of wildlife utilize the forage and seeds. Hyperproteinemia is an excessive level of protein in the blood plasma or serum and has been associated with cattle grazing hairy vetch. Itchy and scaly skin, diarrhea, conjunctivitis, and body wasting are signs of poisoning, but they may not appear for several weeks after the start of grazing. Although rare, it can be fatal. Hairy vetch is pollinated by bumblebees and other long-tongued bees. It is also planted for winter soil cover and as a green manure crop. It has little value for landscaping.

VARIETIES: This variable species has been divided into about 20 varieties and subspecies. SYNONYMS: *Cracca dasycarpa* Alef., *C. varia* Godr. & Gren., *C. villosa* (Roth) Godr. & Gren., *Ervum villosum* (Roth) Trautv., *Vicia dasycarpa* Ten., *V. glabrescens* A. Kern., *V. godronii* A.W. Hill, *V. littoralis* Salzm., *V. plumosa* Martrin-Donos, *V. pseudocracca* Bertol., *V. reuteriana* Boiss. & Buhse, *V. varia* Host, and about 15 additional synonyms. OTHER COMMON NAMES: cow vetch, winter vetch, woollypod vetch.

SIMILAR SPECIES: *Vicia exigua* Nutt., little vetch, and *Vicia leavenworthii* Torr. & A. Gray, Leavenworth vetch, are winter annuals growing in the southern Great Plains. They are highly variable and difficult to separate from *Vicia villosa*.

III. *MIMOSACEAE* R. BR.

Mostly trees, shrubs, woody vines, or less commonly herbs; stems armed or unarmed; leaves alternate, usually even-bipinnately compound or less commonly pinnately compound; pinnae and leaflets few to many, leaflets sometimes touch-sensitive; petiole and rachis bearing 1 to several glands, glands depressed or stalked; inflorescences densely flowered globose heads, cylindrical spikes, umbels, or rarely racemes, pedunculated from axils; flowers regular, perfect or rarely unisexual, usually small; calyx usually gamosepalous; lobes 5, minute; corolla gamopetalous or polypetalous, valvate in bud, petals 5 (rarely 4), inconspicuous; stamens usually 10 or multiples thereof, sometimes 5 (rarely 1), distinct or united basally, filaments usually long-exserted; fruits legumes, straight to curved or spirally twisted, valves usually 2, dehiscent, occasionally transversely separated between the seeds.

A family of about 40 genera and at least 2,000 species worldwide. They are found chiefly in the tropics and warm temperate areas. The family is represented in the Great Plains by 5 genera and 12 species.

Key to the Genera

1. Shrubs or small trees
   2. Stamens many; terminal pair of leaflets shaped like the others. .... *Acacia*
   2. Stamens 10 or fewer; terminal pair of leaflets curved. .... *Prosopis*
1. Herbs
   3. Stems unarmed; corollas white to greenish-white; petals free. .... *Desmanthus*
   3. Stems armed or prickly; corollas yellow or pink; petals partially fused at the base
      4. Corollas pink; anthers eglandular; legumes narrowly oblong. .... *Mimosa*
      4. Corollas yellow; anthers topped with a caducous gland; legumes broadly oblong.
      .... *Neptunia*

Acacia Mill.

acacia (Lat. from Gk.): a thorny Egyptian tree.

Shrubs or small trees, armed or unarmed with stipular prickles or spines; leaves alternate, bipinnate (in North American species); terminal pair of leaves shaped like the others; inflorescences racemes, spikes, or globose heads, terminal or axillary; calyx campanulate, variably lobed; corolla creamy-white, yellow, or orange, often fragrant; petals united into a tube or not; stamens many, distinct or united at the base into fascicles; style linear; stigma punctiform; fruits legumes, diverse in texture, laterally compressed to turgid, curved or straight, dehiscent or indehiscent; seeds few to several or many. x=13.

A genus of more than 800 species. Most are tropical in both hemispheres. Four grow in the extreme southern Great Plains, and 2 are common.

1. Inflorescences globose heads or headlike spikes; pinnae 7–12 pairs; prickles slightly curved to nearly straight (1–3 mm long). .... *Acacia berlandieri*
1. Inflorescences oblong spikes or racemes; pinnae 1–3 pairs; prickles recurved like cats' claws (5–9 mm long). .... *Acacia greggii*

Guajillo
Acacia berlandieri Benth.

London Journal of Botany 1:522. 1842.

berlandieri: named after Jean-Louis Berlandier (1803–51), French Mexican naturalist, physician, and anthropologist.

GROWTH FORM: shrub, semievergreen, flowers February to April, reproduces from seeds. LIFE-SPAN: perennial. ORIGIN: native. HEIGHT: to 5 m. TWIGS: ascending, several, branching from the base, sparingly branched above, flaring out widely near the top, gray to white, internodes striate, generally armed with prickles (1–3 mm long); prickles scattered on the internodes, slightly curved to nearly straight. LEAVES: alternate, even-bipinnately compound (9–18 cm long), delicate, nearly fernlike; pinnae 7–12 pairs; leaflets 15–80 pairs per pinna, crowded, linear to oblong (2–6 mm long, 0.5–1 mm wide), slightly asymmetric; apex acute; base rounded to obtuse; margins entire; surfaces tomentose when immature, glabrate when mature, veins prominent; petioles bearing a sessile gland; stipules small, usually caducous, not spinescent. INFLORESCENCES: globose heads or headlike spikes (8–15 mm in diameter), rarely elongated, axillary, solitary or in clusters; flowers many; peduncles 2–5 cm long, pubescent. FLOWERS: perfect, regular, fragrant; calyx campanulate, lobes 5; petals 5; corolla creamy-white to yellow, pubescent; stamens numerous, exserted, distinct. FRUITS: legumes laterally compressed (8–15 cm long, 1.5–2.5 cm wide), thin, straight or somewhat curved, apex obtuse or apiculate, surfaces velvety-tomentose when mature, green when immature and becoming dark brown, margins thickened, one margin generally straighter than the other, tardily dehiscent; seeds several. SEEDS: oval to spherical, slightly compressed, reddish-brown to brown, smooth, indurate. $2n=26$.

HABITAT: Guajillo grows on limestone ridges and caliche hills of rangelands and prairies. It is most abundant in sandy soils and is uncommon in deep soils. It is more widespread south of the Great Plains. USES AND VALUES: In most cases, the herbage is fair for livestock and wildlife. However, it contains more than 30 alkaloids and trace amounts of several amphetamines. Total alkaloid content of dried leaves has been recorded to be in the range of 0.3–0.7 percent. Wilted leaves may cause hydrocyanic poisoning of livestock. Seeds furnish important food for small mammals and birds. It pollinated by various bees and is an important honey plant. Guajillo adds a soft, lush quality to landscapes and mixes well with a variety of desert trees and shrubs. It is semievergreen, making it a complement to deciduous trees. It can be planted with taller trees to create a landscape screen, and its smaller stature makes it a good accent tree. ETHNOBOTANY: Gums and dyes have been extracted from guajillo.

SYNONYMS: *Senegalia berlandieri* (Benth.) Britton & Rose. OTHER COMMON NAMES: Berlandier acacia, guajillo acacia, huajillo, catclaw mimosa, round-flowered catclaw.

SIMILAR SPECIES: *Acacia angustissima* (Mill.) Kuntze, prairie acacia, grows in the southern Great Plains as far north as Kansas. It has heads (1–1.5 cm in diameter) of white flowers and leaves with 10–20 pairs of pinnae. Its stems have no prickles. *Acacia roemeriana* Scheele, roundflower catclaw, has heads (about 1 cm in diameter) of creamy-yellow flowers, leaves with 1–4 pinnae, and internodal prickles. It is scattered in the southwestern Great Plains.

Catclaw acacia
Acacia greggii A. Gray

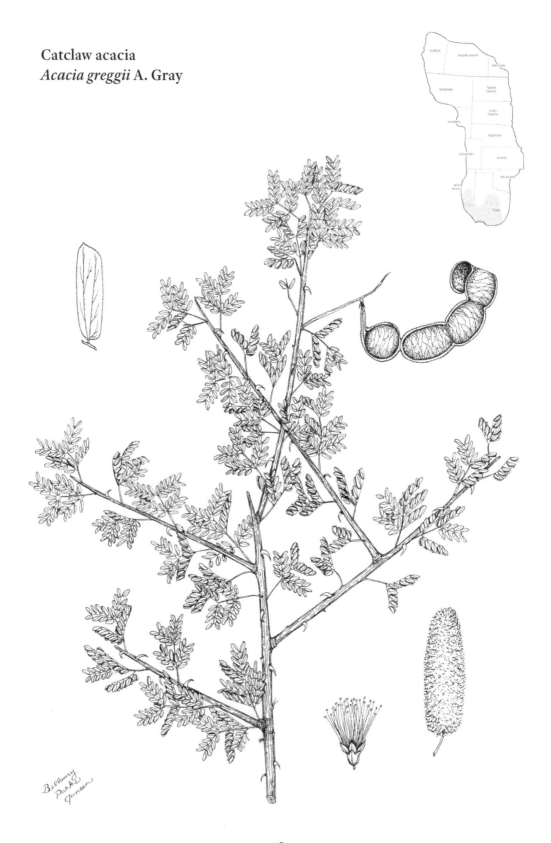

Smithsonian Contributions to Knowledge 3(5):65. 1852.

greggii: named after Josiah Gregg (1806–50); author, explorer, and amateur naturalist who worked in the American Southwest and northern Mexico.

GROWTH FORM: shrub or small tree, flowers April to October, reproduces from seeds. LIFE-SPAN: perennial. ORIGIN: native. HEIGHT: usually 1–2 m (rarely to 10 m). TWIGS: ascending, much-branched, creating thickets, forming rounded crowns, pale or reddish-brown to gray; prickles 5–9 mm long, usually along the internodes, occasionally at a node; prickles recurved, similar in appearance to cats' claws, stout, dark brown or gray. LEAVES: alternate, even-bipinnately compound; pinnae 1–3 pairs (2–3 cm long), most commonly 2 pairs with the lower pair nearly perpendicular to the petiolule and the upper pair forming a V; leaflets 3–7 pairs per pinna, crowded, obovate to narrowly oblong (2–7 mm long, 0.8–1.8 mm wide); apex obtuse; base unequal, asymmetrical; margins entire; surfaces lightly pubescent, lightly reticulate, green or grayish-green. INFLORESCENCES: spikes or racemes, oblong (2–6 cm long, 1 cm wide), axillary, solitary or paired; flowers many. FLOWERS: perfect, regular, fragrant; calyx campanulate; lobes 5, obscure (2–3 mm long), puberulent; petals 5, creamy-white; stamens numerous, long-exserted. FRUITS: legumes thin (5–8 cm long, 1.5–2 cm wide), curved or often curled and contorted, usually flexible, tardily becoming rigid, margins thickened, light brown to reddish-brown, compressed, constricted between the seeds; seeds usually 3–5. SEEDS: nearly oval, compressed, indurate. $2n=26$.

HABITAT: Catclaw acacia grows in dry arroyos, ravines, mesas, washes, and canyon slopes of rangelands. It is most abundant in dry, coarse-textured soils. USES AND VALUES: Catclaw acacia provides poor forage for livestock and wildlife. The foliage is seldom browsed. The prickles may cause injury to soft tissues of browsing animals. It contains a potentially dangerous cyanogenic glycoside. Seeds are an important food for many species of wildlife, and it is one of the main foods for some species of quail and small mammals. It provides important cover for many kinds of wildlife. Catclaw acacia is an important nectar source for honeybees. Plants are available from garden centers either as single-trunked or multiple-trunked forms. Catclaw acacia can be planted as a drought-tolerant accent or perimeter tree. It is the host plant for several species of insects that produce lac, a material formerly used in varnish and shellac. ETHNOBOTANY: Great Plains Indians avoided the potentially dangerous ripe catclaw acacia beans and collected unripe beans to be boiled or eaten raw. Paiutes, Cahuillas, Pimas, and others dried the beans and ground them for mush and cakes, while the Havasupais ground the beans to make flour for bread. Wood was used for construction, firewood, and bows. Twigs were split and used for basketry and brooms.

SYNONYMS: *Acacia durandiana* Buckley, *Senegalia greggii* (A. Gray) Britton & Rose. OTHER COMMON NAMES: catclaw, devilsclaw acacia, Gregg catclaw, long-flowered catclaw, paradise flower, Texas mimosa, wait-a-minute bush.

Desmanthus Willd.

desme (Gk.): a bundle; + *anthos* (Gk.): flower, referring to the dense heads.

Perennial herbs (or occasionally shrubs); stems numerous from a woody caudex; erect or spreading, unarmed, leaves bipinnately compound, glands present between the pinnae pairs or only at the junction of the lowermost pinnae; pinnae and leaflets small, few to many; stipules acicular to subulate, generally persistent; inflorescences heads, solitary, axillary, long-peduncled; calyx tube campanulate to short-cylindric, lobes 5; corolla white to greenish-white; petals 5, free; stamens 5 (or 10), usually long-exserted, free; fruits legumes, oblong to linear, straight or curved, laterally compressed, valves thinly to heavily coriaceous, dehiscent, separating along both sutures, not twisting, seeds few to several. x=14.

About 30 species have been described in tropical and subtropical America. One species is common in the Great Plains; 4 others are occasionally collected.

Illinois bundleflower
Desmanthus illinoensis (Michx.) MacMill.
ex B.L. Rob. & Fernald

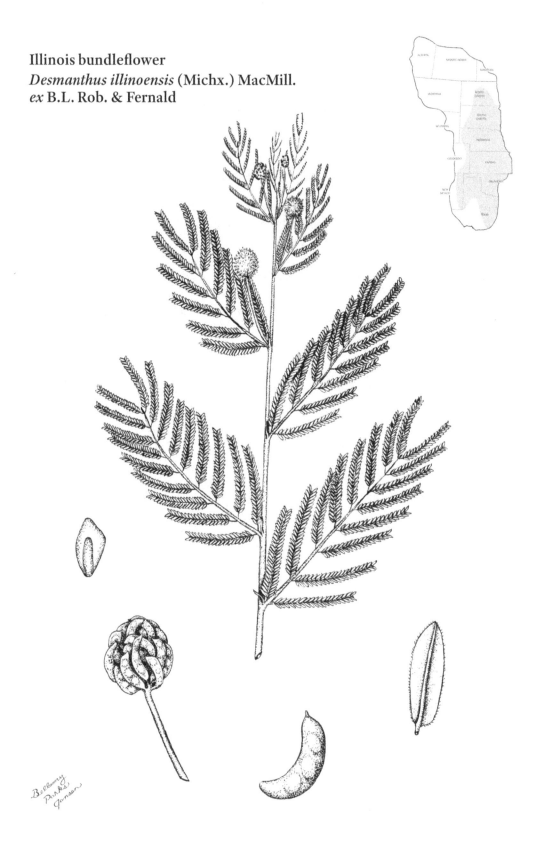

Gray's Manual of Botany, 7th ed.: 503. 1908.
illinoensis: of or from Illinois.

GROWTH FORM: forb, flowers June to August, reproduces from seeds. LIFE-SPAN: perennial.
ORIGIN: native. HEIGHT: 0.8–2 m. STEMS: herbaceous, erect to ascending or spreading,
clustered, from a woody caudex and taproot, slightly grooved, glabrous or nearly so.
LEAVES: alternate, bipinnately compound (3–10 cm long); pinnae 6–16 pairs (2–4 cm
long), gland present between the pinnae pairs or only between the lower pair; leaflets
15–30 pairs, linear to oblong (2–5 mm long, to 1 mm wide); apex obtuse; base oblique;
margins entire, ciliate; surfaces glabrous; midvein prominent; petiole 2–10 mm long;
stipules setaceous to filiform (4–10 mm long), usually persistent. INFLORESCENCES:
heads (8–16 mm wide) axillary; flowers 20–70; peduncles ascending (2–8 cm long).
FLOWERS: perfect, regular; calyx tube campanulate; lobes 5 (1 mm long); corolla white
to greenish; petals 5, oblanceolate (to 2 mm long), united to the middle, becoming
separate; stamens 5, usually long-exserted. FRUITS: legumes thin and strongly curved
(1–2.5 cm long, 4–7 mm wide), laterally compressed, numerous in globose clusters,
reddish-brown; seeds few to several. SEEDS: rhombic (3–5 mm long, nearly as wide),
compressed, yellowish-red to brown. $2n=28$.

HABITAT: Illinois bundleflower is common in dry or moist soils of prairies, rangelands,
open woodlands, ravines, waste places, and roadsides. USES AND VALUES: It is readily
eaten by deer, pronghorn, and all classes of livestock. It decreases with heavy grazing.
It is one of the most important legumes for livestock and wildlife in the Great Plains.
Its seeds are eaten by small mammals and birds. It has limited value for erosion control
and landscaping. ETHNOBOTANY: Members of the Pawnee tribe made a decoction
from the leaves to apply as a remedy for itching. The roots contain an alkaloid and
were ground and mixed with other plant components to produce a stimulating drink.

SYNONYMS: *Acuan illinoense* (Michx.) Kuntze, *Desmanthus brachylobus* Benth., *Mimosa
illinoensis* Michx. OTHER COMMON NAMES: atikatsatsiks (Pawnee), bundle-flower,
Illinois acacia, Illinois mimosa, kitsitaris (Pawnee), pezhe gagtho (Omaha-Ponca),
prairie bundleflower, prairie mimosa, prickleweed, spider bean, wild sensitiveplant.

SIMILAR SPECIES: *Desmanthus cooleyi* (Eaton) Branner & Coville, Cooley desmanthus,
has small stipules (2 mm long or less), heads 1.5–2 cm in diameter, white corollas
with yellow anthers, straight legumes (4–7 cm long), and rhombic seeds. *Desmanthus
leptolobus* Torr. & A. Gray, slenderlobed bundleflower, has longer stipules (4–6 mm
long), inconspicuous heads (less than 1 cm in diameter), whitish corollas, and straight
or slightly curved legumes (3.5–6 cm long) with a filiform beak. *Desmanthus obtusus*
S. Watson, bluntpod bundleflower, has stipules 2.5–4 mm long, heads about 1 cm in
diameter with about 10 flowers, off-white corollas, and straight legumes (2.5–4 cm
long) that are slightly constricted between the seeds. *Desmanthus velutinus* Scheele,
velvet bundleflower, stems and leaves are initially villous and subsequently glabrate.
All four grow in the southern part of the region.

Mimosa L.

mimus (Lat.): mimic, in reference to the touch-sensitive leaves.

Perennial herbs (or shrubs), stems arched, decumbent, sprawling, much-branched, glabrous with recurved prickles; prickles internodal to subnodal, paired, flattened, broadened at the base; leaves in clusters from spurs or alternate, even-bipinnately compound, petioled; pinnae 1–16 pairs; leaflets numerous, symmetric or asymmetric; stipules acicular to linear-lanceolate, inconspicuous, persistent; inflorescences heads, spikes, or racemes, axillary; peduncles 1–9 cm long, flowers many; calyx tubular to campanulate; petals connate basally or distinct, usually pink; stamens 10 or fewer, anthers eglandular; fruits legumes, narrowly oblong to linear, straight or variously curved, angled, laterally compressed, or quadrangular in cross section, dehiscent, sparsely to densely prickly or not prickly; seeds several to many. x=13.

A genus with about 500 species. Circumtropical in distribution but mainly in North and South America. One is common in the Great Plains.

Sensitive briar
Mimosa quadrivalis L.

Species Plantarum 1:522. 1753.

quadri (Lat.): having four; + *valvis* (Lat.): valves, in reference to the legumes.

GROWTH FORM: forb, flowers May to September, reproduces from seeds, forming large patches. LIFE-SPAN: perennial. ORIGIN: native. LENGTH: 0.5–2 m. STEMS: herbaceous, arched to decumbent or scrambling, glabrous, armed with numerous recurved prickles (2–4 mm long); prickles generally internodal. LEAVES: alternate, even-bipinnately compound (6–15 cm long); pinnae 4–8 pairs (2–5 cm long), eglandular; leaflets 18–30, oblong or elliptic (2–9 mm long, 1–1.5 mm wide); apex cuspidate to acute, mucronate; base rounded to truncate, asymmetric; margins entire, lightly ciliate; upper surface glabrous, midvein prominent, lateral veins anchor-shaped or arching; lower surface reticulate; stipules linear-lanceolate (3–6 mm long), glabrous, persistent. INFLORESCENCES: heads (1.5–3 cm in diameter) globose, dense, axillary; flowers many; peduncles 3–9 cm long. FLOWERS: usually perfect, sometimes uppermost staminate, regular, sessile; calyx campanulate (0.1–0.2 mm long), lobes united; corolla pink (rarely rose or purple), petals basally connate, funnelform (2.5–4 mm long), lobed at the apex; stamens 10 or fewer; filaments slender, elongate (7–10 mm long), pink, distinct or united at the base; anthers eglandular. FRUITS: legumes oblong to linear (4–12 cm long, 4–7 mm wide), clustered, valves 4, strongly ribbed, densely prickly on the ribs, beaked, not laterally compressed, not constricted between the seeds, brown to golden-brown, dehiscent; seeds 6–32. SEEDS: ovate to rhombic (3–5 mm long), variable, slightly compressed, smooth, brown to black, indurate. $2n=26$.

HABITAT: Sensitive briar grows on sterile, dry hillsides and ravines of open woodlands and rangelands. It is most abundant in sandy or gravelly soils. USES AND VALUES: Cattle and horses graze the foliage and tender branches before the prickles harden. Deer and wild turkeys eat the leaves. Ground-foraging birds and small mammals eat the seeds. It is pollinated by bees and attracts many kinds of insects. It is not used for erosion control. Plants are sometimes grown as a curiosity because the leaves fold and droop when touched. ETHNOBOTANY: The seeds have been used in laxatives.

VARIETIES: *Mimosa quadrivalvis* has been divided into about 15 varieties. Var. *nuttallii* (DC. *ex* Torr. & A. Gray) Beard *ex* Barneby is most common in the Great Plains. Var. *occidentalis* (Wooton & Standl.) Barneby, western sensitivebriar, is found in the southwestern Great Plains. It is similar, but only the midrib of each leaflet is prominent, and its seeds are quadrate (more than 5 mm long). Other varieties in the Great Plains, all growing in the southern part of the region, include *hystricina* (Small *ex* Britton & Rose) Barneby, *latidens* (Small) Barneby, *nelsonii* (Britton & Rose) Barneby, and *platycarpa* (A. Gray) Barneby. SYNONYMS: *Leptoglottis hamata* (Humb. & Bonpl. *ex* Willd.) Standl., *Mimosa tetragona* Poir., *Schrankia hamata* Humb. & Bonpl. *ex* Willd., *Schrankia quadrivalvis* (L.) Merr. OTHER COMMON NAMES: bashful briar, catclaw mimosa, catclaw sensitivebriar, Nuttall sensitivebriar, prairie mimosa, sensitive rose.

Neptunia Lour.

neptunium (Lat.): uncountable, in reference to the difficulty of counting the many flowers in a raceme.

Perennial herbs; stems decumbent from an orangish taproot; armed with soft spines; leaves alternate, bipinnately compound, leafstalk without glands; pinnae and leaflets few to numerous; stipules lanceolate, striate, persistent; inflorescences heads, short-cylindric to capitate, axillary, solitary or paired; flowers dimorphic, upper perfect, lower staminate or sterile; calyx campanulate, lobes shorter than the tube; corolla yellow to greenish-yellow, petals partially fused or nearly free; stamens 10, free anthers each topped with a caducous gland; fruits legumes, stipitate, broadly oblong, laterally compressed, somewhat asymmetric, valves thinly coriaceous to membranous, dehiscent through one or both sutures, not curving or twisting; seeds few to several. $x=14$.

About 10 species have been described in the tropics and warm regions of North and South America, Australia, Asia, and Africa. One species is common in the extreme southern Great Plains.

Yellow puff
Neptunia lutea (Leavenw.) Benth.

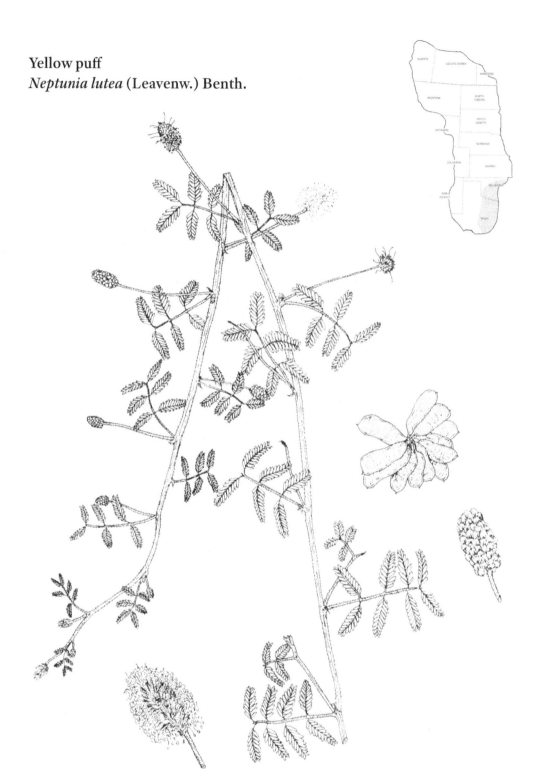

Reprinted from *Texas Range Plants* (Stephan L. Hatch and Jennifer Pluhar) by permission of Texas A & M University Press.

Journal of Botany, Being a Second Series of the Botanical Miscellany 4 (31):356. 1841.
luteus (Lat.): yellow, in reference to the color of the corollas.

GROWTH FORM: forb, flowers May to July, reproduces from seeds. LIFE-SPAN: perennial. ORIGIN: native. LENGTH: to 1.5 m. STEMS: herbaceous (sometimes appearing as slightly woody), prostrate to decumbent, few to several from a woody, orangish taproot, branching freely, hirsutulous to glabrate with soft spines. LEAVES: alternate, even-bipinnately compound; pinnae 2–10 pairs; leaflets 9–18 pairs, shortly oblong (3–4 mm long), nearly symmetric, rapidly folding to the touch, at night, and on cloudy days; apex obtuse to acute; base acute to rounded; margins entire, ciliate; upper surface glabrous to glabrate, weakly veined; lower surface glabrous to glabrate, raised reticulate venation; petioles and rachises glandless; stipules lanceolate (2–4 mm long), membranous, free, persistent. INFLORESCENCES: heads ovoid to short-cylindric (to 2.5 cm long, to 1.5 cm wide), axillary; flowers up to 60; peduncles (1–2 cm long) usually with 2 bracts; bracts subulate (1–3 mm long, 1–2 mm wide), caducous. FLOWERS: dimorphic; upper perfect, regular; lower staminate; calyx about 1 mm long, glabrous; lobes 5, longer than the tube, not conspicuously nerved; corolla yellow, petals 5 (2–3 mm long), united, not striate; stamens 10, separate, long-exserted. FRUITS: legumes broadly oblong (2.5–5 cm long, 1–1.7 cm wide), distally apiculate or rounded, somewhat asymmetrical, laterally compressed, coriaceous or becoming woody, unilocular, exserted-stipitate (4–14 mm long), freely dehiscent; seeds usually 3–10. SEEDS: ovoid but elongated, marked with an elliptical line or depression, smooth, olive to brown or black, indurate. $2n=28$.

HABITAT: Yellow puff grows in moist or dry soils of prairies, grasslands, rangelands, savannas, meadows, open brushlands, open woodlands, and roadsides. It can grow in a broad range of soils but is most common in sandy soils. It is tolerant of disturbed soils. USES AND VALUES: Palatability is low, and it furnishes only fair forage for cattle, sheep, and wildlife. Palatability and forage quality are higher for goats. It is not a toxic species. Legumes are eaten by small mammals, and the seeds are eaten by ground-foraging birds and small mammals. It is pollinated by various bees and other insects. The flowers of yellow puff are visited by many kinds of songbirds. It serves as a host to hairstreak (*Theclinae*) and other butterfly larvae. This drought-tolerant species has been used for erosion control on sandy sites. It is an attractive wildflower, but it has few applications in landscaping. It spreads quickly, and each plant can cover an area 2 m by 2 m. It is sometimes grown as a curiosity because the leaves exhibit rapid plant movement and fold when touched.

VARIETIES: Two varieties are var. *multipinnata* B.L. Turner and var. *tenuis* (Benth.) B.L. Rob. Var. *multipinnata* is most common in the Great Plains. SYNONYMS: *Acacia lutea* Leavenw., *Neptunia tenuis* Benth. OTHER COMMON NAMES: lemon acacia, neptunia, sensitive briar, yellow neptunia, yellow sensitivebriar.

Prosopis L.

prosopis (Gk.): an ancient name for burdock, an unrelated spiny plant.

Low woody shrubs or small trees; stems usually armed with straight, stout nodal spines; leaves bipinnately compound, mostly clustered from spurs; petiole with an obscure gland between the lower pinnae pairs, pinnae 1 to several pairs; leaflets several to many, narrow, terminal pair of leaflets usually curved; inflorescences cylindric spikes, spikelike racemes, or globose heads; calyx lobes 5, usually small; corollas yellow or yellowish-brown, petals 5; stamens 10 or fewer, anthers bearing apical glands; fruits legumes, long, nearly straight, usually constricted between seeds, valves thick and succulent, becoming fibrous and woody; seeds few to several. x=14.

A genus of about 35 species in drier, subtropical regions of North and South America, Africa, and Asia. Five species occur in North America, and 1 species is important in the southern Great Plains.

Honey mesquite
Prospois glandulosa Torr.

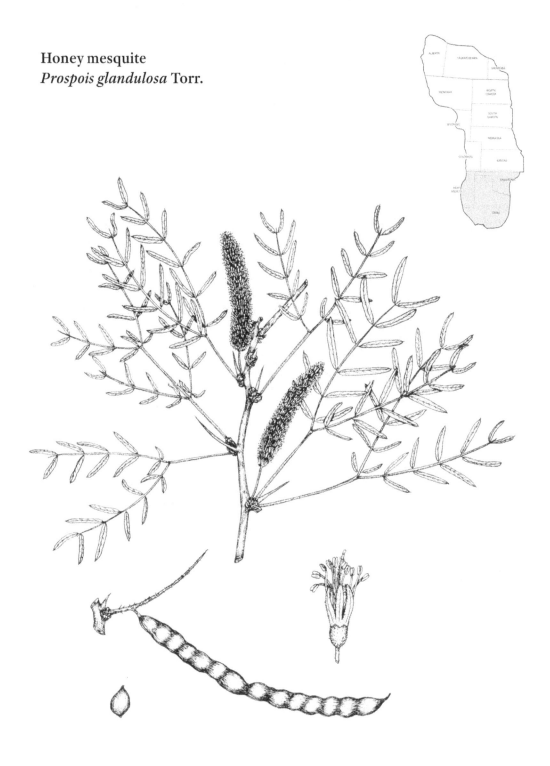

Annals of the Lyceum of Natural History of New York 2:192–93, pl. 2. 1827.
glandulosa (Lat.): with glands, in reference to glands at several locations on the plants.

GROWTH FORM: shrub or small tree, crown rounded, flowers May to July, reproduces from basal sprouts and seeds. LIFE-SPAN: perennial. ORIGIN: native. HEIGHT: to 6 m. TWIGS: zigzag, rigid, glabrous, reddish-brown or grayish-brown, usually armed with 1–2 stout spines at the nodes. TRUNKS: much-branched, single or multiple. LEAVES: deciduous, alternate, even-bipinnately compound, pinnae usually 1 pair per leaf; pinnae 6–15 cm long, each with 12–36 leaflets; leaflets linear to oblong (2–6 cm long, 2–3 mm wide), terminal pair usually curved, sessile or nearly so; apex acute, mucronate; base obtuse; margins entire; surfaces glabrous, lateral veins indistinct; petiole 5–8 cm long, circular gland on the upper side at the base of the pinnae; stipules modified into spines (rarely spineless) to 5 cm long, rigid, straight. INFLORESCENCES: racemes 7–9 cm long, spikelike, axillary, pendulous; flowers many; peduncle 1–3 cm long, glabrous; bracts 1–3, caducous. FLOWERS: perfect, regular, fragrant; calyx tube campanulate (about 1 mm long); lobes 5, triangular (0.3–0.4 mm long), finely pubescent at the tips; corolla greenish-yellow, petals 5; petals elliptic to obovate (about 3 mm long), distinct, pubescent within; stamens 10; pedicels glandular (to 0.5 mm long). FRUITS: legumes linear (10–20 cm long, about 1 cm wide), straight or nearly so, terete, slightly constricted between the seeds, valves thick, succulent and becoming woody, indehiscent; seeds 2–14. SEEDS: ovoid (6–7 mm long, 4–5 mm wide), compressed, tan to brown, indurate. $2n=28$.

HABITAT: Honey mesquite is common on dry, sandy or gravelly pastures, rangelands, and prairies. It increases following control of fires and abusive grazing, and it is considered a pernicious weed on abused rangeland. USES AND VALUES: It furnishes poor to good forage for cattle, horses, sheep, deer, and peccaries. It is good to excellent forage for goats. Legumes and seeds are important to many species of wildlife and serve as an emergency feed for livestock. Ingestion of large quantities of foliage may cause rumen stasis and impaction. Honey mesquite wood is used for fuel, flooring, posts, and furniture. The flowers attract bees and butterflies and are an important resource for honey. It is used occasionally as an ornamental, and it is a good soil binder. ETHNOBOTANY: The seeds, usually called beans, were pounded into flour and important in the diets of tribes in the southern Great Plains. The legumes were eaten raw or cooked. The flour was mixed with water and fermented.

VARIETIES: The three varieties described are var. *glandulosa*, var. *prostrata* Burkart, and var. *torreyana* (L.D. Bensen) M.C. Johnst. Var. *glandulosa* is most common in the Great Plains. SYNONYMS: *Algarobia glandulosa* (Torr.) Cooper, *A. glandulosa* (Torr.) Torr. & A. Gray, *Neltuma constricta* (Sarg.) Britton & Rose, *N. glandulosa* (Torr.) Britton & Rose, *N. neomexicana* Britton, *Prosopis chilensis* var. *glandulosa* (Torr.) Standl., *P. juliflora* var. *glandulosa* (Torr.) Cockerell, *P. odorata* Torr. & Frém. OTHER COMMON NAMES: glandular mesquite, mesquite, mezquite.

GLOSSARY

a- prefix meaning "without"

abaxial on the side away from the axis; lower surface

abrupt changing sharply or quickly, rather than gradually

absent not present; never developing

acaulescent stemless; without an aboveground stem or apparently so

accent plant that adds special interest to a garden or landscape by virtue of its size, shape, or color

acicular needle-shaped; acerose

acidic soil with a low pH (less than 5.5), normally interferes with the growth of most species

acuminate gradually tapering to a sharp point and forming concave sides along the apex; compare with acute

acute tapering to a pointed apex with more or less straight sides; sharply pointed; angle less than ninety degrees

adaxial on the side nearest the axis; upper surface

addicted having a compulsive need for a habit-forming substance

adnate attached or grown together; fusion of unlike parts, such as petioles and stipules

adventitious roots growing directly from an organ other than a root, such as the base of the stem; developing irregularly or accidentally

alkaline soil with a high pH (8.5 or higher) and high exchangeable sodium content (15 percent or more), normally interferes with the growth of most species

alkaloid nitrogen containing organic chemical that behaves as a base and is usually of plant origin; many alkaloids are toxic to animals

alluvial soils that have been deposited by flowing water

alternate located singly at each node; not opposite or whorled

amplexicaul said of leaves, bracts, or stipules surrounding or clasping the stem

-angled suffix meaning "a corner"

angular forming an angle; with one or more angles

annual within one year; applied to plants that do not live more than one year

anorexia loss of appetite for food

anther part of a stamen in which pollen develops

anthesis period of flowering; period during which pollination occurs

antidote remedy to counter the effects of a toxic substance

antioxidants substances that inhibit oxidation or reactions promoted by oxygen

antrorse directed upward or forward; opposed to retrorse

apetalous without petals

apex, apices tip or distal end

apical relating to the apex

apiculate ending in an abrupt, sharp point

appendage adjunct to something larger or more important; a subordinate or derivative plant part

appressed lying against an organ; flatly pressed against

arching curved like an arch

arcuate curved like a bow or arch

aristate awned or tapering; a very long, narrow apex

armed having sharp thorns or spines

aromatic fragrant or having an odor; bearing essential oils

arroyo gully or channel; a watercourse in an arid region

articulate jointed, provided with nodes; separating clearly at maturity

ascending growing or angled upward; obliquely upward

astringent able to draw together soft tissues; styptic

asymmetrical not symmetrical; not divisible into equal halves

attenuate gradually narrowing to a slender apex or base

auricle ear-shaped lobes, such as those that occur at the base of leaf blades of some legumes

auriculate shaped like an ear or earlobe

axil angle between an organ and its axis

axillary growing in an axil

axis, axes central or main longitudinal support upon which parts are attached

badlands extensive tracts of untillable, heavily eroded land with little vegetation

banner upper petal (standard) of the papilionaceous flower in the *Fabaceae*

bark exterior covering of a woody stem or root; tissues lying outside the cambium

barren unproductive sites, usually with shallow soils

basal located at or near the base of a structure, such as leaves arising from the base of the stem

base lower part

basifixed attached by the base (compare with dolabriform)

beak narrow or prolonged tip; hard point or projection (frequently the remnants of the style base)

bearded furnished with long, stiff hairs

bi- prefix meaning "two"

bicolored having two colors

bidentate having two teeth

biennial plant that lives for two years

bifid one-cleft or two-toothed; applied to the summit of petals, stipules, or leaflets

bilabiate having two lips, as in irregular flowers

bilocular having two compartments or locules in the ovary, anther, or fruit

bipinnate twice pinnate

bivalved having two valves

bladder thin-walled, inflated structure

blade the part of the leaf above the petiole or petiolule

bloat digestive disturbance of ruminants (especially cattle) marked by abdominal swelling due to a buildup of gas; potentially fatal

blotched spot or mark in an irregular shape

blowout depression in the surface of sand or sandy soil caused by wind erosion

bluff steep cliff, bank, or promontory

blunt having a point or edge that is not sharp

borne attached to or carried by

bottomland low-lying land, typically near a stream; land subject to occasional overflow

bracteate having bracts

bracteole bract borne on a secondary axis

bractlet second-order bract, usually highly reduced

bracts reduced leaves (frequently subtending flowers)

branch lateral stem

branchlet small lateral stem

bristle stiff, slender appendage

bristly covered with bristles

browse twigs, leaves, and other parts of woody plants consumed by herbivores; the act of consuming portions of woody plants

brushland area covered with shrubs

bulbous bulb-shaped; bearing bulbs

bulge swelling; an abrupt expansion

cache stockpile, hoard, or collection

caducous early deciduous; falling off early

calcareous soil containing sufficient calcium carbonate (often with magnesium carbonate) to effervesce when treated with hydrochloric acid

caliche sedimentary rock; hardened natural cement of calcium carbonate bound with other materials such as gravel, sand, silt, or clay

calyx, calyces the sepals of a flower considered collectively, usually green bracts

campanulate shaped like a bell

canescent pale or gray-colored because of a dense, fine pubescence

canyon deep cleft between cliffs or hills resulting from erosion and weathering

capillary extremely slender; hairlike

capitate aggregated into a dense cluster; headlike; caplike

carinate having one or more keels

carpel a single pistil; one unit of a compound pistil

cathartic substance causing the evacuation of the bowels

catkin flexible raceme or spike, often pendant; inflorescence crowded with imperfect flowers

caudate bearing a slender taillike projection or appendage

caudex, caudices short, usually somewhat woody, vertical stem located just below the soil surface

caulescent having a stem

cauline pertaining to the stem or belonging to the stem

cespitose tufted; several or many stems in a close tuft

chartaceous texture of writing paper and usually not green

chasmogamous flower that is open at the time of anthesis

ciliate fringed with hairs on the margins

ciliolate fringed with minute hairs on the margins

clasping one organ or tissue partially or totally wrapped around a second

claw narrowed base of some sepals and petals

cleft divided into teeth or divisions that extend halfway or more to the midvein

cleistogamous applied to flowers or florets fertilized without opening

clump dense cluster

cluster number of similar tissues or organs growing together; a bunch

coarse composed of relatively large parts; not fine-textured or fine-structured

coil one or more loops

colic abdominal pain

colony group of plants of the same species growing in close association with each other; all members of the group may have originated from a single plant

compact having a small, dense structure

compound composed of several parts united into a single structure

compressed flattened, typically laterally

comprised made up of

congested overcrowded; full

concave hollowed inward, like the inside of a bowl

conjunctivitis inflammation or swelling of the conjunctiva of the eye; pink eye

connate fusion of like parts, such as petals, to form a corolla tube

connivent converging but not actually fused

conspicuous obvious; easy to notice

constricted drawn together; appearing to be tightly held

contorted bent; twisted

contracted decreased in size

convex rounded on the surface like the bottom or exterior of a bowl

copious abundant

cordiform, cordate heart-shaped, with rounded lobes and a sinus at the base

coriaceous leathery in texture

corolla all of the petals considered collectively

corrugated surface with alternate ridges and grooves

counterirritant substance or act that overcomes an irritation

cover crop crop grown for protecting or enriching the soil

craving to want greatly; powerful desire for something

creeping continually spreading; describing a shoot or horizontal stem (usually a rhizome or stolon) that roots at the nodes

crenate having rounded teeth; scalloped margins

crenulate diminutive of crenate

crescent shape having a convex edge and a concave edge

cross rugose surface horizontally wrinkled

crowded pressed close together; a number of structures in a small space

crown persistent base of an herbaceous perennial; the shape of the foliage of a shrub or tree

crude protein measure of how much protein (expressed as a percent) is in the herbage; one measure of forage quality

cumulative increasing by successive additions

cuneate wedge-shaped; narrowly triangular with the narrow end at the point of attachment

cure to dry, as in standing herbage or hay; cause of the relief or recovery from a disease

curled formed in the shape of curves or spirals

cushion pad-shaped or pillow-shaped

cuspidate bearing a sharp, firm, and elongated point at the apex

cuttings portions of stems used to root new plants

cyanogenic capable of producing cyanide

cylindric, cylindrical shaped like a cylinder

deciduous not persistent but falling away in less than one year; caducous

declined bent downward; drooping

decoction extract obtained by boiling the plant material in water

decumbent curved upward from a horizontal or inclined base, with only the end ascending

decurrent extending downward from the point of attachment

decurved curved downward

deflexed bent or curving downward or backward

dehiscent opening at maturity along a definite suture

delicate fine structure or texture

deltate, deltoid triangular; shaped like the Greek letter delta

dense crowded

dentate pointed, coarse teeth spreading at right angles to the margin

denticulate diminutive of dentate

depressed flattened from above; pressed down

dermatitis inflammation of the skin

descending sloping downward

diadelphous stamens arranged in two groups, often unequal in number; compare to monadelphous

diffuse open and much-branched, loosely branching

digitate several members arising from one point at the summit of a support, like the fingers arising from the hand as a point of origin

dilated enlarged; expanded; widened

dimorphic two types (forms) of leaves, flowers, or other structures on the same plant

dioecious unisexual flowers on separate plants; pistillate and staminate flowers on separate plants

disseminate to disperse or spread

distal remote from the place of attachment

distinct clearly evident; separate; apart

disturbance alteration or destruction of the vegetative cover

disturbed areas in which the vegetative cover has been altered or destroyed

divaricate very widely spreading

divergent widely spreading

diverse showing variety or difference

divided separated or cut into distinct parts by inclusions extending to near the base or midvein

division one of the parts of the whole

dolabriform hatchet- or T-shaped; attached in the middle (compare with basifixed)

dorsal relating to the back of an organ; opposite the ventral side

dorsiventral having both dorsal and ventral surfaces; compressed and having distinct upper and lower surfaces (as do most leaves)

dotted marked with small spots

downy soft, fine pubescence

droop to hang downward; pendulous

dull lacking brilliance or luster; not shiny

dune mound or ridge of sand formed by the wind

dwarf less than normal size

dye pigment or other substance used to color other items such as cloth

edible fit to be eaten; consumed as food

eglandular without glands

ellipsoid circular in cross section and elliptic in long section

elliptic, elliptical shaped like an ellipse; narrowly pointed at the ends and widest in the middle

elongate narrow, the length many times the width or thickness

emarginate having a shallow notch at the tip

embedded enclosed in a supporting structure or organ

emetic substance causing vomiting

enfold to cover with or as if with folds

entire whole; with a continuous margin

erect upright; not reclining or leaning

erose irregularly notched at the apex; appearing gnawed or eroded

erosion undesirable process of soil materials being moved by wind or water

even- prefix meaning "number of structures divisible by two"

evergreen woody plants that retain their leaves throughout the year

evident obvious; distinct; easily seen

exceed greater than; larger than

expanded increased; extended

exposed open to view

exserted protruding or projecting beyond; not included

extensive wide or considerable range or spread

extract to separate or remove; material that has been separated

fading gradually growing faint

falcate sickle-shaped

fascicle small bundle or cluster

fencerow uncultivated strip of land on each side and under a fence

fertile capable of producing fruit

filament stalk of a stamen supporting the anther; threadlike structure

filiform threadlike; long and very slender

firm hard; indurate

fissure deep groove

flake thin, flattened piece or layer

flat, flattened having the major surfaces essentially parallel and distinctly greater than the minor surfaces; compressed; without a slope

fleshy pulpy; succulent; plant material with high water content

flexible capable of being easily bent or flexed; pliant

flexuous bent alternately in opposite directions; wavy form

floriferous flower-bearing

folded part or organ that is doubled over or laid over; V-shaped in cross section

foliaceous leaflike

foliage plant material that is mainly leaves

-foliate suffix meaning "pertaining to or consisting of leaflets" (i.e., *trifoliate* means that the leaves are made up of three leaflets)

foothill region at the base of a mountain range

forage herbage usually consumed by animals

forb herbaceous plants other than grasses and grasslike plants

fractured cracked; broken

fragrant having a sweet or delicate odor

free not attached to other organs

fringed having a border consisting of hairs or other structures

fruit ripened ovary (pistil); the seed-bearing organ

funnelform shaped like a funnel

furrowed bearing longitudinal grooves or channels; sulcate

fused attached

fusiform shaped like a spindle

gamopetalous petals at least partially united

gamosepalous sepals at least partially united

gaping open; wide open

geniculate bent abruptly, like a knee (stems may be bent in this manner)

gibbous swollen on one side

glabrate, glabrescent nearly glabrous or becoming so with age

glabrous without hairs

glade open space in a wooded area

gland protuberance or depression that secretes a fluid or appears to secrete a fluid

glandular supplied with glands

glaucous covered with a waxy coating that gives a bluish-green color; possessing a waxy surface that easily rubs off

globose nearly spherical in shape

glossy having a surface luster; shiny

glycoside, glucoside organic compounds that yield a sugar and another substance upon hydrolysis; may be found in plants and may be toxic to animals

graze consume growing or standing grass or forb herbage; to place animals on pastures to enable them to consume the herbage

groove long, narrow channel or depression; sulcus

gully trench eroded in the land surface by running water

habituated having become accustomed to

hastate shaped like an arrowhead with the narrow basal lobes standing out at wide angles

head dense cluster of sessile or nearly sessile flowers on a short axis; an inflorescence type

hemispheric shaped like half of a sphere

herb, herbaceous not woody; dying each year or dying back to the crown

herbage aboveground material produced by herbaceous plants; vegetation that is available for consumption by grazing animals

hirsute with straight, rather stiff hairs

hirsutulous minutely hirsute

hispid with stiff or rigid hairs; bristly hairs

hoary covered with fine gray or white pubescence

hollow unfilled space; empty

honey plant plants that furnish nectar suitable for bees to make honey

hooked curved or bent like a hook

horizontal parallel to the plane of the earth

humistrate laid flat on the surface of the soil

hyaline thin and translucent or transparent

hydrochloric acid a solution of hydrogen chloride, a gas, dissolved in water

hydrocyanic acid aqueous solution of hydrogen cyanide that is poisonous

hypanthium ring or cup around the ovary formed by a fusion of the bases of sepals, stamens, and petals

imbricate overlapping (like shingles on a roof)

impaction fixed firmly as if by packing

improper grazing animal utilization of herbaceous material and impact on the site that causes a significant decline in the plant community

included not exserted nor protruding

inconspicuous not easily seen; not evident

incurved curved toward the applicable center axis

indehiscent not opening, staying closed at maturity; not splitting

indistinct not easily seen; not sharply outlined or separable

indurate hard

inflated swollen or expanded; puffed up; bladdery

inflorescence arrangement of flowers on an axis

infra- prefix meaning "below"

infrastipular below the stipules

infructescence arrangement of fruits on an axis

infusion liquid resulting from soaking or steeping material in water without boiling

internode part of a stem between two successive nodes

interrupt to break the uniformity; to come between two similar objects or structures

intricate having many complex parts or elements

introduced not native to North America

intrude to place, thrust, or force between

inversely inverted position; upside down

involucre whorl or circles of bracts below the flower or spikelet cluster

involute rolled inward from the edges, the upper surface within

irregular asymmetrical; not equal in similar parts

joint node or point of articulation

junction place at which two structures or organs join

juvenile young; not mature or fully developed

keel sharp fold or ridge at the back of a compressed blade; the united two lower petals of a papilionaceous flower

lac resinous substance secreted by an insect that was used as glue

lacerate appearing torn at the edge or irregularly cleft several times

lanate woolly with intertwined, curly hairs

lance-linear shaped like a narrow lance head

lanceolate rather narrow, pointed and tapering to both ends, widest below the middle

landscaping process of making land more attractive usually by adding plants and other features

lateral belonging to or borne on the side

laterally compressed flattened from side to side

lax loose; open and spreading

leafless without leaves

leaflet division of a compound leaf

leafy with many leaves

legume *Fabaceae* fruit composed of a single carpel but with two sutures and dehiscing at maturity along the sutures; a pod

lenticel slightly raised, lens-shaped area on the surface of a woody stem

lenticular lens-shaped

life-span length of time a plant will live; usually expressed as *annual, biennial,* or *perennial*

linear long and narrow with parallel sides

lobe projecting part of an organ with divisions less than half the distance to the base or midvein, usually rounded or obtuse

locular having locules

locule compartment of an ovary, fruit, or anther

loment jointed, dry fruit, constricted and breaking apart between the seeds

lomentaceous having fruits like loments

longitudinal placed or occurring lengthwise

loose not arranged tightly together

lunate crescent-shaped

lustrous reflecting light evenly without glitter or sparkle

maculate blotched or mottled

margin edge; border

marsh area of wet soil characterized by herbaceous vegetation, primarily grasses and grasslike plants

mat tangled mass of plants growing close to the soil surface and generally rooting at the nodes

mature fully developed, usually including seed production

meadow moist, level lowland on which herbaceous plants dominate

membranous, membranaceous thin, opaque, not green; like a membrane

-merous suffix referring to the number of parts

mesa relatively flat-topped elevated land surface

mesic an environment with a moderate amount of moisture; not wet or dry

midvein, midrib central or principal vein of a leaf or bract

minute small

monadelphous stamens and filaments united into one group; compare to diadelphous

mottled marked with spots or blotches

mound small rounded mass projecting above the surface

mucro short, sharply pointed tip

mucronate tipped with a short, slender sharp point

mucronulate tipped with a very small mucro

mush ground meal boiled in water

naked uncovered; lacking pubescence; lacking enveloping structures

native occurring in North America before settlement by Europeans

nectar sugary fluid, usually within flowers, which encourages pollination by insects; basic material used by bees to make honey

nerve vein or rib

neuter lacking stamens and pistil; without functional sexual parts

nitrate compound of nitrogen accumulated by some plants that can cause poisoning if consumed by animals

nodding inclined somewhat from the vertical; drooping

node joint along a stem from which leaves are borne; a joint in a stem or inflorescence

nodulose minute knobs or nodules, frequently associated with infection of *Fabaceae* roots with rhizobia

notch gap; V-shaped indentation

ob- prefix meaning "inversely"

obconic, obconical inversely cone-shaped with the attachment at the broad end rather than the narrow end

obcordate inversely heart-shaped or cordate with the attachment at the point

obdeltoid inversely triangular-shaped with the attachment at the point of the triangle rather than along the side

oblanceolate inversely lanceolate with the broadest portion nearer to the apex

oblique having the axis not perpendicular to the base; neither perpendicular nor parallel

oblong longer than broad, with sides nearly equal; rounded on both sides with parallel sides

obovate opposite of ovate, with the widest part toward the far end; egg-shaped with the widest part above the middle

obovoid opposite of ovoid, with the attachment at the narrower end

obscure inconspicuous; not easily seen

obsolescent nearly obsolete

obsolete not apparent; missing; rudimentary

obtriangular inversely triangular; triangular leaf with the attachment at a pointed end

obtuse shape of an apex, with an angle greater than ninety degrees

obvious easily seen

ochroleucous yellowish-white

odd- prefix referring to a number not evenly divisible by two

opposite borne across from one another at the same node

orbiculate, orbicular nearly circular in outline

origin place where the species originally occurred

ornamental plant cultivated for its beauty rather than agronomic use

oval broadly elliptic with round ends; shorter than oblong

ovary expanded basal part of the pistil that contains the ovules

ovate, ovoid shaped like an egg, with the broadest portion closer to the base

overlap to extend over and cover part of an adjacent structure

oxalate group of chemical compounds that may cause poisoning in animals; salts of oxalic acid

paired two; together

palatable acceptable in taste and texture for consumption

pale dim; not bright; deficient in color; pallid

pallid pale

palmate three or more lobes, veins, or leaflets arising from a common point

panicle, paniculate inflorescence with a main axis and rebranched branches

papery having the texture of writing paper

papilionaceous flower type in the *Fabaceae* having a banner petal (or standard), two wing petals, and two partially fused to fused keel petals

partition structure dividing a space or organ into two parts

pasture fenced area containing standing forage harvested by grazing animals

pedicel the stalk of a spikelet or single flower in an inflorescence

pedicellate having a pedicel

pedicelled borne on a pedicel

peduncle stalk of a flower cluster or spikelet cluster

pendant, pendulous suspended or hanging downward; drooping

perennial lasting more than two years; applied to plants or plant parts that live more than two years

perfect applied to flowers having both functional stamens and pistil

perianth floral envelope consisting of the calyx and corolla (when both are present)

pernicious harmful, especially in a gradually increasing manner

perpendicular at a ninety-degree angle to another surface

persistent remaining attached

petal part or member of the corolla, often brightly colored

petalous having petals

petiolar growing from the petiole; pertaining to the petiole

petiolate with a petiole

petiole stalk of a leaf blade

petiolule stalk of a leaflet of a compound leaf

pH measure of the acidity (below 7) or alkalinity (above 7) of the soil; pH 7 is neutral

photosensitization hypersensitivity of the skin to sunlight following ingestion of photodynamic compounds from certain plants

phyllode leaflike petiole serving as the blade

pigmented colored with a substance

pilose with long, soft, straight hairs

pilosulous minutely pilose

pinna, pinnae primary division of a pinnate leaf

pinnate having two rows of lateral divisions along a main axis (like barbs of a feather)

pioneer plant first plant to colonize an area

pistil combination of the stigma, style, and ovary; the female reproductive organ of a flower

pistillate applied to flowers bearing pistils only; unisexual flowers

plains flat to rolling land usually covered with herbaceous plants and dominated by grasses

plateau extensive area of land with a relatively flat surface raised sharply above the adjacent land at least on one side; tableland

plumose having fine, featherlike elongated hairs

plump rounded; full

pod dry, dehiscent fruit with a single carpel splitting along two sutures

pollen grains borne in the anther containing the male gametophyte

pollination process of the transfer of pollen from the anther to the stigma

poultice moist, soft mass of plant material

potherb plants that are boiled before being eaten

polygamous bearing some flowers with stamens only, some with pistils only, and some with both on the same or different plants

prairie extensive tract of level to rolling land with vegetation dominated by grasses together with forbs, shrubs, and grasslike plants

predominantly mainly; for the most part

prickle small, sharp outgrowth of the bark or epidermis

primary first

primary branch branch arising directly from the main inflorescence axis or stem axis

procumbent prostrate; lying flat on the ground; trailing but not taking root

prominent readily noticeable; projecting out beyond the surface

prostrate lying flat on the ground; procumbent

pseudo- prefix meaning "false"

puberulent diminutive of "pubescent"

pubescent covered with short, soft hairs

pulp soft, succulent portion of a fruit

pulvinate cushion-shaped

punctate having dots, usually with small glandular pits

punctiform marked by or composed of dots or points

pungent sharp and penetrating odor; firm- or sharp-pointed

pustular, pustulate having small eruptions or blisters

quadrate nearly square

raceme inflorescence in which all of the spikelets or flowers are pedicelled on a rachis

racemose having the shape of a raceme without being a true raceme

rachis axis of a spike, spicate raceme, or raceme inflorescence or pinnately compound leaf

rangeland land on which the native vegetation is made up of herbaceous plants and shrubs suitable for grazing and browsing; primary resource of a ranching operation

ranked arranged in one or more vertical rows

-ranked suffix indicating rows or series

ravine narrow, steep-sided valley typically eroded by running water

recurved curved away from the apex

reduced smaller than normal; not functional

reflexed bent or turned downward abruptly

regular having structures of the flower, especially the corolla, of similar shape and equally spaced about the center of the flower; flowers with two or more planes of symmetry

reniform kidney-shaped

resinous producing any of numerous viscous substances such as resin or amber

restoration returning the contour of the land and the vegetation to its original condition

reticulate in the form of a network; netted with many leaf veins

reticulum fine network; netlike

retrorse pointing backward toward the base

retuse with a slight notch at a rounded apex

revegetation replacing current vegetation or starting vegetation on denuded land

rhizobium genus of bacteria associated with the formation of root nodules on plants

rhizomatous having rhizomes

rhizome underground stem with nodes, scalelike leaves, and short internodes

rhombic having the shape of a four-sided figure with opposite sides parallel and equal but with two of the angles oblique; diamond-shaped

rib prominent nerve or vein

rich soils fertile soils

ridge a narrow, raised band

right-of-way usually vegetated land along roads, highways, railroad tracks, pipelines, or transmission lines

rigid firm; not flexible

roadcut cut through a hill made to build a road

robust healthy; full-sized

rock garden artificial mound of soil and rocks that is planted with species requiring little moisture

root sprouts adventitious roots

rotenone substance obtained from plants that is often toxic to insects and fish

rotund nearly circular

rough not smooth; surface marked by inequalities

rudimentary underdeveloped

rugose wrinkled surface

rugulose somewhat wrinkled surface

runner stolon

sac pouch or baglike structure

saccate bag-shaped

sagittate arrowhead-shaped with the lobes turned downward

saline a nonsodic soil containing sufficient soluble salts to impair plant productivity

scaberulous slightly roughened

scabridulous minutely roughened

scabrous rough to the touch; with short, angled hairs requiring magnification for observation

scales reduced leaves at the base of a shoot or a rhizome; a thin chafflike portion of the bark of woody plants; thin, flat structures

scaly having scales

scape leafless flowering stem arising from near the surface of the soil

scapiform resembling a scape

scapose bearing a flower or flowers on a scape or on a plant part resembling a scape

scarious thin, dry, membranous, not green

scurfy covered with minute scales or specialized mealy hairs

scythe hand tool with a long, curved blade used for cutting down plants

season-long throughout one season; frequently used to describe grazing during the growing season

secund pointed or directed to one side

seed ripened ovule

seep place where water oozes slowly to the land surface

segment part of a structure which may be separated from the other parts

selenium nonmetallic element that is frequently extracted from the soil by plants, relatively large quantities may be poisonous to animals

semi- prefix meaning "half"

senescence process of aging; decline following maturity

sepal member of the calyx bracts, usually green and subtending the petals

septate divided by one or more wall or membrane

sericeous silky with soft, appressed hairs

serrate saw-toothed margins, with teeth pointing toward the apex

serrulate minutely serrate

sessile without a pedicel or stalk

setaceous with bristlelike hairs

setiform in the form of a bristle

setose covered with hairs; hispid

sheathing structure that encloses or wraps around another structure, such as the base of a stipule when it surrounds the petiole

shelterbelt linear planting of shrubs and trees to reduce wind flow and wind erosion

shiny lustrous; possessing a sheen

shoot young stem or branch

showy attractive, such as a large colorful flower; striking appearance

shrub low-growing woody plant; bush with one or more trunks

sickle tool with a semicircular blade used for cutting down plants

silage herbage harvested, compacted, and stored in airtight conditions without drying and usually used as animal feed

silky fine, lustrous, long hair; resembling silk in appearance or texture

silvery lustrous and gray or white; having the luster of silver

simple not branched; not compound; single

sinuate, sinuous strongly wavy margins

sinus indentation between two lobes or teeth segments, such as on leaf blade margins

slough place in which shallow water stands for most or all of the year

sod-forming creating a dense mat, usually associated with rhizomatous plants

solitary single; alone; one by itself

sparingly meager; not dense

sparse scattered; opposite of dense

spatulate shaped like a spatula, being broader above than below

spherical, spheroid having the shape of a sphere or ball; a globular body

spicate spikelike

spiciform shaped like a spike

spike unbranched inflorescence in which the spikelets or flowers are sessile on a rachis (central axis)

spine stiff, pointed outgrowth that is usually woody

spinescent bearing a spine; terminating in a spine

spinose having spines

split divided lengthwise

splotch blotch or spot; blending of a spot and blotch

sprout shoot of a plant, especially the first from a root or a germinating seed

spur hollow, tubular projection from a sepal or petal; outgrowth of woody tissue

stalk supporting structure of an organ

stamen pollen-producing structure of a flower; typically an anther borne at the apex of a filament

staminate flower containing only stamens; unisexual flowers

standard upper petal of the papilionaceous flower in *Fabaceae*; banner

stasis slowing down or cessation of flow or of a process

steep to soak in warm water at a temperature under the boiling point

stem portion of the plant bearing nodes, internodes, leaves, and buds

sterile without functional pistils, may or may not bear stamens

stiff not easily bent; rigid

stigma apical portion of the pistil that receives the pollen

stipe in general, a stalk or stem that supports an organ

stipel stipule of a leaflet

stipitate borne on a stalk or stipe

stipules appendages, usually leaflike, occurring in pairs, one on either side of the petiole base; may be leaflike, glands, or modified into spines

stolon horizontal, aboveground, modified propagating stem with nodes, internodes, and leaves

stoloniferous bearing stolons

stout sturdy; strong; rigid

stramineous straw-colored; strawlike in texture

striate marked with slender longitudinal grooves or lines; appearing striped

strigose rough, with short, stiff hairs or bristles

strigulose minutely strigose

style slender, elongated portion of the pistil which bears the stigma at its apex

sub- prefix that denotes somewhat, slightly, or in less degree

substrate underlying layer

subtend to underlie; located below

subterranean belowground

subulate shaped like an awl

succulent fleshy and juicy

suffrutescent slightly shrubby; having a slightly woody or shrubby base

suffuse spread over; spread throughout

sulcate having grooves or furrows

sulcus groove or furrow

summit top or apex

superior higher; above; more elevated in position

surmount directly on top of

surpass exceed

suture line or seam marking the union of two parts; the line of dehiscence of a fruit or capsule

swale low-lying depression in the land

swamp seasonally flooded area with more woody plants than a marsh

swollen enlarged

symmetrical regular in number and size of parts

synonym scientific name of a plant that has been replaced with a more current name

tannin any member of a large group of chemicals which can precipitate proteins and are derived from plants

taproot primary root of a plant that grows directly downward and gives rise to lateral branches

tardily occurring late

teeth pointed divisions, as on calyces and leaf margins

tendril twisting leaf modification by which a plant clings to a support

teratogenic substance halting pregnancy or producing a congenital malformation

terete cylindric and slender; circular in cross section

terminal borne at or belonging to the extremity or summit

tetrahedral geometric form with four faces

texture classification of soil based on the relative content of sand, silt, and clay particles

thicket dense growth of shrubs or small trees

throat opening; the orifice of a corolla or calyx

tinged slightly colored

tip apex; distal end

tomentose surface covered with matted, tangled, densely woolly hairs

tomentulose finely or slightly tomentose

tomentum covering of dense, woolly hairs

tonic beverage that refreshes, stimulates, restores, or invigorates

tooth pointed projection or division

trailing prostrate and creeping but not rooting

translucent semitransparent; transmitting light rays only partially

transverse at right angles to the long axis; crosswise; cross section

tri- prefix meaning "three"

triangular having three edges and three angles

trifoliate having three leaflets

triquetrous three-angled with concave faces between the angles

truncate ending abruptly; appearing to be cut off at the end

trunk main woody stem of a tree or shrub

tuber much-thickened underground part of a stem or rhizome; belowground storage organ

tubular having the shape of a tube, such as the corolla of some flowers

tuft cluster; bunch; not rhizomatous

tumble to roll over and over as when blown by the wind

turbinate top-shaped; inversely conical

turgid swollen by pressure from within

twig small branch of a tree or shrub

twining said of a plant spiraling about a support

umbel simple flat-topped or rounded inflorescence with pedicels or rays radiating from a common point

umbellate resembling an umbel

unarmed without thorns, prickles, or spines

uncinate hooked at the tip

uncinulate minutely uncinate

undulate strongly wavy in a perpendicular plane

uni- prefix meaning "one"

unilocular having one locule

united two or more wholly or partially fused parts

unpalatable not desirable for food; not readily eaten

valvate having valves; said of perianth parts that do not overlap in bud

valve one portion of a compound ovary; part of a legume

variety category or taxonomy below the species and subspecies level

vein single branch of the vascular system of a plant

velvety, velutinous soft and smooth, like velvet

venation pattern formed by the veins in a leaf or other plant organ

ventral relating to the underside of an organ; opposite the dorsal side

verrucose covered with wartlike projections

vestiture any covering on a surface making it other than glabrous

villosulous minutely villous

villous with long, soft macrohairs; similar to pilose but with a higher density of hairs

vine plant with flexible stems supported by climbing, twining, or creeping along a surface

virgate long, straight, and slender

wanting not present

washes areas of active water erosion

wavy margin with small, regular lobes; undulating surface or margin

weak frail; not stout nor rigid

whorl cluster of several branches or leaves around the axis arising from a common node

wilt limp from loss of water

wing thin projection or border; either of the two side petals in some legume flowers

wiry being thin and resilient

woodland vegetation consisting of primarily woody plants

woolly covered with long, entangled soft hairs

wrinkle small ridge or furrow on a surface

zigzag series of short, sharp bends

zygomorphic said of a flower having parts unequal in size or form; capable of division into essentially symmetrical halves by only one longitudinal plane passing through the axis on the side away from the axis; lower surface

ABBREVIATIONS FOR NOMENCLATURE AUTHORITIES

Abrams LeRoy Abrams (1874–1956), American botanist with Stanford University.

Agardh, J. Jakob Georg Agardh (1813–1901), Swedish botanist and professor of botany with Lund University.

Aiton William Aiton (1731–93), Scottish botanist and director of the Royal Botanic Gardens, Kew, England.

Alef. Friedrich Georg Christoph Alefeld (1820–72), German botanist, author, and medical practitioner.

Almeida Joaquim de Almeida Pinto (1813–61), Brazilian botanist.

Almeida, M.R. Marselein Rusario Almeida (1939–2017), Indian botanist.

Almeida, S.M. Sarah M. Almeida (born 1940), Indian botanist.

Amman Johann Amman (1707–41), Swiss-Russian botanist and professor with the Russian Academy of Sciences, Saint Petersburg.

Arcang. Giovanni Arcangeli (1840–1921), Italian botanist with the University of Turin and director of the Botanical Garden of Pisa.

Arechav. José Arechavaleta (1838–1912), Spanish-born pharmacist, geologist, naturalist, and professor with the University of the Republic, Montevideo, Uruguay.

Barneby Rupert Charles Barneby (1911–2000), British-born botanist with the New York Botanical Garden.

Bartal. Biagio Bartalini (1746–1822), Italian professor of botany and director of the botanical garden in Siena.

Barton, W.P.C. William Paul Crillon Barton (1786–1856), American medical botanist, surgeon, and professor of botany with the University of Pennsylvania.

Bates John Mallory Bates (1846–1930), American botanist and clergyman in Nebraska.

Beard John Stanley Beard (1916–2011), British-born ecologist and forester who worked in Australia.

Bensen, L.D. Lyman David Bensen (1909–93), American botanist with the University of Arizona and Pomona College, California.

Benth. George Bentham (1800–84), English botanist at the Royal Botanic Gardens, Kew, England, and secretary of the Horticultural Society of London.

Bertol. Antonio Bertoloni (1775–1869), Italian physician and professor of botany with the University of Bologna.

Besser Wilibald Swibert Joseph Gottlieb von Besser (1784–1842), Austrian-born botanist who worked in Ukraine.

Bicknell, E.P. Eugene Pintard Bicknell (1859–1925), American banker, botanist, and ornithologist who worked in New York.

Bieb., M. Baron Friedrich August Marschall von Bieberstein (1768–1826), German explorer in Russia and the Caucasus.

Bigelow Jacob Bigelow (1787–1879), American physician and botanist with Harvard University.

Bisch. Gottlieb Wilhelm T. G. Bischoff (1797–1854), German professor of botany.

Blake, S.F. Sidney Fay Blake (1892–1959), American botanist with the United States Department of Agriculture.

Blank. Joseph William Blankenship (1862–1938), American botanist with Montana State University.

Bobrov Evgenij Grigorievicz Bobrov (1902–83), Russian botanist.

Boiss. & Buhse Pierre Edmond Boissier (1810–85), Swiss botanist, mathematician, and explorer, and Friedrich Alexander Buhse (1821–98), German botanist.

Boivin, B., & Raymond Joseph Robert Bernard Boivin (1916–85), Canadian botanist, and Louis-Florent-Marcel Raymond (1915–72), Canadian botanist with the Montreal Botanical Garden.

Boreau Alexandre Boreau (1803–75), French pharmacist, botanist, and author of *Flora of the Center of France* (1840).

Boriss. Antonina Georgievna Borissova (1903–70), Russian botanist.

Boynton, F.E. Frank Ellis Boynton (1859–1942), American botanical collector for the herbarium and superintendent of the Biltmore Estates, North Carolina.

Br., R. Robert Brown (1773–1858), Scottish botanist, paleobotanist, and first keeper of botany with the British Museum, London.

Branner & Coville John Casper Branner (1850–1922), American geologist, botanist with Indiana University, and president of Stanford University (1913–15), and Frederick Vernon Coville (1867–1937), American chief botanist for the United States Department of Agriculture and founder of the United States National Arboretum.

Braun, P.J. Pierre Josef Braun (born 1959), German botanist.

Brewer William Henry Brewer (1828–1910), American botanist and geologist, California State Geological Survey.

Britten & Baker f. James Britten (1846–1924), English physician and botanist with the Royal Botanic Gardens, Kew, England, and with the British Museum, London, and Edmund Gilbert Baker (1864–1949), British plant collector.

Britton Nathaniel Lord Britton (1859–1934), American botanist and taxonomist who co-founded the New York Botanical Garden.

Britton & Kearney Nathaniel Lord Britton (1859–1934), American botanist and taxonomist who co-founded the New York Botanical Garden, and Thomas Henry Kearney (1874–1956), American botanist and agronomist with the United States Department of Agriculture.

Britton & Rose Nathaniel Lord Britton (1859–1934), American botanist and taxonomist who co-founded the New York Botanical Garden, and Joseph Nelson Rose (1862–1928), American botanist with the United States Department of Agriculture and curator at the Smithsonian Institution.

Britton, Sterns & Poggenb. Nathaniel Lord Britton (1859–1934), American botanist and taxonomist who co-founded the New York Botanical Garden, Emerson Ellick Sterns (1846–1926), American botanist, and Justus Ferdinand Poggenburg I (1840–93), American botanist.

Broich Steven L. Broich (n.d.), American botanist with Oregon State University.

Buckley Samuel Botsford Buckley (1809–84), American botanist, geologist, naturalist, and state geologist of Texas.

Bullock Arthur Allman Bullock (1906–80), English botanist with the Royal Botanic Gardens, Kew, England.

Bunge Alexander Andrejewitsch von Bunge (1803–90), Russian German explorer and professor of botany with the University of Dorpat, Estonia.

Burkart Arturo Erhardo Burkart (1906–75), Argentinean botanist.

Butters & H. St. John Fredrick King Butters (1878–1945), American professor of botany with the University of Minnesota, and Harold St. John (1892–1991), American professor of botany with Washington State University and the University of Hawaii.

Canby William Marriott Canby (1831–1904), American amateur botanist and businessman in Delaware.

Carrière Élie-Abel Carrière (1818–96), French botanist.

Cav. Antonio José Cavanilles (1745–1804), Spanish clergyman, botanist, taxonomist, and director of the Royal Botanical Garden, Madrid.

Chapm. Alvan Wentworth Chapman (1809–99), American physician and botanist specializing in plants from the southeastern United States.

Clayton, J. John Clayton (1685–1773), physician and botanist in Virginia.

Clem. & E.G. Clem. Frederic Edward Clements (1874–1945), American plant ecologist with the University of Nebraska and Carnegie Institute, and Edith Gertrude (Schwartz) Clements (1877–1971), American botanist and wife of Frederic, ecologist, and author.

Clos Dominique Clos (1821–1908), French physician and professor of botany with the University of Toulouse.

Cockerell Theodore Dru Alison Cockerell (1866–1948), English-born American zoologist and botanist.

Cooper Daniel Cooper (c. 1817–42), English botanist, physician, and curator for the Botanical Society of London.

Cory Victor Louis Cory (1880–1964), American botanist with Southern Methodist University.

Coult., J.M. John Merle Coulter (1851–1928), American botanist with Lake Forest College and the University of Chicago, founder of the *Botanical Gazette*, and president of Indiana University.

Darl. William Darlington (1782–1863), American botanist, physician, and member of the United States House of Representatives from Pennsylvania.

DC. Augustin Pyramus de Candolle (1778–1841), Swiss botanist, chair of botany with the College of Medicine in Montpellier, France, and later director of botany of the Academy of Geneva.

DC., A. Alphonse Louis Pierre Pyrame de Candolle (1806–93), French Swiss botanist and son of Augustin Pyramus de Candolle.

Debeaux Jean Odon Debeaux (1826–1910), French military pharmacist and botanist.

Desr. Louis Auguste Joseph Desrousseaux (1753–1838), French botanist.

Desv. Nicaise Auguste Desvaux (1784–1856), French botanist and director of the botanical garden in Angers, France.

Dietr., D. David Nathaniel Dietrich (1799–1888), German botanist and curator of the botanical garden in Jena, Germany.

Don, D. David Don (1799–1841), Scottish botanist and professor of botany with King's College, London.

Don, G. George Don (1798–1856), Scottish botanist who collected plants for the Royal Horticultural Society, London.

Dorn Robert D. Dorn (born 1942), American botanist with the Rocky Mountain Herbarium at the University of Wyoming.

Dorthes Jacob Anselme Dorthes (1759–94), French botanist and naturalist.

Douglas David Douglas (1798–1834), Scottish botanist who collected plants in the northwestern United States for the Royal Horticultural Society, London.

Dulac Joseph Dulac (1827–97), French botanist, theologian, and archaeologist.

Dum. Cours. Georges Louis Marie Dumont de Courset (1746–1824), French botanist and agronomist.

Eastw. Alice Eastwood (1859–1953), Canadian American botanist and curator of the herbarium at the California Academy of Sciences.

Eaton Amos Eaton (1776–1842), American botanist, geologist, and co-founder of the Rensselaer School, Troy, New York.

Eaton & Wright Amos Eaton (1776–1842), American botanist, geologist, and co-founder of the Rensselaer School, Troy, New York, and John Wright (1811–46), American physician, botanist, and student of Amos Eaton.

Eifert Imre János Eifert (1934–2020), American botanist with the University of Texas.

Elliott Stephen Elliott (1771–1830), American banker, legislator, and botanist in Charleston, South Carolina.

Engelm. Georg Engelmann (1809–84), German-born American physician and botanist in Saint Louis, Missouri.

Fabr. Philipp Conrad Fabricius (1714–74), German physician and botanist.

Fernald Merritt Lyndon Fernald (1873–1950), American plant taxonomist, plant geographer, and director of the Gray Herbarium, Harvard University.

Fisch. Friedrich Ernst Ludwig von Fischer (1782–1854), German-born Russian botanist and director of the Imperial Botanical Garden in Saint Petersburg, Russia.

Forbes, F.B., & Hemsl. Francis Blackwell Forbes (1839–1908), American botanist, merchant, and opium trader, and William Botting Hemsley (1843–1924), English botanist and Keeper of the Herbarium and Library, Royal Botanic Gardens, Kew, England.

Forst., E. Edward Forster (1765–1849), English banker and botanist.

Fraser John Fraser (1750–1811), Scottish botanist who collected plants in many parts of the world, including North America.

Gaertn. Joseph Gaertner (1732–91), German physician and botanist in Tübingen, Germany, and Saint Petersburg, Russia.

Gagnep. François Gagnepain (1866–1952), French botanist who specialized in Asian plants.

Gand. Michel Gandoger (1850–1926), French botanist and mycologist who collected throughout the Mediterranean region.

Gandhi & L.E. Br. Kancheepuram Natarajan Gandhi (born 1948), Indian-born nomenclature registrar with Harvard University, and Larry E. Brown (born 1937), American plant taxonomist and professor in Texas.

Gentry Howard Scott Gentry (1903–93), American botanist with the United States Department of Agriculture.

Gérard Louis Gérard (1733–1819), French naturalist.

Gleason Henry Allan Gleason (1882–1975), American ecologist, botanist, and taxonomist with the University of Chicago, University of Illinois, University of Michigan, and New York Botanical Garden.

Gmel., J.F. Johann Friedrich Gmelin (1748–1804), German botanist and naturalist with the University of Göttingen and University of Tübingen, Germany.

Godr. & Gren. Dominique Alexandre Godron (1807–80), French physician, botanist, and geologist, and Jean Charles Marie Grenier (1808–75), French botanist and naturalist.

Gray, A. Asa Gray (1810–88), professor of botany with Harvard University and author of *Gray's Manual of Botany*.

Greene Edward Lee Greene (1843–1915), American clergyman and botanist with the University of California–Berkeley and Catholic University.

Greenm. Jesse More Greenman (1867–1951), American botanist with the University of Chicago, Washington University, and Missouri Botanical Garden.

Gren. & Godr. Jean Charles Marie Grenier (1808–75), French botanist and naturalist, and Dominique Alexandre Godron (1807–80), French physician, botanist, and geologist.

Grimes, J.W. James Walter Grimes (born 1953), American botanist with the Royal Botanic Gardens in Melbourne, Australia.

Gron. Johannes Fredericus Gronovius (1690–1762), senator in the Netherlands, amateur botanist, and friend of Linnaeus.

Heller, A. Amos Arthur Heller (1867–1944), American plant collector for the New York Botanical Garden and professor of botany with the University of Minnesota and California Academy of Sciences.

Herm., F.J. Frederick Joseph Hermann (1906–87), American scientist with the Forest Service, United States Department of Agriculture.

Heynh. Gustav Heynhold (1800–1860), German botanist with the botanic gardens in Frankfurt and Dresden.

Hill, A.W. Arthur William Hill (1875–1941), English botanist, taxonomist, and director of the Royal Botanic Gardens, Kew, England.

Hook. William Jackson Hooker (1785–1865), English botanist, and director of the Royal Botanic Gardens, Kew, England.

Hook. & Arn. William Jackson Hooker (1785–1865), English botanist, and director of the Royal Botanic Gardens, Kew, England, and George Arnott Walker-Arnott (1799–1868), Scottish botanist at the University of Glasgow.

Hopkins, M. Milton Hopkins (1906–83), American botanist and professor of biology with the University of Oklahoma.

Hornem. Jens Wilken Hornemann (1770–1841), Danish botanist with the University of Copenhagen and director of the botanical garden.

Hosok. Takahide Hosokawa (1909–81), Japanese botanist with Kyushu University and Kumamoto University.

House Homer Doliver House (1878–1949), American botanist who worked at several institutions in New York, Washington DC, and North and South Carolina.

Huds. William Hudson (1730–93), British botanist and apothecary in London who published *Flora Angilica*.

Humb. & Bonpl. Friedrich Wilhelm Heinrich Alexander von Humboldt (1769–1859), Prussian naturalist, explorer, and geographer, and Aimé Jacques Alexandre Bonpland (1773–1858), French explorer and botanist.

Irwin, H.S., & Barneby Howard Samuel Irwin (1928–2019), American botanist and taxonomist with the New York Botanical Garden and Brooklyn Botanic Garden, and Rupert Charles Barneby (1911–2000), British-born botanist with the New York Botanical Garden.

Isely Duane Isely (1918–2000), American taxonomist with Iowa State University.

Jacq. Nikolaus Joseph von Jacquin (1727–1817), Danish-born chemist, botanist, and director of the botanical gardens at the University of Vienna.

Johnst., M.C. Marshall Corning Johnston (born 1930), American botanist with the University of Texas.

Jones, M.E. Marcus Eugene Jones (1852–1934), American mining engineer, geologist, and botanist in Utah and California.

Judz. Emmet J. Judziewicz (born 1953), American botanist and professor of biology with the University of Wisconsin–Stevens Point.

Juz. Sergei Vasilievich Juzepczuk (1893–1959), Russian botanist.

Kartesz John T. Kartesz (born 1948), American botanist and founder of the Biota of North America Program with the University of North Carolina.

Kartesz & Gandhi John T. Kartesz (born 1948), American botanist and founder of the Biota of North America Program with the University of North Carolina, and Kancheepuram Natarajan Gandhi (born 1948), Indian-born nomenclature registrar with Harvard University.

Kearney Thomas Henry Kearney (1874–1956), American botanist and agronomist with the United States Department of Agriculture and California Academy of Sciences.

Kearney & Peebles Thomas Henry Kearney (1874–1956), American botanist and agronomist with the United States Department of Agriculture and California Academy of Sciences, and Robert Hibbs Peebles (1900–1955), American botanist with the United States Department of Agriculture.

Kelso, E.H. Estelle H. Kelso (1911–87), botanist in Colorado.

Ker Gawl. John Bellenden Ker Gawler (1764–1842), English botanist.

Kern., A. Anton Joseph Kerner von Marilaun (1831–98), Austrian botanist with the University of Innsbruck and University of Vienna.

Killip & J.F. Macbr. Ellsworth Paine Killip (1890–1968), American botanist with the Smithsonian Institution, and James Francis Macbride (1892–1976), American botanist who worked primarily in Peru for the Field Museum of Natural History, Chicago.

Koch, K. Karl Heinrich Emil Koch (1809–79), German botanist with the University of Jena.

Krock. Anton Johann Krocker (1744–1823), German physician and botanist in Breslau, Germany (Wrocław, Poland).

Kuntze Carl Ernst Otto Kuntze (1843–1907), German physician and botanist who collected plants in many parts of the world.

Kuprian. Lyudmila Andreyeva Kuprianova (1914–87), Russian scientist with the Komarov Botanical Institute, Saint Petersburg.

L. Carolus Linnaeus (after 1761 Carl von Linné) (1707–78), Swedish zoologist, botanist, and physician; author of *Species Plantarum* (1753) upon which botanical nomenclature is based; the "Father of Taxonomy."

Lam. Jean-Baptiste Antoine Pierre de Monnet de Lamarck (1744–1829), French botanist and naturalist and chair of botany with the Jardin des Plantes.

Larisey Mary Maxine Larisey (1909–2000), American botanist with the School of Pharmacy, Medical College of South Carolina.

Lassen Per Lassen (born 1942), Swedish botanist.

Leavenw. Melines Conklin Leavenworth (1796–1832), American physician, soldier, and plant collector.

Ledeb. Carl Friedrich von Ledebour (1785–1851), German Estonian botanist and professor with the University of Tartu, Estonia.

Lehm. Johann Georg Christian Lehmann (1792–1860), German physician, botanist, and founder of the Hamburg Botanical Garden.

Lindl. John Lindley (1799–1865), English botanist, gardener, and professor.

Link Johann Heinrich Friedrich Link (1767–1851), German naturalist and botanist; professor of natural history, curator of the herbarium, and director of the botanic garden in Berlin.

Lodd. Joachim Conrad Loddiges (1738–1826), German botanist.

Loudon John Claudius Loudon (1783–1843), Scottish botanist and landscape designer.

Lour. João de Loureiro (1717–91), Portuguese botanist and Jesuit missionary who worked in Asia.

Lundell Cyrus Longworth Lundell (1907–94), American botanist with the Tropical Plant Research Foundation, Washington DC.

Lunell Joël Lunell (1851–1920), Swedish-born physician and botanist who immigrated to the United States and worked in North Dakota.

Macbr., J.F. James Francis Macbride (1892–1976), American botanist who worked primarily in Peru for the Field Museum of Natural History, Chicago.

Mack. & Bush Kenneth Kent Mackenzie (1877–1934), American attorney and amateur botanist, and Benjamin Franklin Bush (1858–1937), American botanist and ornithologist who collected plants for the Arnold Arboretum and the Missouri Botanical Garden.

MacMill. Conway MacMillan (1867–1929), American botanist who worked in Minnesota.

Maesen & S.M. Almeida Laurentius Josephus Gerardus van der Maesen (born 1944), Dutch botanist, and Sarah M. Almeida (born 1940), Indian botanist.

Mahler William F. Mahler (1930–2013), professor of botany and director of the herbarium at Southern Methodist University.

Makino Tomitarô Makino (1862–1957), Japanese botanist and taxonomist; "Father of Japanese Botany."

Marsh Charles Dwight Marsh (1855–1932), American botanist who worked for the United States Department of Agriculture and on the Biological Survey of the Panama Canal Zone.

Martrin-Donos Julien Victor de Martrin-Donos (1800–1870), French botanist.

Maxim. Carl Johann Maximovich (1827–91), Russian botanist and curator of the herbarium of the Saint Petersburg Botanical Gardens.

McCoy, S. Scott McCoy (1897–1980), American botanist who collected in Indiana.

Medik. Friedrich Kasimir Medikus (1736–1808), German physician and botanist and curator of the botanical garden in Mannheim.

Mciklc Robert Desmond Meikle (1923–2017), Irish botanist who worked at the Royal Botanic Gardens, Kew, England.

Mendenh., M.G. Meghan G. Mendenhall (born 1960), American botanist with the University of Texas.

Merr. Elmer Drew Merrill (1876–1956), American botanist and taxonomist who worked in the Philippines and later was director of the New York Botanical Garden and Arnold Arboretum, Harvard University.

Michx. André Michaux (1746–1803), French botanist and explorer in North America.

Mill. Philip Miller (1691–1771), English botanist and chief gardener at the Chelsea Physic Garden.

Miq. Friedrich Anton Wilhelm Miquel (1811–71), Dutch botanist, and professor of botany with the University of Amsterdam and Utrecht University.

Miyabe & Tatew. Kingo Miyabe (1876–1964), Japanese botanist and professor with the University of Tokyo, and Misao Tatewaki (1899–1976), Japanese botanist.

Moench Conrad Moench (1744–1805), German botanist and professor of botany with Marburg University.

Moldenke Harold Norman Moldenke (1909–96), American botanist and taxonomist with the New York Botanical Garden and later with the Trailside Nature and Science Center, Mountainside, New Jersey.

Moric. Moïse Étienne Moricand (1779–1854), Swiss botanist, naturalist, and explorer.

Muhl. Gotthilf Heinrich Ernst Muhlenberg (1753–1815), German-educated Lutheran minister and pioneer botanist in Pennsylvania.

Nees & Schwein. Christian Gottfried Daniel Nees von Esenbeck (1776–1858), German botanist physician, zoologist, and professor at the Leopoldina Academy and University of

Breslau, and Lewis David von Schweinitz (1780–1834), German-born American clergyman and botanist who worked in Pennsylvania and North Carolina.

Nelson, A. Aven Nelson (1859–1952), American botanist, one of the founding professors of the University of Wyoming, and later president of the University of Wyoming.

Neumann Joseph Henri François Neumann (1800–1858), French botanist.

Nutt. Thomas Nuttall (1786–1859), English-born botanist who lived and worked in North America.

Ortega Casimiro Gómez Ortega (1741–1818), Spanish physician, botanist, and the first professor of the Royal Botanical Garden, Madrid.

Osterh. George Everett Osterhout (1858–1937), American botanist and businessman who worked in Colorado.

Pall. Peter Simon von Pallas (1741–1811), Prussian botanist who worked in Russia.

Palmer, E.J. Ernest Jesse Palmer (1875–1962), American botanist in Missouri.

Payson Edwin Blake Payson (1893–1927), American botanist who collected in Colorado.

Peck, M. Morton Eaton Peck (1871–1959), American botanist and professor of botany with Willamette University, Oregon.

Pépin Pierre Denis Pépin (c. 1802–76), French botanist and agronomist with the National Museum of Natural History, Paris.

Pers. Christiaan Hendrik Persoon (1761–1836), South African–born botanist and physician who worked in France.

Phil. Rodolfo Amandus Philippi (1808–1904), German-born Chilean paleontologist and zoologist who became a professor of botany and zoology and director of the natural history museum in Santiago.

Phillips, L.Ll. Lyle Llewellyn Phillips (1923–2000), American botanist who worked in North Carolina.

Piper Charles Vancouver Piper (1867–1926), Canadian-born botanist and agriculturist with Washington State College (later Washington State University) and the United States Department of Agriculture.

Poir. Jean Louis Marie Poiret (1755–1834), French clergyman, botanist, and explorer of North Africa.

Pollard Charles Louis Pollard (1872–1945), American plant collector and librarian in Vermont.

Presl, C. Carl Bořivoj Presl (1794–1852), Czech botanist and professor with the University of Prague.

Pursh Frederick Traugott Pursh (1774–1820), German-born botanist who immigrated to the United States in 1799 and collected plants in Canada and the United States.

Raf. Constantine Samuel Rafinesque-Schmaltz (1783–1840), Constantinople-born pioneer botanist and naturalist who traveled and collected in Kentucky and Ohio.

Richardson John Richardson (1787–1865), Scottish naval surgeon, explorer, and naturalist.

Rob., B.L. Benjamin Lincoln Robinson (1864–1935), American botanist with the Gray Herbarium, Harvard University.

Roem. Johann Jacob Roemer (1763–1819), Swiss physician and professor of botany in Zurich.

Rose Joseph Nelson Rose (1862–1928), American botanist with the United States Department of Agriculture and the Smithsonian Institution.

Roth Albrecht Wilhelm Roth (1757–1834), German-born physician and botanist with the botanical institute at the University of Jena.

Rouy Georges Rouy (1851–1924), French botanist.

Rydb. Per Axel Rydberg (1860–1931), Swedish-born botanist with the University of Nebraska and New York Botanical Garden.

Salisb. Richard Anthony Salisbury (1761–1829), British botanist.

Salzm. Philipp Salzmann (1781–1851), German naturalist and botanist.

Sanjappa & Pradeep Munivenkatappa Sanjappa (born 1951), Indian botanist, and A. K. Pradeep (n.d.), Indian botanist with the University of Calcutta.

Sarg. Charles Sprague Sargent (1841–1927), American botanist who was the first director of the Arnold Arboretum, Harvard University.

Savi Gaetano Savi (1769–1844), Italian naturalist with the University of Pisa.

Scheele George Heinrich Adolf Scheele (1808–64), German botanist and explorer.

Schindl. Anton Karl Schindler (1879–1964), German dentist and botanist.

Schischk. Boris Konstantinovich Schischkin (1886–1963), Russian botanist and professor at Leningrad University.

Schleich. Johann Christoph Schleicher (1768–1834), German-born Swiss botanist.

Schneid., C.K. Camillo Karl Schneider (1876–1951), German botanist, dendrologist, and landscape architect.

Schrank Franz von Paula von Schrank (1747–1835), German priest, botanist, entomologist, professor with the University of Ingolstadt and director of the botanical gardens in Munich.

Schreb. Johann Christian Daniel von Schreber (1739–1810), German naturalist, professor in Erlangen, and editor of the eighth edition of Linnaeus's *Genera Plantarum*.

Schub., B.G. Bernice Giduz Schubert (1913–2000), American botanist and curator of the herbarium at Harvard University.

Schum., K. Karl Moritz Schumann (1851–1904), German botanist and curator of the Botanischen Museum, Berlin.

Scop. Giovanni Antonio Scopoli (1723–88), Italian physician, naturalist, and professor at the University of Pavia.

Senn, H. Harold Archie Senn (1912–97), Canadian botanist who worked in the Americas.

Ser. Nicolas Charles Seringe (1776–1858), French physician, botanist, first director of the de Candolle Herbarium, and professor with the University of Lyon.

Shafer John Adolph Shafer (1863–1918), American botanist with the Carnegie Museum of Natural History and later with the New York Botanical Gardens.

Sheld., E. Edmund Perry Sheldon (1869–1913), American forester who worked in Minnesota and Oregon.

Shinners Lloyd Herbert Shinners (1918–71), Canadian-born professor of botany with Southern Methodist University.

Shuttlew. Robert James Shuttleworth (1810–74), English botanist.

Sibth. John Sibthorp (1758–96), English botanist and professor of botany at the University of Oxford.

Siebold & Zucc. Philipp Franz von Siebold (1796–1866), German physician and botanist, and Joseph Gerhard Zuccarini (1797–1848), German botanist and professor of botany with the University of Munich.

Sinskaya Eugeniya Nikolayevna Sinskaya (1889–1965), Russian botanist and professor.

Širj. Grigorij Ivanović Širjaev (1882–1954), Russian botanist.

Small John Kunkel Small (1869–1938), American botanist and curator at the New York Botanical Garden.

Sm., C.P. Charles Piper Smith (1877–1955), American botanist in San Jose, California, and contributor to the United States National Herbarium.

Sm., J.G. Jared Gage Smith (1866–1957), American botanist with the United States Department of Agriculture.

Sm., J.G., & Rydb. Jared Gage Smith (1866–1957), American botanist with the United States Department of Agriculture, and Per Axel Rydberg (1860–1931), Swedish-born botanist with the University of Nebraska and New York Botanical Garden.

Spreng. Kurt Polycarp Joachim Sprengel (1766–1833), German botanist, physician, and professor of botany with the University of Halle.

St.-Hil., J. Jean Henri Jaume Saint-Hilaire (1772–1845), French naturalist.

Standl. Paul Carpenter Standley (1884–1963), American botanist with the United States National Museum and later the Field Museum of Natural History, Chicago.

Steud. Ernst Gottlieb von Steudel (1783–1856), German physician and botanist.

Stokes Jonathan S. Stokes (1755–1831), English physician and botanist.

Sw. Olof Swartz (1760–1818), Swedish botanist and taxonomist.

Sweet Robert Sweet (1783–1835), English botanist and horticulturalist.

Tausch Ignaz Friedrich Tausch (1793–1848), Bohemian botanist and professor of economic and technical botany with the University of Prague.

Ten. Michele Tenore (1780–1861), Italian botanist and director of the Botanical Garden of Naples.

Tharp & F.A. Barkley Benjamin Carroll Tharp (1885–1964), American botanist at the University of Texas, and Fred Alexander Barkley (1908–89), American professor of botany with Northeastern University and the University of Texas.

Thiret John William Thiret (1926–2005), American plant taxonomist.

Thunb. Carl Peter Thunberg (1743–1828), Swedish botanist who collected in South Africa and Japan.

Tidestr. Ivar Tidestrom (1864–1956), Swedish-born botanist with the United States Department of Agriculture and later with Catholic University of America.

Torr. John Torrey (1796–1873), American botanist, physician, chemist, and professor with several universities.

Torr. & A. Gray John Torrey (1796–1873), American botanist, physician, chemist, and professor with several universities, and Asa Gray (1810–88), American professor of botany at Harvard University and author of *Gray's Manual*.

Torr. & Frém. John Torrey (1796–1873), American botanist, physician, chemist, and professor with several universities, and John Charles Frémont (1813–90), American soldier, explorer, and presidential candidate.

Trautv. Ernst Rudolf von Trautvetter (1809–89), Russian botanist and taxonomist.

Trew Christoph Jakob Trew (1695–1769), German physician and botanist.

Turner, B.L. Billie Lee Turner Sr. (1925–2020), American botanist with the University of Texas.

Ulbr. Oskar Eberhard Ulbrich (1879–1952), German botanist, and curator and professor with the Botanical Museum, Berlin.

Uspensky Konstantin Aleksandrovich Uspensky (1804–85), Russian philosopher, theologian, orientalist, and archaeologist.

Vail Anna Murray Vail (1863–1955), American botanist and first librarian of the New York Botanical Garden.

Vassilcz. Ivan Tikhonovich Vassilczenko (1903–95), Russian botanist.

Vent. Étienne Pierre Ventenat (1757–1808), French botanist.

Vict. & Rousseau Joseph Lewis Conrad Kirouac Marie-Victorin (1885–1944), Canadian botanist and member of the Brothers of the Christian Schools in Quebec, and Jacques Rousseau (1905–70), Canadian botanist.

Waldst. & Kit. Franz de Paula Adam von Waldstein (1759–1823), Austrian soldier, explorer, and naturalist & Pál Kitaibel (1757–1817), Hungarian botanist and chemist.

Wallr. Karl Friedrich Wilhelm Wallroth (1792–1857), German botanist and physician.

Walter Thomas Walter (1740–89), British-born American botanist who lived in South Carolina.

Watson, P. Peter William Watson (1761–1830), English botanist and merchant.

Watson, S. Sereno Watson (1826–92), American botanist and curator of the Gray Herbarium, Harvard University.

Welsh, S.L. Stanley Larson Welsh (born 1928), American botanist and professor at Brigham Young University.

Wemple Don Kimberly Wemple (born 1929), American botanist at Iowa State University.

Willd. Carl Ludwig von Willdenow (1765–1812), German botanist, taxonomist, pharmacist, and director of the Berlin Botanical Garden.

Wood, Alph. Alphonso Wood (1810–81), American botanist and theology instructor.

Wooton & Standl. Elmer Otis Wooton (1865–1945), American botanist and professor with New Mexico State College, and Paul Carpenter Standley (1884–1963), American botanist, curator of the United States National Herbarium and National History Museum, Chicago.

Yakovlev Gennady Pavlovic Yakovlev (1938–2011), Russian botanist and director of the Saint Petersburg State Chemical-Pharmaceutical Academy.

Zabel Hermann Zabel (1832–1912), German botanist and director of the forestry botanical garden in Hannoversch Münden.

Zuccagni Attilio Zuccagni (1754–1807), Italian botanist.

Aandahl, A. R. *Soils of the Great Plains*. Lincoln: University of Nebraska Press, 1982.

Agricultural Research Service. *Selected Weeds of the United States*. Agricultural Handbook 366. Washington DC: United States Department of Agriculture, 1970.

Agricultural Research Service, United States Department of Agriculture. *Germplasm Resources Information Network (GRIN) Taxonomy for Plants*. https://www.ars-grin.gov/.

Agricultural Research Service, United States Department of Agriculture. National Plants Database. https://www.plants.usda.gov.

Allen, O. N., and E. K. Allen. *The Leguminosae*. Madison: University of Wisconsin Press, 1981.

Bailey Hortorium. *Hortus Third*. New York: Macmillan, 1976.

Bare, J. E. *Wildflowers and Weeds of Kansas*. Lawrence: Regents Press of Kansas, 1979.

Barkley, T. M. *Field Guide to the Common Weeds of Kansas*. Lawrence: University Press of Kansas, 1983.

Barneby, R. C. "Atlas of North American *Astragalus*." *Memoirs of the New York Botanical Garden* 13 (1954): 1–1188.

Barneby, R. C. "A Revision of the North American Species of *Oxytropis* DC." *Proceedings of the California Academy of Sciences* 27 (1952): 117–312.

Barnes, R. F., C. J. Nelson, K. J. Moore, and M. Collins, eds. *Forages*. Ames: Iowa State University Press, 2007.

Bartgis, R. L. "Rediscovery of *Trifolium stoloniferum* Muhlenberg *ex* A. Eaton." *Rhodora* 87 (1985): 425–29.

Beauchamp, R. M. "New Names in American Acacias." *Phytologia* 46 (1980): 5–9.

Biota Program of North America Project. *North American Plant Atlas*. https://www.bonap.org.

Blackwell, W. H. *Poisonous and Medicinal Plants*. Englewood Cliffs NJ: Prentice-Hall, 1990.

Brady, N. C., and R. R. Weil. *The Nature and Properties of Soils*. Englewood Cliffs NJ: Prentice-Hall, 2002.

Brooks, R. E. "*Trifolium stoloniferum*, Running Buffalo Clover: Description, Distribution, and Current Status." *Rhodora* 85 (1983): 343–54.

Brummitt, R. K., and C. E. Powell, eds. *Authors of Plant Names*. Kew, England: Royal Botanic Gardens, 1992.

Bryson, C. T., and M. S. DeFelice, eds. *Weeds of the Midwestern United States and Central Canada*. Athens: University of Georgia Press, 2010.

Budd, A. C. *Wild Plants of the Canadian Prairies*. Publication 1662. Quebec: Research Branch, Agriculture Canada, 1979.

Bugwood Center for Invasive Species and Ecosystem Health. *Bugwood Image Database System*. https://www.bugwood.org.

Burrows, G. E., and R. J. Tyrl. *Toxic Plants of North America*. Hoboken NJ: Wiley-Blackwell, 2013.

Campbell, J. J. N., M. Evans, M. E. Medley, and N. L. Taylor. "Buffalo Clovers in Kentucky (*Trifolium stoloniferum* and *T. reflexum*): Historical Records, Presettlement Environment, Rediscovery, Endangered Status, Cultivation, and Chromosome Number." *Rhodora* 90 (1988): 339–418.

Christiansen, P., and M. Müller. *An Illustrated Guide to Iowa Prairie Plants*. Iowa City: University of Iowa Press, 1999.

Clewell, A. F. "Native North American Species of *Lespedeza* (Leguminosae)." *Rhodora* 68 (1966): 359–405.

Coon, N. *Using Plants for Healing*. Emmaus PA: Rodale Press, 1979.

Correll, D. S., and M. C. Johnson. *Manual of the Vascular Plants of Texas*. Renner: Texas Research Foundation, 1970.

Cronquist, A. *An Integrated System of Classification of Flowering Plants*. New York: Columbia University Press, 1981.

Cronquist, A., A. H. Holmgren, N. H. Holmgren, J. K. Reveal, and P. K. Holmgren. *Intermountain Flora: Vascular Plants of the Intermountain West*. Bronx: New York Botanical Garden, 1989.

Currah, R., A. Smreciu, and M. Van Dyk. *Prairie Wildflowers*. Edmonton: University of Alberta, 1983.

Cusick, A. W. "*Trifolium stoloniferum* (Fabaceae) in Ohio: History, Habits, Decline, and Rediscovery." *Sida* 13 (1989): 467–80.

Dean, C. C. *Flora of Indiana*. Indianapolis: Department of Conservation, 1940.

Densmore, F. *How Indians Use Wild Plants for Food, Medicine, and Crafts*. New York: Dover, 1974.

Dewey, K. E. *Great Plains Weather*. Lincoln: University of Nebraska Press, 2019.

Diffendal, R. F. *Great Plains Geology*. Lincoln: University of Nebraska Press, 2017.

———. "Plate Tectonics, Space, Geologic Time, and the Great Plains: A Primer for Non-Geologists." *Great Plains Quarterly* 11 (1991): 83–102.

Diggs, G. M., B. L. Lipscomb, and R. J. O'Kennon. *Illustrated Flora of North Central Texas*. Fort Worth: Botanical Research Institute of Texas, 1999.

Dirr, M. A. *Manual of Woody Landscape Plants*. Champaign IL: Stipes, 1983.

Dorn, R. D. *Vascular Plants of Wyoming*. Cheyenne WY: Mountain West, 2001.

Duke, J. A. *Handbook of Edible Weeds*. Boca Raton FL: CRC Press, 1992.

Dunn, D. B., and J. M. Gillett. *The Lupines of Canada and Alaska*. Monograph No. 2. Ottawa: Canada Department of Agriculture, 1966.

Dunn, C. D., M. B. Stephenson, and J. Stubbendieck. *Common Forbs and Shrubs of Nebraska*. Extension Circular 118. Lincoln: University of Nebraska, 2017.

Elias, T. S. *Trees of North America*. New York: Van Nostrand Reinhold, 1980.

Elisens, W. J., and J. G. Packer. "A Contribution to the Taxonomy of the *Oxytropis campestris* Complex in Northwestern North America." *Canadian Journal of Botany* 58 (1980): 1820–31.

Evers, R. A., and R. P. Link. *Poisonous Plants of the Midwest and Their Effects on Livestock*. Special Publication 24. Urbana-Champaign: College of Agriculture, University of Illinois, 1972.

Farrar, J. *Field Guide to Wildflowers of Nebraska and the Great Plains*. Iowa City: University of Iowa Press, 2011.

Farrar, J. L. *Trees of the Northern United States and Canada*. Ames: Iowa State University Press, 1995.

Fassett, N. C. *The Leguminous Plants of Wisconsin*. Madison: University of Wisconsin Press, 1939.

Fernald, M. L. *Gray's Manual of Botany*. New York: Macmillan, 1950.

Gambill, W. G., Jr. *The Leguminosae of Illinois*. Illinois Biological Monographs. Volume 22. Urbana: University of Illinois Press, 1953.

Gay, C. W., D. Dwyer, C. Allison, S. Hatch, and J. Schickedanz. *New Mexico Range Plants*. Extension Circular 374. Las Cruces: New Mexico State University, 1980.

Gilmore, M. R. *Uses of Plants by Indians of the Missouri River Region*. Lincoln: University of Nebraska Press, 1991.

Glad, J. B., and R. R. Halse. "Invasion of *Amorpha fruticosa* L. (Leguminosae) along the Columbia and Snake Rivers in Oregon and Washington." *Madroño* 40 (1993): 62–63.

Great Plains Flora Association. *Atlas of the Flora of the Great Plains*. Ames: Iowa State University Press, 1977.

——. *Flora of the Great Plains*. Lawrence: University Press of Kansas, 1986.

Greuter, W., ed. *International Code of Botanical Nomenclature (Tokyo Code)*. Regnum Vegetabile No. 131. Königstein, Germany: Koeltz Scientific Books, 1994.

Grimes, J. W. "A Revision of the New World Species of Psoraleeae (Leguminosae: Papilionoideae)." *Memoirs of the New York Botanical Garden* 61 (1990): 1–114.

Grinnell, G. B. *The Cheyenne Indians: Their History and Ways of Life*. Volume 2. New York: Cooper Square, 1962.

Haddock, M. J. *Wildflowers and Grasses of Kansas*. Lawrence: University Press of Kansas, 2005.

Haddock, M. J., C. C. Freeman, and J. Bare. *Kansas Wildflowers and Weeds*. Lawrence: University Press of Kansas, 2015.

Harris, J. G., and M. W. Harris. *Plant Identification Terminology*. Spring Lake UT: Spring Lake, 2001.

Hart, J., and J. Moore. *Montana: Native Plants and Native People*. Helena: Montana Historical Society, 1976.

Hatch, S. L., and J. Pluhar. *Texas Range Plants*. College Station: Texas A&M University Press, 1993.

Hermann, F. J. *A Botanical Synopsis of the Clovers (Trifolium)*. Monograph No. 22. Washington DC: United States Department of Agriculture, 1953.

——. *Notes on Western Range Forbs*. Agriculture Handbook 293. Washington DC: Forest Service, United States Department of Agriculture, 1966.

Howard, A. D. "Drainage Evolution in Northeastern Montana and Northwestern North Dakota." *Geological Society of America Bulletin* 69 (1958): 575–88.

International Plant Name Index. Royal Botanic Gardens, Kew, England, Harvard University Herbarium, and the Australian National Herbarium. https://www.ipni.org.

Isely, D. "*Astragalus* (Leguminosae: Papilionoideae), I: Keys to United States Species." *Iowa State Journal of Research* 58 (1983): 1–172.

——. "*Astragalus* (Leguminosae: Papilionoideae), II: Species Summary A–E." *Iowa State Journal of Research* 59 (1984): 97–216.

———. "*Astragalus* (Leguminosae: Papilionoideae), III: Species Summary F–M." *Iowa State Journal of Research* 60 (1985): 179–322.

———. "*Astragalus* (Leguminosae: Papilionoideae), IV: Species Summary N–Z". *Iowa State Journal of Research* 61 (1986): 153–296.

———. "Classification of Alfalfa (*Medicago sativa* L.) and Relatives." *Iowa State Journal of Research* 57 (1983): 207–20.

———. "*Desmodium paniculatum* (L.) DC. and *D. viridiflorum* (L.) DC." *American Midland Naturalist* 49 (1953): 920–33.

———. "Keys to Sweet Clovers (*Melilotus*)." *Proceedings of the Iowa Academy of Sciences* 61 (1954): 119–31.

———. "Legumes of the United States, I: Native *Acacia*." *Sida* 3 (1969): 365–86.

———. "Legumes of the United States, II: *Desmanthus* and *Neptunia*." *Iowa State Journal of Science* 44 (1970): 495–511.

———. "Legumes of the United States, III: *Schrankia*." *Sida* 4 (1971): 232–45.

———. "Legumes of the United States, IV: *Mimosa*." *American Midland Naturalist* 85 (1971): 410–24.

———. "Legumes of the United States, VI: *Calliandra, Pithecellobium, Prosopis*." *Madroño* 21 (1972): 273–98.

———. "Leguminosae of the North-Central States, IV: Subfamily Psoraleae." *Iowa State Journal of Science* 37 (1962): 103–62.

———. "Leguminosae of the United States, I: Subfamily Mimosoideae." *Memoirs of the New York Botanical Garden* 25, no. 1 (1973): 1–152.

———. "Leguminosae of the United States, II: Subfamily Caesalpiniodeae." *Memoirs of the New York Botanical Garden* 25, no. 2 (1975): 1–228.

———. "Leguminosae of the United States, III: Subfamily Papilionoideae, Tribes Sophoreae, Podalyrieae, Loteae." *Memoirs of the New York Botanical Garden* 25, no. 3 (1981): 1–264.

———. *Native and Naturalized Leguminosae (Fabaceae) of the United States*. Provo: Monte L. Bean Life Science Museum, Brigham Young University, 1998.

———. "New Combinations and Two New Varieties in *Astragalus, Orophaca*, and *Oxytropis* (Leguminosae)." *Systematic Botany* 8 (1983): 420–26.

James, L. F. "Syndromes of *Astragalus* Poisoning in Livestock." *Journal of the American Veterinary Medical Association* 158 (1981): 614–18.

Johnson, J. R., and G. Larson. *Grassland Plants of South Dakota and the Northern Great Plains*. Agricultural Experiment Station Bulletin 566R. Brookings: South Dakota State University, 1999.

Johnston, M. C. "The North American Mesquites: *Prosopis* Section *Algarobia* (Leguminosae)." *Brittonia* 14 (1962): 72–90.

Kannowski, P. B. *Wildflowers of North Dakota*. Grand Forks: University of North Dakota Press, 1989.

Kaul, R. B., D. Sutherland, and S. Rolfsmeier. *The Flora of Nebraska*. Lincoln: School of Natural Resources, University of Nebraska, 2011.

Kearney, T. H., and R. H. Peebles. *Arizona Flora*. Berkeley: University of California Press, 1960.

Keeler, R. F., K. R. Van Kampen, and L. F. James, eds. *Effects of Poisonous Plants on Livestock*. New York: Academic Press, 1978.

Kindscher, K. *Edible Wild Plants of the Prairie: An Ethnobotanical Guide*. Lawrence: University Press of Kansas, 1987.

———. *Medicinal Wild Plants of the Prairie: An Ethnobotanical Guide*. Lawrence: University Press of Kansas, 1992.

Kingsbury, J. M. *Poisonous Plants of the United States and Canada*. Englewood Cliffs NJ: Prentice-Hall, 1964.

Kirk, D. R. *Wild Edible Plants of the Western United States*. Healdsburg CA: Naturegraph, 1975.

Lackschewitz, K. *Vascular Plants of West-Central Montana—Identification Guidebook*. General Technical Report INT-277. Ogden UT: Forest Service, United States Department of Agriculture, 1991.

Larisey, M. M. "A Monograph of the Genus *Baptisia*." *Annals of the Missouri Botanical Garden* 27 (1940): 119–244.

Larson, G. E., and J. R. Johnson. *Plants of the Black Hills and Bear Lodge Mountains*. B732. Brookings: Agricultural Experiment Station, South Dakota State University, 2007.

Lavin, S. J., F. M. Shelley, and J. C. Archer. *Atlas of the Great Plains*. Lincoln: University of Nebraska Press, 2011.

Ledingham, G. F. "Chromosome Numbers in *Astragalus* and *Oxytropis*." *Canadian Journal of Genetics and Cytology* 2 (1960): 119–28.

———. "Chromosome Numbers of Some Saskatchewan Leguminosae with Particular Reference to *Astragalus* and *Oxytropis*." *Canadian Journal of Botany* 35 (1957): 657–66.

Leininger, W. C., J. E. Taylor, and C. L. Wambolt. *Poisonous Range Plants in Montana*. Bulletin 348. Bozeman: Cooperative Extension Service, Montana State University, 1977.

Looman, J., and K. F. Best. *Budd's Flora of the Canadian Prairie Provinces*. Publication 1662. Quebec: Research Branch, Agriculture Canada, 1979.

Luckow, M. "Monograph of *Desmanthus* (Leguminosae–Mimosoideae)." *Systematic Botany Monographs* 38 (1993): 1–166.

Majak, W., B. M. Brooke, and R. T. Ogilvie. *Stock-Poisoning Plants of Western Canada*. Alberta: Agriculture Canada, 2008.

Martin, A. C. *Weeds*. New York: St. Martin's Press, 2001.

Martin, W. C., and C. R. Hutchins. *Flora of New Mexico*. Vol. 1. Lehre, Germany: Koeltz Scientific Books, J. Cramer Verlag, 1980.

———. *Flora of New Mexico*. Vol. 2. Lehre, Germany: Koeltz Scientific Books, J. Cramer Verlag, 1981.

McVaugh, R. *Leguminosae: Flora Novo-Galiciana*. Vol. 5. Ann Arbor: University of Michigan Press, 1987.

Merck Veterinary Manual. Merck. https://www.merckvetmanual.com.

Miller, R. W., and D. T. Gardner. *Soils in Our Environment*. Upper Saddle River NJ: Prentice-Hall, 2001.

Minnesota Wildflowers. https://www.minnesotawildflowers.info.

Missouri Botanical Garden. Tropicos. https://www.tropicos.org/.

Moerman, D. E. *Native American Ethnobotany*. Portland: Timber Press, 1998.

Mohlenbrock, R. H. "A Revision of the Genus *Stylosanthes*." *Annals of the Missouri Botanical Garden* 44 (1957): 299–355.

Mosquin, T., and J. M. Gillett. "Chromosome Numbers in American *Trifolium* (Leguminosae)." *Brittonia* 17 (1965): 136–43.

Nowick, E. *Historical Common Names of Great Plains Plants*. Vol. 2. Lincoln: Zea Books, 2015.

Ode, D. J. *Dakota Flora*. Pierre: South Dakota State Historical Society Press, 2006.

Owensby, C. E. *Kansas Prairie Wildflowers*. Ames: Iowa State University Press, 1980.

Peterson, G. A., and C. V. Cole. "Productivity of Great Plains Soils: Past, Present, and Future." In *Conservation of Great Plains Ecosystems: Current Science, Future Options, of Ecology, Economy and Environment*, vol. 5, edited by S. R. Johnson and A. Bouzaher, 325–42. Dordrecht, Netherlands: Kulwer Academic Publishers, 1995.

Phillips, J. *Wild Edibles of Missouri*. Jefferson City: Missouri Department of Conservation, 1979.

Plaster, E. J. *Soil Science and Management*. Albany: Delmar, 1996.

Polhill, R. M., and P. H. Raven, eds. *Advances in Legume Systematics*, Part 1. Kew, England: Royal Botanic Gardens, 1981.

Pool, R. J. *Handbook of Nebraska Trees*. Conservation Bulletin 32. Lincoln: Conservation and Survey Division, University of Nebraska, 1971.

Porter, C. L. "*Astragalus* and *Oxytropis* in Colorado." *Wyoming University Publications* 16 (1951): 1–49.

Rogers, D. J. *Edible, Medicinal, Useful, and Poisonous Wild Plants of the Northern Great Plains–South Dakota region*. Saint Francis SD: Buechel Memorial Lakota Museum, 1980.

—— . *Lakota Names and Traditional Uses of Native Plants by Sicangu (Brule) People in the Rosebud Area, South Dakota*. Saint Francis SD: Rosebud Educational Society, 1980.

Rossum, S., and S. J. Lavin. "Where Are the Great Plains? A Cartographic Analysis." *Professional Geographer* 52 (2000): 543–52.

Roy, M., P. U. Clark, R. W. Barendregt, J. R. Glassman, and R. J. Enkin. "Glacial Stratigraphy and Paleomagnetism of the Late Cenozoic Deposits in the North-Central United States." *Geological Society of America Bulletin* 116 (2004): 30–41.

Rydberg, P. A. *Flora of Colorado*. Bulletin No. 100. Fort Collins: Colorado Agricultural Experiment Station, 1906.

—— . *Flora of the Prairies and Plains of Central North America*. New York: New York Botanical Gardens, 1932.

—— . "Genera of North American Fabaceae: V. *Astragalus* and Related Genera." *American Journal of Botany* 15 (1928): 584–95.

Saunders, C. F. *Edible and Useful Wild Plants*. New York: Dover, 1976.

Scifres, C. J., R. W. Bovey, C. E. Fisher, G. O. Hoffman, and R. D. Lewis, eds. *Mesquite: Growth and Development, Management, Economics, Control, and Uses*. College Station: Agricultural Experiment Station, Texas A&M University, 1973.

Scoggan, H. J. *The Flora of Canada, Part 3*. National Museum of Natural Sciences Publications in Botany, No. 7. Ottawa: National Museums of Canada, 1978.

Seabrook, J. A., and L. A. Dionne. "Studies on the Genus *Apios*, I: Chromosome Number and Distribution of *Apios americana* and *A. priceana*." *Canadian Journal of Botany* 54 (1976): 2567–72.

Sibley, D. A. *The Trees of North America: Michaux and Redoute's American Masterpiece*. New York: Abbeville, 2017.

Small, E., and B. S. Brookes. "Taxonomic Circumscription and Identification in the *Medicago sativa-falcata* (Alfalfa) Continuum." *Economic Botany* 38 (1984): 83–96.

Small, E., and M. Jomphe. "A Synopsis of the Genus *Medicago* (Leguminosae)." *Canadian Journal of Botany* 67 (1989): 3260–94.

Stephens, H. A. *Poisonous Plants of the Central United States.* Lawrence: University Press of Kansas, 1980.

——. *Woody Plants of the North Central Plains.* Lawrence: University Press of Kansas, 1973.

Steyermark, J. A. *Flora of Missouri.* Ames: Iowa State University Press, 1963.

Stroh, J. R., A. E. Carleton, and W. J. Seamands. *Management of Lutana Cicer Milkvetch for Hay, Pasture, Seed, and Conservation Uses.* Research Journal 66. Laramie: University of Wyoming, 1972.

Stubbendieck, J., M. P. Carlson, and C. D. Dunn. *Nebraska Plants Toxic to Livestock.* Extension Circular 3037. Lincoln: University of Nebraska, 2018.

Stubbendieck, J., M. J. Coffin, and C. D. Dunn. *Weeds of the Great Plains.* Lincoln: Nebraska Department of Agriculture, 2019.

Stubbendieck, J., and E. C. Conard. *Common Legumes of the Great Plains.* Lincoln: University of Nebraska Press, 1989.

Stubbendieck, J., S. L. Hatch, N. M. Bryan, and C. D. Dunn. *North American Wildland Plants.* Lincoln: University of Nebraska Press, 2017.

Swinehart, J. B. "Geology." In *Encyclopedia of the Great Plains*, edited by D. J. Wishart, 629–30. Lincoln: University of Nebraska Press, 2004.

Trimble, D. E. *Geologic History of the Great Plains.* Bulletin 1493. Washington DC: United States Geological Survey, 1980.

Turner, B. L. *The Leguminosae of Texas.* Austin: University of Texas Press, 1959.

——. "Texas Species of *Desmanthus* (Leguminosae)." *Field and Laboratory* 18 (1950): 54–65.

Turner, B. L., and O. S. Fearing. "Chromosome Numbers in the Leguminosae, III: Species of the Southwestern United States and Mexico." *American Journal of Botany* 47 (1959): 603–8.

Turner, B. L., H. Nichols, G. C. Denny, and O. Doron. *Atlas of the Vascular Plants of Texas.* Vol. 2. Fort Worth: Botany Research Institute of Texas, 2003.

United States Government. Integrated Taxonomic Information System (ITIS). https://www.itis.gov.

Van Bruggen, T. *The Vascular Plants of South Dakota.* Ames: Iowa State University Press 1976.

Vance, F. R., J. R. Jowsey, and J. S. McLean. *Wildflowers of the Northern Great Plains.* Minneapolis: University of Minnesota Press, 1984.

Vogel, V. J. *American Indian Medicine.* Norman: University of Oklahoma Press, 1970.

Wasowski, S. *Gardening with Prairie Plants.* Minneapolis: University of Minnesota Press, 2002.

Waterfall, U. T. *Keys to the Flora of Oklahoma.* Stillwater: Oklahoma State University Press, 1962.

Weaver, J. E. *Native Vegetation of Nebraska.* Lincoln: University of Nebraska Press, 1965.

——. *North American Prairie.* Lincoln: Johnsen, 1954.

Weaver, J. E., and F. W. Albertson. *Grasslands of the Great Plains.* Lincoln: Johnsen, 1956.

Weber, W. A., and R. C. Wittmann. *Catalog of the Colorado Flora: A Biodiversity Baseline.* Boulder: University Press of Colorado, 1992.

Wedel, W. R. "Notes on the Prairie Turnip (*Psoralea esculenta*) among the Plains Indians." *Nebraska History* 59 (1978): 155–79.

Welsh, S. L. "Legumes of the North-Central States: Galegeae." *Iowa State Journal of Science* 35 (1960): 111–250.

———. "*Oxytropis* DC.—Names, Basionyms, Types, and Synonyms—Flora North America Project." *Great Basin Naturalist* 51 (1991): 377–96.

Whitson, T. D., ed. *Weeds and Poisonous Plants of Wyoming and Utah*. Laramie: Cooperative Extension Service, University of Wyoming, 1987.

Whitson, T. D., L. C. Burrill, S. A. Dewey, D. W. Cudney, B. E. Nelson, R. D. Lee, R. Parker. *Weeds of the West*. Laramie: University of Wyoming, 1991.

Wishart, D. J., ed. *Encyclopedia of the Great Plains*. Lincoln: University of Nebraska Press, 2004.

Yatskievych, G. *Flora of Missouri*. Saint Louis: Missouri Botanical Garden Press, 1999.

INDEX

Page references for the scientific names of species appear in boldface.